T0140359

Studies in Fuzziness and Soft Computing

Volume 365

Series editor

Janusz Kacprzyk, Polish Academy of Sciences, Warsaw, Poland
e-mail: kacprzyk@ibspan.waw.pl

The series "Studies in Fuzziness and Soft Computing" contains publications on various topics in the area of soft computing, which include fuzzy sets, rough sets, neural networks, evolutionary computation, probabilistic and evidential reasoning, multi-valued logic, and related fields. The publications within "Studies in Fuzziness and Soft Computing" are primarily monographs and edited volumes. They cover significant recent developments in the field, both of a foundational and applicable character. An important feature of the series is its short publication time and world-wide distribution. This permits a rapid and broad dissemination of research results.

More information about this series at http://www.springer.com/series/2941

John N. Mordeson · Sunil Mathew
Davender S. Malik

Fuzzy Graph Theory with Applications to Human Trafficking

 Springer

John N. Mordeson
Department of Mathematics
Creighton University
Omaha, NE
USA

Davender S. Malik
Department of Mathematics
Creighton University
Omaha, NE
USA

Sunil Mathew
Department of Mathematics
National Institute of Technology
Calicut, Kerala
India

ISSN 1434-9922   ISSN 1860-0808   (electronic)
Studies in Fuzziness and Soft Computing
ISBN 978-3-030-09494-2   ISBN 978-3-319-76454-2   (eBook)
https://doi.org/10.1007/978-3-319-76454-2

Printed on acid-free paper

This Springer imprint is published by the registered company Springer International Publishing AG part of Springer Nature
The registered company address is: Gewerbestrasse 11, 6330 Cham, Switzerland

In memory of Lotfi A. Zadeh, who inspired so many.

Foreword

When Lotfi A. Zadeh introduced "Fuzzy sets" in 1965 in his seminal paper [1], he amplified the usual mathematical concepts of sets and relations. He borrowed from Paul R. Halmos' Nave Set Theory [2] the definition of a relation as a set of ordered pairs (x, y) such that $x, y \in X$. A fuzzy relation he defined as a fuzzy set in the product space $X \times X$, or just as generally: an n-ary fuzzy relation is a fuzzy set A in the product space $X \times X \times \cdots \times X$, with the membership function $\mu(x_1, \ldots, x_n)$, where $x_i \in X$, $i = 1, \ldots, n$.

The first Ph.D. thesis related to Fuzzy Sets and supervised by Zadeh was "Fuzzy Sets and Pattern Recognition" written in 1967 by the Chinese student Chang [3], and in the following year he acted on a suggestion about topologically motivated questions regarding fuzzy sets which Zadeh had received from mathematician Richard W. Hamming. Zadeh later recalled this: "And then I showed this to C.-L. Chang. He was a student at that time. And he had good mathematical background and I told him, "Why don't you look into that?" And so then he looked into that and then he wrote a short paper on "Fuzzy Topological Spaces," on fuzzy topology, and that paper started this whole field with fuzzy topology" [4]. Chang wrote this paper to apply the concepts of fuzzy sets and fuzzy relations to generalize some basic concepts of the mathematical theory of topology "such as open set, closed set, neighborhood, interior set, continuity, and compactness" [5].

Among others, the paper was read by the then leading researcher of computer image analysis Azriel Rosenfeld, who felt inspired to write an analogous paper on "Fuzzy groups" to apply the concepts of fuzzy sets and fuzzy relations to generalize some basic concepts of the mathematical theory of groups, e.g., fuzzy sub-groupoids, fuzzy ideals, and their lattices, their homomorphic preimages, and fuzzy subgroups [6]. Rosenfeld's paper appeared in 1971 and it "became the starting point of an entire literature on fuzzy algebraic structures" [7].

Based on Zadeh's fuzzy relations, the first definition of fuzzy graphs was given in 1973 by Arnold Kauffman, the French engineer and professor of Applied Mechanics and Operations Research, in the world's first textbook on fuzzy sets and systems [8]. An English version of this book appeared in 1975, the same year as Rosenfeld also elaborated a concept of a fuzzy graph. He introduced "fuzzy analogs" of several basic graph-theoretic concepts. This volume contained the proceedings of the US–Japan Seminar on Fuzzy Sets and their applications that was held at the University of California in Berkeley, California, on July 1–4, 1974 [9]. In his paper, entitled "Fuzzy Graphs," Rosenfeld defined a fuzzy graph as a graph that consists of vertices and edges with membership value in the interval $[0, 1]$. More specifically, he defined a fuzzy graph as a pair $G = (\sigma, \mu)$ of functions $\sigma : S \rightarrow [0, 1]$ and $\mu : S \times S \rightarrow [0, 1]$ where for all $x, y \in S$ we have $\mu(x, y) = \mu(y, x)$ and $\mu(x, y) \leq \sigma(x) \wedge \sigma(y)$ with $\wedge$ denoting the minimum. Rosenfeld's paper presented the concepts of subgraphs, paths, connectedness, cliques, bridges, trees, and forests and established some of their properties.

At this US–Japan Seminar, an alternative analysis of fuzzy graphs from the viewpoint of connectedness was also presented by the computer scientists Raymond T. Yeh and S. Y. Bang. The two defined a fuzzy graph as a pair $G = (V, R)$ of a set V of vertices and a fuzzy relation R on V. In their concept of a fuzzy graph, the vertices are crisp and the edges connecting the vertices are fuzzy with a membership function $\mu_R(V \times V) \rightarrow [0, 1]$ This definition was more suitable for clustering analysis and later also for database theory rather than the more general definition given by Rosenfeld. The door was now wide open for new considerations, e.g., fuzzy trees, cycles and cocycles of fuzzy graphs, and bipartite graphs.

Concluding his paper, Rosenfeld hoped that "much more could be done along this line" and that his paper "will serve to stimulate further work of fuzzified graph theory." He trod an innovative path from fuzzy relations to fuzzy graphs: Every fuzzy relation can be regarded as defining a weighted graph where the arc has a weight (a number between 0 and 1).

Fuzzy groups and fuzzy graphs were combined in 1987 by computer scientist Prabir Bhattacharya to now consider fuzzy groups of fuzzy graphs [10]. He obtained "a fuzzy analog of a basic result in graph theory that given a graph one can associate a group with it in a natural way." Bhattacharya also used Rosenfeld's definition of a metric in a fuzzy graph, called "m-distance," to define the eccentricity of a vertex as the maximum of all m-distances between this vertex and the others, the center of a connected fuzzy graph to be the vertex with minimum eccentricity, and the radius of a connected fuzzy graph as the minimum of the eccentricities of its vertices. This author ended his paper with the hopeful expectation that "Much more work could be done to investigate the structure of fuzzy graphs which would be useful since fuzzy graphs have applications in pattern clustering. Our examples indicate that not all properties of fuzzy graphs could be expected to be analogs of properties of (crisp) graphs."

In 1989, the mathematician Kiran Bhutani was able to show "that every fuzzy group can be imbedded in a fuzzy group of the group of automorphisms of some fuzzy graph" [11]. She examined ordered fuzzy graphs, and she defined a fuzzy graph to be complete on a finite set S if the membership value of two of its elements, $\mu(x, y)$, equals the minimum of fuzzy group values of x and y.

Since the 1990s, the field of fuzzy graphs has developed at a dramatic pace. This is in large part due to the research work in the Center for Mathematics of Uncertainty at Creighton University that was headed by the professor of mathematics John Mordeson. Also, Sunil Mathew has been a leader in the field of fuzzy graph theory. Davender Malik from this university's Mathematics Department had shown Mordeson Rosenfeld's paper on fuzzy groups and Mordeson wrote later: "This paper opened a whole new area in fuzzy mathematics, namely fuzzy abstract algebra. Dr. Malik and I began a long-lasting research association in fuzzy abstract algebra [12]." Shortly after this encounter, the two created the center that was then named "Creighton's Center for Research in Fuzzy Math and Computer Science," [13]. Mordeson wrote in 2012, "the center hosted visiting scholars from China, India, Korea, Japan, and Saudi Arabia." The center collaborated with medical and psychological departments from various universities and research centers that led to new joint research work and an increasing number of publications in fuzzy theory and its applications.

Following Lotfi A. Zadeh's introduction of fuzzy sets in 1965, he did not expect applications of his theory in the field of technology as he said 1994 in an interview: "I expected people in the social sciences-economics, psychology, philosophy, linguistics, politics, sociology, religion, and numerous other areas to pick up on it. It's been somewhat of a mystery to me why even to this day, so few social scientists have discovered how useful it could be. Instead, fuzzy logic was first embraced by engineers and used in industrial process controls and in "smart" consumer products such as hand-held camcorders that cancel out jittering and microwaves that cook your food perfectly at the touch of a single button. I didn't expect it to play out this way back in 1965" [14]. In the 1960s, he had already written the following: "What we still lack and lack rather acutely, are methods for dealing with systems which are too complex or too ill-defined to admit of precise analysis. Such systems pervade life sciences, social sciences, philosophy, economics, psychology, and many other "soft" fields." [15].

Applications of fuzzy theory in social sciences and humanities appeared later and the investigations of professor Mordeson and his colleagues have principally undertaken this pioneering and important research work. In 2005, he contacted both Dr. Terry Clark of Creigthon's Political Science Department and Dr. Mark Wierman of the Computer Science Department so as to join "formal theoretical work with empirical research. We felt that standard mathematics is too precise to model human thinking and action" as Mordeson wrote in 2012. The center's new approach focused on collaborations "with students from political science, mathematics, and economics" and "using fuzzy mathematics to model global issues, e.g., nuclear stability, children with special needs, economic freedom, smart power, political stability, cooperative threat reduction, failed states, economic stability, creative economy, quality of life, solar energy, wind energy, college freshman weight gain, population management of Sub-Saharan Africa, safe skies, remittances, health care, nuclear deterrence, developmental disorders in children, human development, and globalization." [12].

Published here now is a new product of this interdisciplinary research work on Fuzzy Set Theory in the social sciences and humanities: Mordeson and Malik, together with Indian mathematician Sunil Mathew, present a two-volume set on

Fuzzy Graph Theory and the second volume here contains applications to human trafficking.

Fuzzy Set Theory provides a very good tool to handle real-world problems or real-life problems. This often enunciated word has already been proven in technological applications, but it is also true for much more important problems in our real world in which we live together. We can all see that human trafficking is not fake but a fact of real life. Millions of victims are currently "modern slaves," who are exploited for labor or for sex.

Because of war, poverty, oppression, hunger and thirst, more and more refugees are coming to the "first world" that is our real world. Over the last years, the clash of cultures, climate change, and many misunderstandings have resulted in enormous real-life problems which are continually developing and are complexly interwoven with each other.

This is the factual (not "fake") news of our time! We need to understand these "social networks," and we urgently need to find solutions to solve the problems that they produce for the improvement of all our lives. Fuzzy graphs are appropriate models for depicting such networks of sources, transits, and destinations, and Mordeson, Malik, and Mathew have successfully used fuzzy graphs to study human trafficking networks.

PS: During the time that I started to write this foreword Mark Wierman passed away very suddenly, and some weeks later, after I had already finished it, very sadly Lotfi A. Zadeh also died. They both were eminent and important members of the "Fuzzy community," and their work will continue with books like these.

Berkeley, CA, USA   Rudolf Seising
and Munich, Germany
September 2017

References

1. L.A. Zadeh, Fuzzy sets. Inf. Control **8**, 338–353 (1965)
2. P.R. Halmos, *Naive Set Theory* (Van Nostrand, New York, 1960)
3. C.-L. Chang, Fuzzy sets and pattern recognition. Ph.D. Thesis, Department of Electrical Engineering, University of California, Berkeley, December 1967
4. L.A. Zadeh in an interview with R. S. in the year 2000
5. C.L. Chang, Fuzzy topological spaces. J. Math. Anal. Appl. **24**, 182–190:182 (1968)
6. A. Rosenfeld, Fuzzy groups. J. Math. Anal. Appl. **35**, 512–517 (1971)
7. A. Rosenfeld, Foreword, in: *Fuzzy Mathematics. An Introduction for Engineers and Scientists*, 2nd edn, ed. by J.N. Mordeson, P.S. Nair (Springer, Berlin, 2001), pp. v–vi:v
8. A. Kauffman, *Introduction a la Theorie des Sous-emsembles Flous* (Masson et Cie Editeurs, Paris 1973) (English translation: Introduction to the Theory of Fuzzy Subsets: Fundamental theoretical elements) (Academic Press, 1975)
9. A. Rosenfeld, Fuzzy graphs, in: *Fuzzy Sets and Their Applications to Cognitive and Decision Processes, New York, San Francisco*, ed. by L.A. Zadeh, K.-S. Fu, K. Tanaka, M. Kimura (Academic Press, London 1975), pp. 77–95

10. P. Bhattacharya, Some remarks on fuzzy graphs, Pattern Recogn. Lett. **6**, 297–302 (1987)
11. K.R. Bhutani, On automorphisms of fuzzy graphs. Pattern Recogn. Lett. **9**, 159–162 (1989)
12. J.N. Mordeson, On fuzziness in mathematics, in *On Fuzziness. A Homage to Lotfi A. Zadeh—Volume II*, ed. by R. Seising, E. Trillas, S. Termini, C. Moraga. Studies in Fuzziness and Soft Computing, vol. 299 (Springer, Berlin, 2013), pp. 455–458:456
13. B. Kokensparger, Creighton enlightens computers with fuzzy math, Creighton University Window, Summer Issue, 12–15:12 (1997)
14. B. Blair, Interview with Lotfi Zadeh, creator of fuzzy logic by Azerbaijada International **2.4** (1994), http://www.azer.com/aiweb/categories/magazine/24_folder/24_articles/24_fuzzylogic.html
15. L.A. Zadeh, Toward a theory of fuzzy systems, electronic research laboratory, College of Engineering, University of California Berkeley, Report ERL-69-2, June 1969. Later printed as: L.A. Zadeh, Toward a theory of fuzzy systems, in *Aspects of Network and Systems Theory, New York*, ed. by R.E. Kalman, N. DeClaris (Holt, Rinehard and Winston, 1971), pp. 469–490

Preface

Lotfi Zadeh introduced the concept of a fuzzy subset of a set in 1965 as a way to represent uncertainty. His work inspired researchers worldwide. Azriel Rosenfeld and Yeh and Bang introduced the notion of a fuzzy graph in 1975. The notion of a fuzzy graph is an area of wide interest and with broad applications.

This book is the second of two books by the authors dealing with fuzzy graphs. This book is strongly motivated by the desire to apply graph-theoretic ideas to the terrible problem of human trafficking. It should be of interest to research mathematicians, computer scientists, and social scientists. We thank the Journal of Fuzzy Mathematics for its support.

In Chap. 1, we examine groups in which some members are seen as exercising a disruptive influence while others appear to hold the group together. The general results developed are applied to the set of countries in which some are highly involved in human trafficking while others try to control trafficking. Results related to group structure that have not appeared in book form before including articulation points, intuitionistic fuzzy graphs, and $(s, t]$-fuzzy graphs.

In Chap. 2, we present results concerning the concept of domination in fuzzy graphs. We apply the results to examine the regional flow of trafficking in persons. We determine destination dominating sets, minimal destination dominating sets, and maximal independent sets for these regions.

In Chap. 3, we use a relatively new concept to study trafficking in persons. This concept is that of the incidence of a vertex and an edge. We determine results concerning bridges, cut vertices, cut pairs, fuzzy incidence paths. We create a mathematical model that can be used to study the effect of states influence on other states.

In Chap. 4, we prove a fuzzy max flow, min cut theorem which can be applied to routes involved in human trafficking.

In Chap. 5, we examine complementary fuzzy incidence graphs. We introduce $(s, t]$-fuzzy incidence graphs and quasi-fuzzy incidence graphs. The results are applied to illegal immigration. We introduce the notion of vague fuzzy incidence graphs and apply the results to determine numerical measures of the susceptibility of flow of particular routes.

Governments, international organizations, and civil society are devoting huge efforts to control trafficking in persons. However, there still lacks accurate information. Due to the lack of accurate information on global sex trafficking, we follow the lead of Trafficking in Persons Global Patterns, Appendices-United Nations Office on Drugs and Crime, Citation Index, 2006, and take a linguistic approach in Chap. 6. We determine an upper bound in linguistic terms on the trafficking from one region to another. We use the terms very low, low, medium, high, and very high to describe the source, transit, and destination designation of countries.

In Chap. 7, we consider causal linkages between commercial sex and human trafficking. A three-link chain of necessary conditions: a population vulnerability to trafficking, a capable trafficking organization, and a sex market is used. Trafficking can be addressed by using policy intervention of any link. Prospects for policy success at these three points of intervention are compared. We use five methods to derive five different linear equations to measure the success of policy intervention.

The book is heavily dependent on papers from the journal New Mathematics and Natural Computation(NM & NC) that involve results on human trafficking. NM & NC is to be commended for its support of the fight against human trafficking.

Acknowledgements

The authors thank George and Sue Haddix for their continued support of mathematics of uncertainty, STEM, and Creighton University. We also thank the journal New Mathematics and Natural Computation for allowing us to use some material published there.

Omaha, USA	John N. Mordeson
Calicut, India	Sunil Mathew
Omaha, USA	Davender S. Malik

Contents

About the Authors

Dr. John N. Mordeson is Professor Emeritus of Mathematics at Creighton University. He received his B.S., M.S., and Ph.D. from Iowa State University. He is a Member of Phi Kappa Phi. He is a President of the Society for Mathematics of Uncertainty. He has published fifteen books and two hundred journal articles. He is on the editorial board of numerous journals. He has served as an external examiner of Ph.D. candidates from India, South Africa, Bulgaria, and Pakistan. He has refereed for numerous journals and granting agencies. He is particularly interested in applying mathematics of uncertainty to combat the problem of human trafficking.

Dr. Sunil Mathew is currently a Faculty Member in the Department of Mathematics, NIT Calicut, India. He has acquired his masters from St. Joseph's College Devagiri, Calicut, and Ph.D. from National Institute of Technology Calicut in the area of Fuzzy Graph Theory. He has published more than seventy-five research papers and written two books. He is a Member of several academic bodies and associations. He is editor and reviewer of several international journals. He has an experience of twenty years in teaching and research. His current research topics include fuzzy graph theory, bio-computational modeling, graph theory, fractal geometry, and chaos.

Dr. Davender S. Malik is a Professor of Mathematics at Creighton University. He received his Ph.D. from Ohio University and has published more than fifty-five papers and eighteen books on abstract algebra, applied mathematics, graph theory, fuzzy automata theory and languages, fuzzy logic and its applications, programming, data structures, and discrete mathematics.

About the Authors

Chapter 1
Strengthening and Weakening Members of a Network

1.1 Fuzzy Sets

In 1965, Lotfi A. Zadeh [34] introduced a new type of set called a fuzzy set and a new logic later known as fuzzy logic. Instead of YES or NO, regarding the existence of an element in a set, he used the degree of membership, which allows an element to exist in a set with a partial grade of membership. The applications of fuzzy logic are profound and widespread.

In the first chapter, we present the basic concepts from fuzzy set theory needed for this book. We assume that the reader is knowledgeable on the basics of set theory. We first must set our notation.

$\mathbb{Z}$ denotes the set of all integers

$\mathbb{R}$ denotes the set of all real numbers

$\mathbb{N}$ denotes the set of positive integers or natural numbers

We let $\wedge$ denote minimum or infimum and $\vee$ denote maximum or supremum.

Let A and B be subsets of a universal set U. We write $A \subseteq B$ if A is a subset of B or equivalently $B \supseteq A$ if B contains A. We write $A \subset B$ if $A \subseteq B$ and there exists $x \in B$ such that $x \notin A$. The intersection of A and B is denoted by $A \cap B$ and the union of A and B is denoted by $A \cup B$. We let $B \setminus A$ denote the set difference of A in B. Thus, $B \setminus A = \{x \in B \mid x \notin A\}$. If $B = U$, then we write A^c for $U \setminus A$ and call A^c the **complement** of A. We let $A \times B$ denote the Cartesian cross product of A and B.

We let $\emptyset$ denote the empty set.

Definition 1.1.1 Let X be a set. A **fuzzy subset** of X is a function from X into the closed interval $[0, 1]$.

We can interpret a fuzzy subset μ of a set X as giving the membership degree of every element of X in some "subset" of X that is given in some descriptive manner. We shall sometimes use the term **fuzzy set**, hereafter to denote a fuzzy subset of a

J. N. Mordeson et al., *Fuzzy Graph Theory with Applications to Human Trafficking*, Studies in Fuzziness and Soft Computing 365, https://doi.org/10.1007/978-3-319-76454-2_1

set, if there is no confusion regarding the underlying set. We let $\mathfrak{FP}(X)$ denote the set of all fuzzy subsets of X. We call $\mathfrak{FP}(X)$ the **fuzzy power set** of X.

A basic concept of fuzzy set theory is the t-cut of a fuzzy subset μ of a set X, where $t \in [0, 1]$. The t-**cut** or t-**level set** of a fuzzy subset μ of X is defined to be $\mu^t = \{x \in X \mid \mu(x) \geq t\}$, where $t \in [0, 1]$. The **strong** t-**cut** of μ is defined as $\mu^{t+} = \{x \in X \mid \mu(x) > t\}$. The support of μ is defined to be $\mathrm{Supp}(\mu) = \{x \in X \mid \mu(x) > 0\}$. We write μ^* for $\mathrm{Supp}(\mu)$ at times. Clearly a t-cut of a fuzzy set is a crisp set.

We next define some set theoretical operations for fuzzy sets. Let μ and ν be fuzzy subsets of a set X. We write $\mu \subseteq \nu$ if for all $x \in X$, $\mu(x) \leq \nu(x)$. If $\mu \subseteq \nu$ and there exists $x \in X$ such that $\mu(x) < \nu(x)$, we write $\mu \subset \nu$. We define $\mu \cap \nu$ by for all $x \in X$, $(\mu \cap \nu)(x) = \mu(x) \wedge \nu(x)$. We define $\mu \cup \nu$ by for all $x \in X$, $(\mu \cup \nu)(x) = \mu(x) \vee \nu(x)$.

The notion of intersection for fuzzy sets can also be defined by use of a variety of t-norms.

Definition 1.1.2 A function $i : [0, 1] \times [0, 1] \to [0, 1]$ is called a t-**norm** if it satisfies the following conditions:
  (i) For all $x \in [0, 1]$, $i(1, x) = x$.
  (ii) For all $x, y \in [0, 1]$, $i(x, y) = i(y, x)$.
  (iii) For all $x, y, z \in [0, 1]$, $i(x, i(y, z)) = i(i(x, y), z)$.
  (iv) For all $w, x, y, z \in [0, 1]$, $w \leq x$ and $y \leq z$ implies $i(w, y) \leq i(x, z)$.

Example 1.1.3 The following are examples of t-norms: for all $x, y \in [0, 1]$,
  (i) Standard intersection: $i(x, y) = x \wedge y$
  (ii) Algebraic product: $i(x, y) = xy$
  (iii) Bounded difference: $i(x, y) = 0 \vee (x + y - 1)$
  (iv) Drastic intersection: $i(x, y) = \begin{cases} x & \text{if } y = 1, \\ y & \text{if } x = 1, \\ 0 & \text{otherwise} \end{cases}$

The notion of union for fuzzy sets can be defined by a variety of t-conorms.

Definition 1.1.4 A function $u : [0, 1] \times [0, 1] \to [0, 1]$ is called a t-**conorm** if it satisfies the following conditions:
  (i) For all $x \in [0, 1]$, $u(0, x) = x$.
  (ii) For all $x, y \in [0, 1]$, $u(x, y) = u(y, x)$.
  (iii) For all $x, y, z \in [0, 1]$, $u(x, u(y, z)) = u(u(x, y), z)$.
  (iv) For all $w, x, y, z \in [0, 1]$, $w \leq x$ and $y \leq z$ implies $u(w, y) \leq u(x, z)$.

Example 1.1.5 The following are examples of t-conorms:
  (i) Standard union: $u(x, y) = x \vee y$
  (ii) Algebraic sum : $u(x, y) = x + y - xy$
  (iii) Bounded sum : $u(x, y) = 1 \wedge (x + y)$
  (iv) Drastic union : $u(x, y) = \begin{cases} x & \text{if } y = 0, \\ y & \text{if } x = 0, \\ 1 & \text{otherwise.} \end{cases}$

Next we introduce the notion of a complement.

Definition 1.1.6 A function $c : [0, 1] \to [0, 1]$ is called a *fuzzy complement* if the following conditions hold:

(*i*) $c(0) = 1$ and $c(1) = 0$.

(*ii*) For all $x, y \in [0, 1]$, $x \le y$ implies $c(x) \ge c(y)$.

A desirous property for a fuzzy complement c to possess is continuity. Another is that it be **involutive**, i.e., for all $x \in [0, 1]$, $c(c(x)) = x$. An example of a fuzzy complement is the standard complement, i.e., $c(x) = 1 - x$ for all $x \in [0, 1]$.

In classical set theory, the operations of intersection and union are dual with respect to complement in the sense that they satisfy De Morgan's laws,

$$(A \cap B)^c = A^c \cup B^c \text{ and } (A \cup B)^c = A^c \cap B^c$$

for subsets A and B of some universe.

For fuzzy subsets, De Morgan's laws become

$$c(i(a, b)) = u(c(a), c(b)) \text{ and } c(u(a, b)) = i(c(a), c(b)),$$

for all $a, b \in [0, 1]$, where i is a t-norm and u is a t-conorm.

It can be easily shown that if c is the standard complement, then the standard intersection and the standard union are duals with respect to c as are many other pairs.

Example 1.1.7 Standard union and intersection are illustrated in this example. Consider the fuzzy subsets σ and μ of $\{1, 2, 3, 4, 5, 6, 7, 8\}$ defined as follows.

$$\sigma = \{(1, 0.1), (2, 0.5), (3, 0.8), (4, 1), (5, 0.8), (6, 0.5), (7, 0.1), (8, 0)\}$$

and

$$\mu = \{(1, 0), (2, 0), (3, 0.2), (4, 0.4), (5, 0.6), (6, 0.8), (7, 1), (8, 1)\}.$$

Then

$$\sigma \cup \mu = \{(1, 0.1), (2, 0.5), (3, 0.8), (4, 1), (5, 0.8), (6, 0.8), (7, 1), (8, 1)\}$$

and

$$\sigma \cap \mu = \{(1, 0), (2, 0), (3, 0.2), (4, 0.4), (5, 0.6), (6, 0.5), (7, 0.1), (8, 0)\}.$$

1.2 Fuzzy Relations

As crisp relations represent the association between elements of two or more sets, a fuzzy relation gives the extent of relationship between elements between two fuzzy sets. Zadeh [35] introduced fuzzy relations in 1971. We provide a formal definition below. Most of the contents of this section are based on Rosenfeld's [27] work in 1975.

If S is a set, then a fuzzy relation μ on S is a fuzzy subset of $S \times S$. When μ takes the values 0 and 1 alone, it becomes the characteristic function of a relation on S. If R is a subset of S and P is a relation on S, then P becomes a relation on R only if $(x, y) \in P$ implies $x \in R$ and $y \in R$. If ζ and η are the characteristic functions of R and P, respectively, then $\eta(x, y) = 1$ implies $\zeta(x) = \zeta(y) = 1$ for all $x, y \in R$. This is equivalent to the expression $\eta(x, y) \leq \zeta(x) \wedge \zeta(y)$ for all $x, y \in R$. Motivated by this, we have the definition of a fuzzy relation on a fuzzy subset as follows.

Definition 1.2.1 Let σ be a fuzzy subset of a set S and μ be a fuzzy relation on S. Then μ is called a **fuzzy relation** on σ if $\mu(x, y) \leq \sigma(x) \wedge \sigma(y)$ for all $x, y \in S$.

Definition 1.2.2 If S and T are two sets and σ and τ are fuzzy subsets of S and T, respectively, then a fuzzy relation μ from the fuzzy subset σ into the fuzzy subset τ is a fuzzy subset μ of $S \times T$ such that $\mu(x, y) \leq \sigma(x) \wedge \tau(y)$ for all $x \in S$ and $y \in T$.

It is interesting to see that for μ to be a fuzzy relation, the degree of membership of a pair of elements never exceeds the degree of membership of either of the elements. Later, while defining a fuzzy graph, this inequality allows us to organize the flow through an edge of a fuzzy graph in such a way that, it never exceeds the capacities of its end vertices. Also, μ^t is a relation from σ^t into τ^t for all $t \in [0, 1]$ and as a consequence, μ^* becomes a relation from σ^* into τ^*.

In Definition 1.2.2, if $\sigma(x) = 1$ for all $x \in S$ and $\tau(y) = 1$ for all $y \in T$, then μ is called a fuzzy relation from S into T. Similarly, if $\sigma(x) = 1$ for all $x \in S$ in Definition 1.2.1, μ is said to be a fuzzy relation on S.

Definition 1.2.3 If σ is a fuzzy subset of a set S, the **strongest fuzzy relation** on σ is the fuzzy relation μ_σ defined by $\mu_\sigma(x, y) = \sigma(x) \wedge \sigma(y)$ for all $x, y \in S$.

Definition 1.2.4 For a fuzzy relation μ on S, the **weakest fuzzy subset** of S, on which μ is a fuzzy relation is σ_μ, defined by $\sigma_\mu(x) = \vee_{y \in S}(\mu(x, y) \vee \mu(y, x))$ for all $x \in S$.

Definition 1.2.5 Let $\mu : S \times T \to [0, 1]$ be a fuzzy relation from a fuzzy subset σ of S into a fuzzy subset τ of T and $\nu : T \times U \to [0, 1]$ is a fuzzy relation from the fuzzy subset ψ of T into a fuzzy subset η of U. Define $\mu \circ \nu : S \times U \to [0, 1]$ by $(\mu \circ \tau)(x, z) = \vee\{\mu(x, y) \wedge \nu(y, z) \mid y \in T\}$ for all $x \in S, z \in U$. Then $\mu \circ \nu$ is called the **max-min composition** of σ and τ.

The composition of any two fuzzy relations as in Definition 1.2.5 is always a fuzzy relation. But in the next result, we only consider two fuzzy relations defined on the same fuzzy set.

Proposition 1.2.6 *If μ and ν are fuzzy relations on a fuzzy subset σ, then $\mu \circ \nu$ is a fuzzy relation on σ.*

Proof Let S be a set and σ be a fuzzy subset of S. Because μ and ν are fuzzy relations on σ, $\mu(x, y) \leq \sigma(x) \wedge \sigma(y)$ and $\nu(y, z) \leq \sigma(y) \wedge \sigma(z)$ for all $x, y, z \in S$. Thus, $\mu(x, y) \wedge \nu(y, z) \leq \sigma(x) \wedge \sigma(y) \wedge \sigma(z) \leq \sigma(x) \wedge \sigma(z)$ for all $y \in S$ and hence, $(\mu \circ \nu)(x, z) = \vee_{y \in S}(\mu(x, y) \wedge \nu(y, z)) \leq \sigma(x) \wedge \sigma(z)$ for all $x, z \in S$. ∎

If μ, ν and τ are fuzzy relations on a fuzzy set σ, then $\mu \circ (\nu \circ \tau)$ and $(\mu \circ \nu) \circ \tau$ are fuzzy relations on σ and

$$\mu \circ (\nu \circ \tau) = (\mu \circ \nu) \circ \tau.$$

Max-min composition is similar to matrix multiplication, where addition is replaced by $\vee$ and multiplication by $\wedge$. We can easily show that the composition of fuzzy relations is associative. Thus, if we denote $\mu \circ \mu$ by μ^2, higher powers of the fuzzy relation μ^2, μ^3, and so on, can be easily defined. Define $\mu^\infty(x, y) = \vee\{\mu^k(x, y) \mid k = 1, 2, ...\}$ for all $x, y \in S$. Also, define $\mu^0(x, y) = 0$ if $x \neq y$ and $\mu^0(x, x) = \mu(x, x)$ otherwise.

Proposition 1.2.7 *If μ and ν are two fuzzy relations on a finite set S, then for all $t \in [0, 1]$, we have $(\mu \circ \nu)^t = \mu^t \circ \nu^t$.*

Proof Let $(x, z) \in (\mu \circ \nu)^t$. Then $(\mu \circ \nu)(x, z) \geq t$. By definition, $\mu(x, y) \wedge \nu(y, z) \geq t$ for some $y \in S$. Therefore, $\mu(x, y) \geq t$ and $\nu(y, z) \geq t$, which implies $(x, y) \in \mu^t$ and $(y, z) \in \nu^t$. Thus, $(x, z) \in \mu^t \circ \nu^t$ by the definition of composition of functions. Thus, $(\mu \circ \nu)^t \subseteq \mu^t \circ \nu^t$.

Let $(x, z) \in \mu^t \circ \nu^t$. Then there exists $y \in S$ such that $(x, y) \in \mu^t$ and $(y, z) \in \nu^t$. Thus, $(\mu \circ \nu)(x, z) \geq t$. Hence $\mu^t \circ \nu^t \subseteq (\mu \circ \nu)^t$.

Consequently, $(\mu \circ \nu)^t = \mu^t \circ \nu^t$. ∎

Proposition 1.2.8 *Suppose μ, ν, λ, ρ are fuzzy relations defined on a fuzzy subset σ of S. If $\mu \subseteq \nu$ and $\lambda \subseteq \rho$, then $\mu \circ \lambda \subseteq \nu \circ \rho$.*

Proof We have $(\mu \circ \lambda)(x, z) = \vee_{y \in S}(\mu(x, y) \wedge \lambda(y, z)) \leq \vee_{y \in S}(\nu(x, y) \wedge \rho(y, z)) = (\nu \circ \rho)(x, z)$ for all $x, z \in S$. ∎

Note that there are several types of composition of fuzzy relations that are available in the literature. Next we have a unary operation on a fuzzy relation.

Definition 1.2.9 Let μ be a fuzzy relation defined on a fuzzy subset σ of a set S. Then the **compliment** μ^c of μ is defined as $\mu^c(x, y) = 1 - \mu(x, y)$ for all $x, y \in S$.

Definition 1.2.10 Let $\mu : S \times T \to [0, 1]$ be a fuzzy relation from a fuzzy subset σ of S into a fuzzy subset ν of T. Then $\mu^{-1} : T \times S \to [0, 1]$, the **inverse** of μ from ν into σ, is defined as $\mu^{-1}(y, x) = \mu(x, y)$ for all $x, y \in T \times S$.

Some of the properties of fuzzy relations are given in the following result. Their proofs are omitted as they are obvious.

Theorem 1.2.11 *Let τ, π, ρ and ν be fuzzy relations on a fuzzy subset σ of a set S. Then the following properties hold.*

(i) $\tau \cup \pi = \pi \cup \tau$.
(ii) $\tau \cap \pi = \pi \cup \tau$.
(iii) $(\tau^c)^c = \tau$.
(iv) $\pi \cup (\rho \cup \nu) = (\pi \cup \rho) \cup \nu$.
(v) $\pi \cap (\rho \cap \nu) = (\pi \cap \rho) \cap \nu$.
(vi) $\pi \circ (\rho \circ \nu) = (\pi \circ \rho) \circ \nu$.
(vii) $\pi \cap (\rho \cup \nu) = (\pi \cap \rho) \cup (\pi \cap \nu)$.
(viii) $\pi \cup (\rho \cap \nu) = (\pi \cup \rho) \cap (\pi \cup \nu)$.
(ix) $(\tau \cup \pi)^c = \pi^c \cap \tau^c$.
(x) $(\tau \cap \pi)^c = \pi^c \cup \tau^c$.
(xi) For every $t \in [0, 1]$, $(\tau \cup \pi)^t = \tau^t \cup \pi^t$.
(xii) For every $t \in [0, 1]$, $(\tau \cap \pi)^t = \tau^t \cap \pi^t$.
(xiii) If $\tau \subseteq \rho$ and $\pi \subseteq \nu$, then $\tau \cup \pi \subseteq \rho \cup \nu$.
(xiv) If $\tau \subseteq \rho$ and $\pi \subseteq \nu$, then $\tau \cap \pi \subseteq \rho \cap \nu$.

Definition 1.2.12 Let μ be a fuzzy relation on σ, where σ is a fuzzy subset of a set S. Then μ is said to be **reflexive** on σ if $\mu(x, x) = \sigma(x)$ for all $x \in S$.

When μ is a reflexive fuzzy relation on σ, it is not hard to see that $\mu(x, y) \le \sigma(x) \wedge \sigma(y) \le \sigma(x) = \mu(x, x)$ and $\mu(y, x) \le \sigma(y) \wedge \sigma(x) \le \sigma(x) = \mu(x, x)$ for all $x, y \in S$. In other words, when we express a fuzzy relation in a matrix form, the elements of any row or any column will be less than or equal to the diagonal element belonging to that row or column. Sometimes we say μ is reflexive on σ. Next we have some interesting properties of reflexive fuzzy relations.

Theorem 1.2.13 *Let μ and ν be fuzzy relations on a fuzzy subset σ of a set S. If μ is reflexive, then $\nu \subseteq \nu \circ \mu$ and $\nu \subseteq \mu \circ \nu$.*

Proof Let $x, z \in S$. Then $(\mu \circ \nu)(x, z) = \vee\{\mu(x, y) \wedge \nu(y, z) \mid y \in S\} \ge \mu(x, x) \wedge \nu(x, z) = \sigma(x) \wedge \nu(x, z)$. But $\nu(x, z) \le \sigma(x) \wedge \sigma(z)$. Therefore, $\sigma(x) \wedge \nu(x, z) = \nu(x, z)$. Thus, $\nu \subseteq \mu \circ \nu$. Similarly, we can prove that $\nu \subseteq \nu \circ \mu$. ∎

Corollary 1.2.14 *If μ is reflexive, then $\mu \subseteq \mu^2$.*

Corollary 1.2.15 *If μ is reflexive, then $\mu^0 \subseteq \mu \subseteq \mu^2 \subseteq \mu^3 \subseteq \cdots \subseteq \mu^\infty$.*

The proofs of Corollaries 1.2.14 and 1.2.15 can be obtained by taking ν as μ, μ^2, and so on, in the proof of Theorem 1.2.13.

Theorem 1.2.16 *Let μ be a fuzzy relation on a fuzzy subset σ of a set S. If μ is reflexive, then $\mu^0(x, x) = \mu(x, x) = \mu^2(x, x) = \mu^3(x, x) = \cdots = \mu^\infty(x, x) = \sigma(x)$ for all $x \in S$.*

Proof We have $\mu(x, x) = \sigma(x)$ for all $x \in S$. Assume that the result is true for $k = n$. That is, $\mu^n(x, x) = \sigma(x)$, for all $x \in S$. Now, for all $x \in S$, we have $\mu^{n+1}(x, x) = \vee\{\mu(x, y) \wedge \mu^n(y, x) \mid y \in S \leq \vee\{\sigma(x) \wedge \sigma(x) \mid y \in S\} = \sigma(x)$. Also, $\mu^{n+1}(x, x) = \vee\{\mu(x, y) \wedge \mu^n(y, x) \mid y \in S\} \geq \mu(x, x) \wedge \mu^n(x, x) = \sigma(x)$. Thus, $\mu^{n+1}(x, x) = \sigma(x)$ for all $x \in S$. ∎

Theorem 1.2.17 *If μ and ν are reflexive fuzzy relations on σ, then $\mu \circ \nu$ and $\nu \circ \mu$ are also reflexive.*

Proof $(\mu \circ \nu)(x, x) = \vee\{\mu(x, y) \wedge \nu(y, x) \mid y \in S\} \leq \vee\{\sigma(x) \wedge \sigma(x) \mid y \in S\} = \sigma(x)$ and $(\mu \circ \nu)(x, x) = \vee\{\mu(x, y) \wedge \nu(y, x) \mid y \in S\} \geq \mu(x, x) \wedge \nu(x, x) = \sigma(x) \wedge \sigma(x) = \sigma(x)$. The proof that $\nu \circ \mu$ is reflexive is similar. ∎

Theorem 1.2.18 *If μ is reflexive on σ, then μ^t is reflexive on σ^t for all $t \in [0, 1]$.*

Proof Suppose μ is reflexive. Let $x \in \sigma^t$. Then $\mu(x, x) = \sigma(x) \geq t$ and thus $(x, x) \in \mu^t$. Hence, μ^t is reflexive on σ^t. ∎

Definition 1.2.19 Let μ be a fuzzy relation on σ, where σ is a fuzzy subset of a set S. Then μ is said to be **symmetric** if $\mu(x, y) = \mu(y, x)$ for all $x, y \in S$.

From the definition, it follows that if μ is symmetric, then the matrix representation of μ is symmetric.

Theorem 1.2.20 *Let μ and ν be fuzzy relations on a fuzzy subset σ of a set S. Then the following properties hold.*
  (i) If μ and ν are symmetric, then $\mu \circ \nu$ is symmetric if and only if $\mu \circ \nu = \nu \circ \mu$.
  (ii) If μ is symmetric, then every power of μ also is symmetric.
  (iii) If μ is symmetric, then μ^t is a symmetric relation on σ^t for all $t \in [0, 1]$.

Proof (i) $(\mu \circ \nu)(x, z) = (\mu \circ \nu)(z, x) \Leftrightarrow \vee\{\mu(x, y) \wedge \nu(y, z) \mid y \in S\} = \vee\{\mu(z, y) \wedge \nu(y, x) \mid y \in S\} \Leftrightarrow \vee\{\mu(x, y) \wedge \nu(y, z) \mid y \in S\} = \vee\{\nu(y, x) \wedge \mu(z, y) \mid y \in S\} \Leftrightarrow \mu \circ \nu = \nu \circ \mu$ because $\nu(y, x) = \nu(x, y)$ and $\mu(z, y) = \mu(y, z)$.
  (ii) Assume that μ^n is symmetric for $n \in \mathbb{N}$. Then $\mu^{n+1}(x, z) = \vee\{\mu(x, y) \wedge \mu^n(y, z) \mid y \in S\} = \vee\{\mu(y, x) \wedge \mu^n(z, y) \mid y \in S\} = \vee\{\mu^n(z, y) \wedge \mu(y, x) \mid y \in S\} = \mu^{n+1}(z, x)$
  (iii) Let $0 \leq t \leq 1$. Suppose $(x, z) \in \mu^t$. Then $\mu(x, z) \geq t$. Because μ is symmetric, $\mu(z, x) \geq t$. Thus, $(z, x) \in \mu^t$. ∎

Definition 1.2.21 Let μ be a fuzzy relation on σ, where σ is a fuzzy subset of a set S. Then μ is said to be **transitive** if $\mu^2 \subseteq \mu$.

From the definition, it is clear that for any fuzzy relation μ, μ^∞ is a transitive fuzzy relation.

Theorem 1.2.22 *Let ν, μ, and τ be fuzzy relations on a fuzzy subset σ of a set S. Then the following properties hold.*

(i) *If μ is transitive and $\nu \subseteq \mu$, $\tau \subseteq \mu$, then $\nu \circ \tau \subseteq \mu$.*

(ii) *If μ is transitive, then so is every power of μ.*

(iii) *If μ is transitive, τ is reflexive and $\tau \subseteq \mu$, then $\mu \circ \tau = \tau \circ \mu = \mu$.*

(iv) *If μ is reflexive and transitive, then $\mu^2 = \mu$.*

(v) *If μ is reflexive and transitive, then $\mu^0 \subseteq \mu = \mu^2 = \mu^3 = \cdots = \mu^\infty$.*

Proof (i) $(\nu \circ \tau)(x, z) = \vee \{\nu(x, y) \wedge \tau(y, z) \mid y \in S\} \leq \vee\{\mu(x, y) \wedge \mu(y, z) \mid y \in S\} = \mu^2(x, z) \leq \mu(x, z)$. Hence, $\nu \circ \tau \subseteq \mu$.

(ii) Assume that μ^n is transitive. Then $\mu^n \circ \mu^n \subseteq \mu^n$ and $\mu^{n+1} \circ \mu^{n+1} = \mu^{2n+2} = \mu^{2n} \circ \mu^2 \subseteq \mu^n \circ \mu = \mu^{n+1}$.

(iii) In (i), take ν to be μ. Then $\mu \circ \tau \subseteq \mu$. Also, $(\mu \circ \tau)(x, z) = \vee\{\mu(x, y) \wedge \tau(y, z) \mid y \in S\} \geq \mu(x, z) \wedge \tau(z, z) = \mu(x, z) \wedge \sigma(z) = \mu(x, z)$. That is, $\mu \circ \tau \supseteq \mu$ and hence $\mu \circ \tau = \mu$. Similarly, we can prove that $\tau \circ \mu = \mu$.

(iv) Follows from (iii).

(v) By (iv), $\mu = \mu^2$. Assume that $\mu^n = \mu^{n+1}$ for $n > 1$. Then $\mu^n \circ \mu = \mu^{n+1} \circ \mu$. Hence, $\mu^{n+1} = \mu^{n+2}$. ∎

Theorem 1.2.23 *Let ν, μ, and τ be fuzzy relations on a fuzzy subset σ of a set S. Then the following properties hold.*

(i) *If μ and τ are transitive and $\mu \circ \tau = \tau \circ \mu$, then $\mu \circ \tau$ is transitive.*

(ii) *If μ is symmetric and transitive, then $\mu(x, y) \leq \mu(x, x)$ and $\mu(y, x) \leq \mu(x, x)$ for all $x, y \in S$.*

(iii) *If μ is transitive, then for any $t \in [0, 1]$, μ^t is a transitive relation on σ^t.*

Proof (i) $(\mu \circ \tau) \circ (\mu \circ \tau) = \mu \circ (\tau \circ \mu) \circ \tau = \mu \circ (\mu \circ \tau) \circ \tau = \mu^2 \circ \tau^2 \subseteq \mu \circ \tau$. Thus, $\mu \circ \tau$ is transitive.

(ii) Because μ is transitive, $\mu \circ \mu \subseteq \mu$. Hence, $(\mu \circ \mu)(x, x) \leq \mu(x, x)$. That is, $\vee\{\mu(x, y) \wedge \mu(y, x) \mid y \in S\} \leq \mu(x, x)$. Because μ is symmetric, $\vee\{\mu(x, y) \wedge \mu(x, y) \mid y \in S\} \leq \mu(x, x)$. Thus, $\mu(x, y) \leq \mu(x, x)$. Because μ is symmetric, $\mu(y, x) \leq \mu(x, x)$.

(iii) Let $0 \leq t \leq 1$. Let $(x, y), (y, z) \in \mu^t$. Then $\mu(x, y) \geq t$ and $\mu(y, z) \geq t$. Therefore, $\mu(x, z) = \vee\{\mu(x, z) \wedge \mu(w, z) \mid w \in S\} \geq \mu(x, y) \wedge \mu(y, z) \geq t$. Thus, $(x, z) \in \mu^t$. ∎

1.3   Definitions and Basic Properties of Fuzzy Graphs

Let V be a nonempty set. Let $\mathcal{E}$ denote the set of all subsets of V with cardinality 2. Let $E \subseteq \mathcal{E}$. A graph is a pair (V, E). The elements of V are thought of as vertices of the graph and the elements of E as the edges. For $x, y \in V$, we let xy denote $\{x, y\}$. Then clearly $xy = yx$. We note that graph (V, E) has no loops or parallel edges.

Definition 1.3.1 A **fuzzy graph** $G = (V, \sigma, \mu)$ is a triple consisting of a nonempty set V together with a pair of functions $\sigma : V \to [0, 1]$ and $\mu : E \to [0, 1]$ such that for all $x, y \in V$, $\mu(xy) \leq \sigma(x) \wedge \sigma(y)$.

The fuzzy set σ is called the **fuzzy vertex set** of G and μ the **fuzzy edge set** of G. Clearly μ is a fuzzy relation on σ. We consider V as a finite set, unless otherwise specified. For notational convenience, we use simply G or (σ, μ) to represent the fuzzy graph $G = (V, \sigma, \mu)$. Also, σ^* and μ^*, respectively, represent the supports of σ and μ, also denoted by $\mathrm{Supp}(\sigma)$ and $\mathrm{Supp}(\mu)$.

Example 1.3.2 Let $V = \{a, b, c\}$. Define the fuzzy set σ on V as $\sigma(a) = 0.5$, $\sigma(b) = 1$ and $\sigma(c) = 0.8$. Define a fuzzy subset μ of $\mathcal{E}$ by $\mu(ab) = 0.5$, $\mu(bc) = 0.7$ and $\mu(ac) = 0.1$. Then $\mu(xy) \leq \sigma(x) \wedge \sigma(y)$ for all $x, y \in V$. Thus, $G = (\sigma, \mu)$ is a fuzzy graph. However, if we redefine $\mu(ab) = 0.6$, then it is no longer a fuzzy graph.

It follows from the Definition 1.3.1 that any unweighted graph (V, E) is trivially a fuzzy graph with $\sigma(x) = 1$ for all $x \in V$ and $\mu(xy) = 0$ or 1 for all $x, y \in V$. Also, we write (V, μ) to denote a fuzzy graph with $\sigma(x) = 1$ for all $x \in V$.

Definition 1.3.3 Let $G = (V, \sigma, \mu)$ be a fuzzy graph. Then a fuzzy graph $H = (V, \tau, \nu)$ is called a **partial fuzzy subgraph** of G if $\tau \subseteq \sigma$ and $\nu \subseteq \mu$. Similarly, the fuzzy graph $H = (P, \tau, \nu)$ is called a **fuzzy subgraph** of G induced by P if $P \subseteq V$, $\tau(x) = \sigma(x)$ for all $x \in P$ and $\nu(xy) = \mu(xy)$ for all $x, y \in P$. We write $\langle P \rangle$ to denote the fuzzy subgraph **induced** by P.

Example 1.3.4 Let $G = (\tau, \nu)$, where $\tau^* = \{a, b, c\}$ and $\mu^* = \{ab, bc\}$ with $\tau(a) = 0.4$, $\tau(b) = 0.8$, $\tau(c) = 0.5$, $\nu(ab) = 0.3$ and $\nu(bc) = 0.2$. Then clearly G is a partial fuzzy subgraph of the fuzzy graph in Example 1.3.2. Also, if $P = \{a, b\}$ and $H = (\tau, \nu)$, where $\tau(a) = 0.5$, $\tau(b) = 1$ and $\nu(ab) = 0.5$, then H is the induced fuzzy subgraph of G in Example 1.3.2, induced by P.

Definition 1.3.5 Let $G = (\sigma, \mu)$ be a fuzzy graph. Then a partial fuzzy subgraph (τ, ν) of G is said to **span** G if $\sigma = \tau$. In this case, we call (τ, ν) a **spanning fuzzy subgraph** of (σ, μ).

In fact a fuzzy subgraph $H = (\tau, \nu)$ of a fuzzy graph $G = (\sigma, \mu)$ induced by a subset P of V is a particular partial fuzzy subgraph of G. Take $\tau(x) = \sigma(x)$ for all $x \in P$ and 0 for all $x \notin P$. Similarly, take $\nu(xy) = \mu(xy)$ if xy is in a set of edges involving elements from P, and 0 otherwise.

Definition 1.3.6 Let $G = (V, \sigma, \mu)$ be a fuzzy graph. Let $0 \leq t \leq 1$. Let $\sigma^t = \{x \in \sigma^* \mid \sigma(x) \geq t\}$ and $\mu^t = \{uv \in \mu^* \mid \mu(uv) \geq t\}$.

Clearly, $\mu^t \subseteq \{xy \mid \sigma(x) \geq t, \sigma(y) \geq t\}$ and hence $H = (\sigma^t, \mu^t)$ is a graph with vertex set σ^t and edge set μ^t. H is called the **threshold graph** of the fuzzy graph G, corresponding to t.

Proposition 1.3.7 *Let $G = (\sigma, \mu)$ be a fuzzy graph and $0 \leq s < t \leq 1$. Then the threshold graph (σ^t, μ^t) is a subgraph of the threshold graph (σ^s, μ^s). Also, if $H = (\nu, \tau)$ is a partial fuzzy subgraph of G and $t \in [0, 1]$, then (ν^t, τ^t) is a subgraph of (σ^t, μ^t).*

1.4   Connectivity in Fuzzy Graphs

We revisit some connectivity concepts in this section. Theorem 1.4.1, by Rosenfeld gives a very strong characterization for a fuzzy bridge.

A **path** P in a fuzzy graph $G = (\sigma, \mu)$ is a sequence of distinct vertices $x_0, x_1, \ldots, x_n$ (except possibly x_0 and x_n) such that $\mu(x_{i-1}x_i) > 0$, $i = 1, \ldots, n$. Here n is called the **length** of the path. The consecutive pairs are called the **edges** of the path. The **diameter** of $x, y \in V$, written diam(x, y), is the length of the longest path joining x to y. The **strength** of P is defined to be $\wedge_{i=1}^n \mu(x_{i-1}x_i)$. In words, the strength of a path is defined to be the weight of the weakest edge. We denote the strength of a path P by $d(P)$ or $s(P)$. The strength of connectedness between two vertices x and y is defined as the maximum of the strengths of all paths between x and y and is denoted by $\mu^\infty(x, y)$ or $CONN_G(x, y)$. A **strongest path** joining any two vertices x, y has strength $\mu^\infty(x, y)$. It can be shown that if (τ, ν) is a partial fuzzy subgraph of (σ, μ), then $\nu^\infty \subseteq \mu^\infty$. We call P a **cycle** if $x_0 = x_n$ and $n \geq 3$. Two vertices that are joined by a path are called **connected**. It follows that this notion of connectedness is an equivalence relation. The equivalence classes of vertices under this equivalence relation are called **connected components** of the given fuzzy graph. They are just its maximal connected partial fuzzy subgraphs.

Let $G = (\sigma, \mu)$ be a fuzzy graph, let x, y be two distinct vertices and let G' be the partial fuzzy subgraph of G obtained by deleting the edge xy. That is, $G' = (\sigma, \mu')$, where $\mu'(xy) = 0$ and $\mu' = \mu$ for all other pairs. We call xy a **fuzzy bridge** in G if $\mu'^\infty(u, v) < \mu^\infty(u, v)$ for some u, v in σ^*. In words, the deletion of the edge xy reduces the strength of connectedness between some pair of vertices in G. Thus, xy is a fuzzy bridge if and only if there exist vertices u, v such that xy is an edge of every strongest path from u to v.

Let (V, E) be a graph. Let $xy \in E$. For a fuzzy graph $G = (\sigma, \mu)$, we write $G \backslash xy$ or $G \backslash \{xy\}$ for the fuzzy subgraph (σ, μ'), where $\mu'(xy) = 0$ and $\mu' = \mu$ otherwise. Let $x \in V$. We write $G \backslash x$ or $G \backslash \{x\}$ for the fuzzy subgraph (σ', μ') obtained by defining $\sigma'(x) = 0$, $\sigma' = \sigma$ otherwise and $\mu'(xy) = 0$ for all $xy \in E$, $\mu' = \mu$ otherwise.

Theorem 1.4.1 ([27]) *Let $G = (\sigma, \mu)$ be a fuzzy graph. Then the following statements are equivalent.*

   (i) xy is a fuzzy bridge.
   (ii) $\mu'^\infty(x, y) < \mu(xy)$.
   (iii) xy is not the weakest edge of any cycle.

Proof $(ii) \Rightarrow (i)$: If xy is not a fuzzy bridge, then $\mu'^\infty(x, y) = \mu^\infty(x, y) \geq \mu(xy)$.

$(i) \Rightarrow (iii)$: If xy is the weakest edge of a cycle, then any path P involving edge xy can be converted into a path P' not involving xy but at least as strong as P, by using the rest of the cycle as a path from x to y. Thus, xy cannot be a fuzzy bridge.

$(iii) \Rightarrow (ii)$: If $\mu'^{\infty}(x, y) \geq \mu(xy)$, then there is a path from x to y not involving xy, that has strength $\geq \mu(xy)$, and this path together with xy forms a cycle of G in which xy is a weakest edge. ∎

1.5 Connectedness in a Directed Graph

We wish to introduce in book form the notion of weakening members in intuitionistic fuzzy graphs. In order to do so it is necessary to review some results from [22, 30, 33].

We consider aspects of group structure, where by a group we mean a directed graph. We base our work on [30] in the next four sections. There are groups in which some members are seen as exercising a disruptive or divisive influence while other individuals appear to help hold the group together. An example is a set of countries in which some countries are highly involved in human trafficking while others attempt to control trafficking. We model this type of situation using various kinds of connectedness of a directed graph (or digraph). In such a directed graph, the vertices represent the members of the group and the directed lines the relationships between vertices. The group structure may be built from many different kinds of relationships, such as sociometric choice, communication, or power. We present a characterization of the way in which individual members contribute to the connectedness of the group.

Let V be a finite set and $E \subseteq V \times V$. Then (V, E) is a **directed graph** . We think of $(u, v) \in V \times V$ as a directed edge from u to v. We sometimes write $u \rightarrow v$ for (u, v). The notation $u \longleftrightarrow v$ is used when $(u, v) \in E$ and $(v, u) \in E$. We assume throughout that for all $(u, v) \in E$, $u \neq v$. We sometimes call a directed graph a **digraph** .

Definition 1.5.1 A digraph is **strongly connected** (or **strong**) if for every pair of distinct points, x and y, there exists a directed path from x to y and one from y to x.

Definition 1.5.2 A digraph is **unilaterally connected** (or **unilateral**) if for every pair of points, x and y, there is a directed path from x to y or one from y to x.

Definition 1.5.3 A digraph is **disconnected** if the points can be divided into two sets with no line joining any point in one set with a point in the other set.

Definition 1.5.4 A digraph is **weakly connected** (or **weak**) if it is not disconnected.

Let U_3 be the set of all strong digraphs, U_2 the set of all unilateral digraphs, and U_1 the set of all weak digraphs. Let U_0 denote the collection of all disconnected digraphs. Then clearly, $U_3 \subseteq U_2 \subseteq U_1$ and $U_0 \cap U_1 = \emptyset$. Let

$$C_3 = U_3, C_2 = U_2 \backslash U_3, \text{ and } C_1 = U_1 \backslash U_2.$$

In order to evaluate the effect of a point on the connectedness of its digraph, we need a precise definition of the removal of a vertex of a digraph.

Definition 1.5.5 Let D be a digraph and x a vertex in it. Then $D \backslash x$ denotes the digraph obtained from D by deleting the vertex x and all edges which are either directed toward x or away from x. We say x is a vertex of type P_{ij} if the digraph D is in class C_i but $D \backslash x$ is in class C_j.

Definition 1.5.6 A point of a digraph is a **strengthening point** if and only if it is of type P_{ij} such that $i > j$; it is a **weakening point** if $i < j$ and the point is called **neutral** if $i = j$.

If a point is of the type P_{ij}, we say that the corresponding individual is an (i, j) member.

When we contrast D and $D \backslash x$, we assume that the edges between pairs of distinct points exist independently.

Whenever a line between two points of any digraph is drawn without any direction displayed, it stands for two directed lines, one in each direction.

1.6 Strengthening and Weakening Members of a Group

By a group, we mean a universal set whose elements are of interest to a particular study. The elements may be people or countries for example. The elements of the group become the vertices of the directed graph.

In this section, we characterize strengthening and weakening group members. We first show that there are no $(1, 3)$ members in a group. We also introduce the concept of the "reachability matrix" of a group. This matrix R is useful for the expression of matrix conditions that characterize the inclusive connectedness categories, strong, unilateral, weak, and disconnected. A straightforward modification of these theorems leads to a description of the exclusive connectedness categories C_3, C_2, C_1, and C_0. We then obtain conditions for each of the (i, j) types. It turns out that weakening members may be identified from direct examination of the reachability matrix.

Example 1.6.1 The graphs in Fig. 1.1 illustrate all the possible types of (i, j) members that can occur except for $(1, 3)$ types.

Theorem 1.6.2 *There are no* $(1, 3)$ *members in any group.*

Proof Suppose that there exists a $(1, 3)$ member x of group G. Then G is in class C_1 and $G \backslash x$ is in class C_3. That is, $G \backslash x$ is strong. However, if we join the point x to the group $G \backslash x$ by at least one line, then any point of G can either reach x or x can reach very point of $G \backslash x$. Hence, G must be at least unilateral, i.e., in either class C_2 or C_3. This contradicts the hypothesis that G is in class C_1. ∎

Fig. 1.1 Types of points

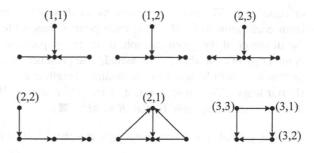

Let $x_1, x_2, \ldots, x_n$ be the n members of the group. Let the matrix $M = [m_{ij}]$ of the group relationship be defined as follows for $i \neq j$:

$$m_{ij} = \begin{cases} 1 & \text{if } (x_i, x_j) \in E. \\ 0 & \text{otherwise,} \end{cases}$$

and $m_{ii} = 1, i = 1, \ldots, n$.

Define the matrix R as follows. The i, j element of R is taken to be 1 if there exists a directed path from x_i to x_j and is 0 otherwise.

The **diameter** d of a digraph D is the greatest distance between any two points of D.

A **loop** is a directed line which begins and ends at the same point. Our convention that all the diagonal elements of M are 1 is equivalent to the existence of a loop at every point.

We treat M as Boolean matrix, i.e., one in which all the elements are either 1 or 0. The usual addition and multiplication operations of matrices are applied to Boolean matrices using Boolean arithmetic. That is, $1 + 1 = 1, 1 + 0 = 1 = 0 + 1, 0 + 0 = 0, 1 \bullet 1 = 1, 0 \bullet 1 = 0 = 1 \bullet 0, 0 \bullet 0 = 0$. In this chapter, all matrices are Boolean, and all the operations are performed as described above.

A **redundant chain** from vertex x_1 to vertex x_k is a sequence of directed edges of the form $x_1 \rightarrow x_2 \rightarrow x_3 \rightarrow \cdots \rightarrow x_{k-1} \rightarrow x_k$, where the vertices from x_1 to x_k are not all distinct.

The following theorem gives a formula for the reachability matrix of a group in terms of its relationship matrix.

Theorem 1.6.3 $R = M^d$.

Proof The matrix M gives all the directed paths of length 1 (or 0) between distinct pairs of points. Hence, if $d = 1$, then clearly $R = M^1$. Assume the matrix M^k gives all paths of length k or less, i.e., if $M^k = [m_{ij}^{(k)}]$ with $m_{ij}^{(k)} = 1$ if and only if there is a path from x_i to x_j of length less than or equal to k. Then $M^{k+1} = M^k M$ and $m_{ij}^{(k)} m_{jk} = 1$ if and only if there is a path from x_{ij} to x_{jk} of length less than or equal to $k + 1$. Hence, by induction it follows that for any positive integer h, $M^h = [m_{ij}^{(h)}]$ is such that $m_{ij}^{(h)} = 1$ if and only if there is a path from x_i to x_j of length less than or equal to h. The matrix M contains all these paths and also a path of length 1 from each point to itself. Thus, each point is reachable from itself. Because d is the diameter of the given digraph, there are no paths of length greater than d that connect points not already connected. The presence of loops at each point assures us that there will be a redundant chain of length d whenever there is a path of any shorter length. Therefore, this last matrix M^d contains all reachability relationships that occur in the digraph. Hence, $R = M^d$. ∎

Corollary 1.6.4 *If k is any integer greater than d, then $M^k = R$.*

We now apply Corollary 1.6.4 to get a working procedure for computing R. After finding M, calculate M^2, M^4, M^8, and so on. As soon as two consecutive matrices of this sequence coincide, we have found R. This procedure usually calculates R in the fewest possible matrix operations.

Corollary 1.6.5 $R = (I - M)^{-1}$.

Proof It is well-known that $(I - M)^{-1} = I + M + M^2 + \cdots + M^d + M^{d+1} + \cdots$. However, $M^d + M^{d+1} + \cdots = M^d$ because we are using Boolean addition. Also, $I + M + M^2 + \cdots + M^d = M^d$. Hence, $(I - M)^{-1} = M^d = R$ by Theorem 1.6.3. ∎

We note that if N is the matrix obtained from M by having 0's on the diagonal, then $R = (-N)^{-1}$.

Example 1.6.6 Let D be a digraph with vertex set $V = \{x, y, z, w\}$. Let the directed edges of D be defined by the graph in Figure 1–2 or the adjacency matrix M.
Thus,

$$
M = \begin{array}{c} \\ x \\ y \\ z \\ w \end{array}
\begin{array}{cccc}
x & y & z & w \\
\end{array}
\left[\begin{array}{cccc}
1 & 1 & 0 & 0 \\
0 & 1 & 0 & 0 \\
0 & 1 & 1 & 0 \\
0 & 0 & 0 & 1
\end{array} \right].
$$

It follows easily that $R = M^2 = M$. We see that D is disconnected and that $D \backslash w$ is weakly connected. Thus, w is of type $(0, 1)$ and so w is a weakening point.

Fig. 1.2 Weakening point

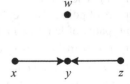

Fig. 1.3 Directed graph D

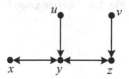

1.7 Inclusive Connectedness Categories

The next four theorems characterize the inclusive connectedness categories. Let W denote the n by n matrix in which every entry is 1, where n is the number of vertices of the digraph. Let $R = [r_{ij}]$ and R^T denote the transpose of R.

Theorem 1.7.1 *The group G is strong if and only if $R = W$.*

Proof The group G is strong if and only if for all vertices x_i and x_j there exists a path in both directions. However, this condition holds if and only if for all values of $r_{ij} = 1 = r_{ji}$. Therefore, G is strong if and only if all the entries in R are equal to 1, i.e., $R = W$. ∎

Example 1.7.2 Let D be a digraph with vertex set $V = \{x, y, z, u, v\}$. Let the directed edges of D be defined by the graph in Fig. 1.3 or the adjacency matrix M.

Thus,

$$M = \begin{array}{c} \\ x \\ y \\ z \\ u \\ v \end{array} \begin{array}{c} \begin{array}{ccccc} x & y & z & u & v \end{array} \\ \left[\begin{array}{ccccc} 1 & 1 & 0 & 0 & 0 \\ 1 & 1 & 1 & 0 & 0 \\ 0 & 1 & 1 & 1 & 0 \\ 0 & 1 & 0 & 1 & 1 \\ 0 & 0 & 1 & 0 & 1 \end{array} \right] \end{array}.$$

It follows that $M^4 = R = W$.

The transpose M^T of a matrix M is obtained by reflecting the entries of M about its diagonal, i.e., by interchanging the rows and columns of M . The next theorem uses the matrix $R + R^T$.

Theorem 1.7.3 *The group G is unilateral if and only if $R + R^T = W$.*

Proof Let x_i and x_j be vertices. Then G is unilateral if and only if there is a path from x_i to x_j or from x_j to x_i, i.e., if and only if $r_{ij} = 1$ or $r_{ji} = 1$. The latter statement is equivalent to $r_{ij} + r_{ji} = 1$. Thus, G is unilateral if and only if $R + R^T = W$. ∎

Example 1.7.4 Let D be a digraph with vertex set $V = \{x, y, z, w\}$. Let the directed edges of D be defined by the graph in Fig. 1.4 or the adjacency matrix M.

Then

Fig. 1.4 Directed graph D

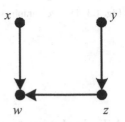

It follows that $d = 2$ and that

$$M = \begin{array}{c} \\ x \\ y \\ z \\ w \end{array} \begin{array}{cccc} x & y & z & w \\ \left[\begin{array}{cccc} 1 & 0 & 0 & 1 \\ 0 & 1 & 1 & 0 \\ 0 & 0 & 1 & 1 \\ 0 & 0 & 0 & 1 \end{array}\right] \end{array} \text{ and } M^T = \begin{array}{c} \\ x \\ y \\ z \\ w \end{array} \begin{array}{cccc} x & y & z & w \\ \left[\begin{array}{cccc} 1 & 0 & 0 & 0 \\ 0 & 1 & 0 & 0 \\ 0 & 1 & 1 & 0 \\ 1 & 0 & 1 & 1 \end{array}\right] \end{array}.$$

It follows that $d = 2$ and that

$$R = \begin{array}{c} \\ x \\ y \\ z \\ w \end{array} \begin{array}{cccc} x & y & z & w \\ \left[\begin{array}{cccc} 1 & 0 & 0 & 1 \\ 0 & 1 & 1 & 1 \\ 0 & 0 & 1 & 1 \\ 0 & 0 & 0 & 1 \end{array}\right] \end{array} \text{ and } R^T = \begin{array}{c} \\ x \\ y \\ z \\ w \end{array} \begin{array}{cccc} x & y & z & w \\ \left[\begin{array}{cccc} 1 & 0 & 0 & 0 \\ 0 & 1 & 0 & 0 \\ 0 & 1 & 1 & 0 \\ 1 & 1 & 1 & 1 \end{array}\right] \end{array}.$$

Hence,

$$R + R^T = \begin{array}{c} \\ x \\ y \\ z \\ w \end{array} \begin{array}{cccc} x & y & z & w \\ \left[\begin{array}{cccc} 1 & 0 & 0 & 1 \\ 0 & 1 & 1 & 1 \\ 0 & 1 & 1 & 1 \\ 1 & 1 & 1 & 1 \end{array}\right] \end{array} \neq W.$$

In order to characterize weak groups, we introduce the digraph D^- obtained from D as follows: First, D^- is a supergraph of D, i.e., all points and lines of D are in D^-. Second, if the directed edge $x \rightarrow y$ is in D, then both of the edges $x \rightarrow y$ and $y \rightarrow x$ are in D^-. This process may be called *symmetrizing* the digraph D. Let M^- and R^- be the relationship matrix and the reachability matrix of D^-, respectively. Then $M^- = M + M^T$.

Theorem 1.7.5 *A group is weak if and only if $R^- = W$.*

Proof Clearly, D is weak if and only if D^- is strong. Therefore, by Theorem 1.7.1, D is weak if and only if $R^- = W$. ∎

Example 1.7.6 Let D be the digraph defined in Example 1.7.4. Then

$$M^- = M + M^T = \begin{array}{c} \\ x \\ y \\ z \\ w \end{array} \begin{array}{cccc} x & y & z & w \\ \left[\begin{array}{cccc} 1 & 0 & 0 & 1 \\ 0 & 1 & 1 & 0 \\ 0 & 1 & 1 & 1 \\ 1 & 0 & 1 & 1 \end{array}\right] \end{array}.$$

It follows that $R^- = (M^-)^3 = W$. Hence, the group is weak.

Theorem 1.7.7 *A group is disconnected if and only if $R^- \neq W$.*

Proof By definition, a group is disconnected if and only if it is not weak. Therefore, by Theorem 1.7.5, it follows that a group is not disconnected if and only if $R^- \neq W$. ∎

Example 1.7.8 Let D be the digraph of Example 1.6.6. Then D is disconnected. We see that

$$M^- = M + M^T = \begin{array}{c} \\ x \\ y \\ z \\ w \end{array} \begin{array}{cccc} x & y & z & w \\ \left[\begin{array}{cccc} 1 & 1 & 0 & 0 \\ 1 & 1 & 1 & 0 \\ 0 & 1 & 1 & 0 \\ 0 & 0 & 0 & 1 \end{array}\right] \end{array} \text{ and}$$

$$(M^-)^2 = \begin{array}{c} \\ x \\ y \\ z \\ w \end{array} \begin{array}{cccc} x & y & z & w \\ \left[\begin{array}{cccc} 1 & 1 & 1 & 0 \\ 1 & 1 & 1 & 0 \\ 1 & 1 & 1 & 0 \\ 0 & 0 & 0 & 1 \end{array}\right] \end{array} = R^- \neq W.$$

1.8 Exclusive Connectedness Categories

We characterize the exclusive categories C_3, C_2, C_1, and C_0 by combining the last four theorems.

Theorem 1.7.7 gives a criterion for a group G to be in C_0. However, for G to be in C_1, the condition of Theorem 1.7.5 holds while that of Theorem 1.7.3 does not. Similarly, G is in C_2 if and only if Theorem 1.7.3 holds while Theorem 1.7.1 does not. Finally, G is in C_3 if and only if Theorem 1.7.1 holds. We summarize these results in the following table.

Exclusive Classes	Criterion
C_3	$R = W$
C_2	$R + R^T = W,\ R \neq W$
C_1	$R^- = W,\ R + R^T \neq W$
C_0	$R^- \neq W$

We next describe further the weakening members of a group. All the possible kinds of weakening members are:

The $(0, j)$ members for $j = 1, 2$, or 3, and the $(1, 2)$, and $(2, 3)$ members.

Recall from Theorem 1.6.2 that there are no $(1, 3)$ members.

There is independent interest in identifying the **isolates** in a group, i.e., the members of the group with no incoming or outgoing edges. The following result is immediate.

Theorem 1.8.1 *The member x_i is an isolate if and only if the only nonzero element in the ith row and the ith column of R is r_{ii}.*

Theorem 1.8.2 *(i) x_i is a $(0, j)$ member for $j = 1, 2$, or 3 if and only if x_i is an isolate and $G \setminus x_i$ is in C_j.*

(ii) x_i is a $(1, 2)$ member if and only if $R + R^T$ has at least one zero and all its zeros occur in the ith row and in the ith column.

(iii) x_i is a $(2, 3)$ member if and only if every element in R is a 1 except in the ith row (or column) of R, where all but the diagonal element are 0.

Proof (i) If x_i is an isolate and $G \setminus x_i$ is in C_j, then x_i is a $(0, j)$ member by definition.

Conversely, suppose x_i is an $(0, j)$ member for $j > 0$ and assume x_i is not an isolate. Then $G \setminus x_i$ is still disconnected. Hence, x_i is a $(0, 0)$ member which is a contradiction.

(ii) If $R + R^T$ has at least one zero and all its zeros occur in the ith row and in the ith column, then by Table 1, G is in C_1 and $G \setminus x_i$ is in C_2.

Conversely, suppose that x_i is a $(1, 2)$ member of G. Then G is in C_1 so that $R + R^T$ must have at least one 0. Assume that there is a 0 which is neither in the ith row nor in the ith column of $R + R^T$. If x_i is eliminated from $R + R^T$, then all 0's not in the ith row or column would remain because the elimination of a member and its bonds cannot add connections to members of G. Therefore, $G \setminus x_i$ would be in C_1, a contradiction.

(iii) The criteria of Table 1 show that under the stated conditions x_i is a $(2, 3)$ member.

Conversely, let x_i be a given $(2, 3)$ member. Then either x_i can reach all other members but no other members can reach x_i or vice versa. In the first case, all the elements in the ith row are 1 and all the nondiagonal elements in the ith column are 0, while every element of R not in the ith row or column is a 1. For the other case, the result follows by interchanging the words "row" and "column." ∎

It follows that $(0, j)$ members can be identified by considering the condition met by the remainder of the matrices R, $R + R^T$, and R^- after the row and column belonging to an isolate are deleted.

Corollary 1.8.3 *The group consisting of exactly two isolates has two $(0, 3)$ members. Any other disconnected group has at most one weakening member.*

Corollary 1.8.4 *A C_1 group has at most two $(1, 2)$ members.*

Corollary 1.8.5 *A C_2 group has at most two (2, 3) members.*

Because weakening members are defined in terms of the kinds of connectedness introduced here. The next result follows from the previous theorem because the possibilities of the three parts listed there are exhaustive and mutually exclusive.

Theorem 1.8.6 *Any group has at most two weakening members.*

There are no analogous theorems describing the strengthening members of a group. However, we can still identify these strengthening members by using the results of Table 1 on the given group G and on $G \backslash x$. In particular, the strengthening members in classes $(i, 0)$ for $i = 1, 2$, or 3 correspond to the liaison persons [29] of the symmetrized group.

1.9   Articulation Points

In [29], Ross and Harary considered articulation points of a crisp undirected graph. These are points whose removal disconnects the graph. The structure of an organization can be expressed by considering whether a given symmetric relationship exists between each pair of members. Some examples of such a relation are mutual influence, two-way communication, and reciprocated sociometric choice. The results in [29] can be applied here for unilaterally crisp directed graphs because these latter graphs are equivalent to crisp undirected graphs. Let $G = (V, E)$ be a connected graph. Recall that the **distance** between two points is the length of any shortest path joining them. We denote the distance between points x and y as $d(x, y)$ in a connected graph. The matrix of distances of a given graph, denoted M_d, is that matrix in which the number appearing in the ith row and jth column is the distance between the ith point and the jth point.

The matrix M_d can be determined in the following manner. Let M denote the adjacency matrix of G. We assume the diagonal entries of the adjacency matrix M to be 0. In this section, we use ordinary addition and multiplication rather than Boolean. Let k be a positive integer and M^k the kth power of M. Let k be the smallest power of M in which for each pair (i, j) there occurs at least one non-zero entry in the ith row and jth column of the matrices $M, M^2, \ldots, M^k$. Then the element in the ith row and jth column of M_d is the exponent of the smallest power of M in which the i, j entry is not zero if $i \neq j$ and is zero if $i = j$.

For every point x of V, let $\alpha(x) = \vee\{d(x, y) \mid y \in V\}$. A **peripheral** point of a graph is a point of maximal associated number, i.e., $\vee\{\alpha(x) \mid x \in V\}$. A **relatively peripheral** point y with respect to x has the property that $d(y, x) = \alpha(x)$, i.e., there is no point farther from x than y.

Theorem 1.9.1 *If y is a relatively peripheral point, then y is not an articulation point.*

Fig. 1.5 A graph

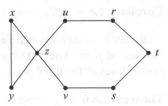

Proof Let y be a relatively peripheral point to x. Then y is not an articulation point else there would exist a point z farther from x than y. ∎

Corollary 1.9.2 *If y is a peripheral point, then y is not an articulation point.*

Proof Every peripheral point is relatively peripheral with respect to some point. ∎

The use of the matrix of distances is useful in applying Theorem 1.9.1. The points given by the columns that contain the maximum numbers appearing in each row are not articulation points.

Example 1.9.3 Consider the graph given in Fig. 1.5.
Then

$$
M_d = \begin{array}{c} \\ x \\ y \\ z \\ u \\ v \\ r \\ s \\ t \end{array}
\begin{array}{c} \begin{array}{cccccccc} x & y & z & u & v & r & s & t \end{array} \\
\left[\begin{array}{cccccccc}
0 & 1 & 1 & 2 & 2 & 3 & 3 & 4 \\
1 & 0 & 1 & 2 & 2 & 3 & 3 & 4 \\
1 & 1 & 0 & 1 & 1 & 2 & 2 & 3 \\
2 & 2 & 1 & 0 & 2 & 1 & 3 & 2 \\
2 & 2 & 1 & 2 & 0 & 3 & 1 & 2 \\
3 & 3 & 2 & 1 & 3 & 0 & 2 & 1 \\
3 & 3 & 2 & 3 & 1 & 2 & 0 & 1 \\
4 & 4 & 3 & 2 & 2 & 1 & 1 & 0
\end{array} \right]
\end{array}.
$$

We see that z is the only articulation point.

Theorem 1.9.4 *If there exists a unique point x at a given distance less than $\alpha(y)$ from any point y, then x is an articulation point.*

Proof Let $S_1 = \{z \in V \mid d(z, y) < d(x, y)\}$ and $S_2 = \{z \in V \mid d(z, y) > d(x, y)\}$. Then $S_1 \neq \emptyset$ and $S_2 \neq \emptyset$ because $y \in S_1$ and S_2 contains a point peripheral with respect to y. The removal of x from the graph separates S_1 and S_2. Hence, x is an articulation point. ∎

Corollary 1.9.5 *If a point x is adjacent to exactly one other point y, then y is an articulation point.*

We now develop a method for the identification of liaison persons, based entirely on distances. We assume the liaison character of points is not determined by preceding methods. Let w be an arbitrary point of the network whose liaison character is not already determined by previous methods. Let x be any point of maximal distance from w. It can be shown that there is more than one such point x because w is taken as a point whose character is not determined. The point x is held fixed in what follows.

Let $S = \{y \in V \mid d(y, x) < d(y, w) + d(w, x)\}$. Then each y in S has the property that there is a shortest path not passing through w which joins y with x.

Let T_0 be the set of all points y_0 such that $d(w, y_0) = 1$ and $d(y_0, x) = d(w, x) + 1$. Then each point of T_0 is adjacent to w and is on the opposite side of w from x. The set T_0 may contain more than one point. Not all points adjacent to w are in T_0. Let $T_{-1} = \{w\}$.

For all $y_0 \in T_0$, let $T_1(y_0) = \{y_1 \in V \mid d(y_1, y_0) = d(y_0, w)$ and $w \neq y_1 \neq w\}$. Thus, each point y_1 in $T_1(y_0)$ is adjacent to y_0, but is different from w and y_0. The following recursive definition of the set $T_n(y_0)$ in terms of the sets $T_k(y_0)$ for $k < n$ is simplified by taking sets $T_0(y_0)$ and $T_{-1}(y_0)$ as consisting of the single points y_0 and w, respectively.

In general, let $T_n(y_0) = \{y_n \in V \mid d(y_n, y_{n-1}) = d(y_{n-1}, w), y_{n-1} \in T_{n-1}(y_0)\}$ $\setminus \{y_n \in V \mid \exists k < n, y_n \in T_k(y_0)\}$. That is, $T_n(y_0)$ is the set of vertices y_n such that $d(y_n, y_{n-1}) = d(y_{n-1}, w)$, where y_{n-1} is in $T_{-1}(y_0)$ and y_n is not in any of the previous $T_k(y_0)$ for $k < n$.

Theorem 1.9.6 *There exists y_0 in T_0 such that the collection of all the points in each of the sets $T_k(y_0)$ is disjoint with the set S if and only if w is an articulation point.*

Proof Assume that w is not an articulation point. Then by definition of articulation point, each point y_0 is joined to x by a path not containing w. But every such path contains at least one point which lies in both S and in some $T_k(y_0)$.

Conversely, for all $x \in V$, the articulation point w must have at least one point y_0 such that any path joining y_0 with x passes through w. ∎

Example 1.9.7 Consider the graph of Example 1.9.3. Consider the point z. The point t is of maximal distance from z. Then the set S consists of those points p such that $d(p, t) < d(p, z) + d(z, t)$. Thus, $S = \{u, v, r, s\}$. Now T_0 is the set of all points y_0 such that $d(z, y_0) = 1$ and $d(y_0, t) = d(z, t) + 1$. Hence, $T_0 = \{x, y\}$. Let $T_{-1} = \{z\}$. Now $T_1(x) = \{y\}$ and $T_1(y) = \{x\}$. Also $T_2(x)$ is the set of points whose distance from y equals $d(y, z) = 1$ and which do not belong to $T_1(x)$. Hence, $T_2(x) = \emptyset$. Because $T_1(x) \cap S = \emptyset$, it follows that z is an articulation point.

Consider the point r. Then x is a point of maximal distance from r. In fact, $d(r, x) = 3$. Those points p such that $d(p, x) < d(p, r) + d(r, x)$ make up the set S. Thus, $S = \{x, y, z, u, v, s\}$. Now T_0 is the set of all points y_0 such that $d(r, y_0) = 1$ and $d(y_0, x) = d(r, x) + 1$. Thus, candidates for S are u and t. Now $d(u, x) \neq d(r, x) + 1$ and $d(t, x) = d(r, x) + 1$. Thus, $T_0 = \{t\}$. Now $y_1 \in T_1(t)$ if and only if $d(y_1, t) = d(t, r) = 1$. Hence, $T_1(t) = \{s\}$. Thus, it follows that r is not an articulation point because $S \cap T_1(t) \neq \emptyset$.

Corollary 1.9.8 *If T_0 is the empty set, then w is not an articulation point.*

Proof Because T_0 is empty, all points of the graph are connected to x by a shortest path that does not go through w. Hence, w is not an articulation point. ∎

1.10   An Application of Fuzzy Graphs to the Problem
Concerning Group Structure

As discussed previously, the concept of a fuzzy graph is often a more relevant mathematical model than a crisp one. In this section, we consider connectedness of a fuzzy graph. The concepts of weakening and strengthening points of a fuzzy graph are introduced and their fundamental properties are investigated. The results in this section are based on [33].

We first review various kinds of connectedness of directed graphs.

We note that a digraph G consisting of exactly one point is strong, because it does not contain two distinct points, the definition is vacuously satisfied. Recall that U_3, U_2, U_1, and U_0 are collections of all strong digraphs, all unilateral digraphs, all weak digraphs, and all disconnected digraphs, respectively. Clearly, we have

$U_3 \subseteq U_2 \subseteq U_1$.

Recall also that

$C_3 = U_3$, $C_2 = U_2 \backslash U_3$, $C_1 = U_1 \backslash U_2$,   and $C_0 = U_0$.

Then each digraph belongs to exactly one of the above categories $C_i, i = 0, 1, 2, 3$.

As previously mentioned, one may be concerned with a group where a class of group members being in relationship with any given member is gradual rather than abrupt. A fuzzy graph may be utilized to represent such a group.

Definition 1.10.1 Let V be a finite set and let Γ be a function of V into the set of all fuzzy subsets of V. Then $G = (V, \Gamma)$ is called a **fuzzy directed graph** .

Definition 1.10.2 Let $G = (V, \Gamma)$ be a fuzzy directed graph. A pair (Y, Γ'), where Y is a subset of V and the function Γ' is defined by $\Gamma' = \Gamma|_Y$ is called a **fuzzy directed subgraph** of G.

Definition 1.10.3 Let σ be a fuzzy subset of V. Define the fuzzy subsets Γ_σ and Γ_σ^{-1} of V as follows: for all $x \in V$,

$$\Gamma_\sigma(x) = \vee\{\sigma(y) \wedge \Gamma_y(x) \mid y \in V\},$$
$$\Gamma_\sigma^{-1}(x) = \vee\{\sigma(y) \wedge \Gamma_x(y) \mid y \in V\}.$$

Suppose $x \in V$. Then $\Gamma(x)$ is a fuzzy subset of V. If $\Gamma(x)(y) > 0$, then we think of (x, y) as a directed edge from x to y with intensity $\Gamma(x)(y)$.

Proposition 1.10.4 *Let σ and τ be fuzzy subsets of V. The following properties hold.*
  (i)   *If $\sigma \subseteq \tau$, then $\Gamma_\sigma \subseteq \Gamma_\tau$.*
  (ii) *If $\sigma \subseteq \tau$, then $\Gamma_\sigma^{-1} \subseteq \Gamma_\tau^{-1}$.*
  (iii) $\Gamma_{\sigma \cap \tau} \subseteq \Gamma_\sigma \cap \Gamma_\tau$.
  (iv) $\Gamma_{\sigma \cap \tau}^{-1} \subseteq \Gamma_\sigma^{-1} \cap \Gamma_\tau^{-1}$.
  (v) $\Gamma_{\sigma \cup \tau} = \Gamma_\sigma \cup \Gamma_\tau$.
  (vi) $\Gamma_{\sigma \cup \tau}^{-1} = \Gamma_\sigma^{-1} \cup \Gamma_\tau^{-1}$.

Proof Properties (i) and (ii) are obvious from Definition 1.10.3. Properties (iii) and (iv) follow directly from (i) and (ii), respectively.

(v)

$$\Gamma_{\sigma \cup \tau}(x) = \vee\{(\sigma(x) \vee \tau(x)) \wedge \Gamma_y(x) \mid y \in V\}$$
$$= (\vee\{\sigma(x) \wedge \Gamma_y(x) \mid y \in V\}) \vee (\vee\{\tau(x) \wedge \Gamma_y(x) \mid y \in V\})$$
$$= \Gamma_\sigma(x) \vee \Gamma_\tau(x)$$
$$= (\Gamma_\sigma \cup \Gamma_\tau)(x).$$

(vi)

$$\Gamma^{-1}_{\sigma \cup \tau}(x) = \vee\{(\sigma(x) \vee \tau(x)) \wedge \Gamma_x(y) \mid y \in V\}$$
$$= (\vee\{\sigma(x) \wedge \Gamma_x(y) \mid y \in V\}) \vee (\vee\{\tau(x) \wedge \Gamma_x(y) \mid y \in V\})$$
$$= \Gamma^{-1}_\sigma(x) \vee \Gamma^{-1}_\tau(x)$$
$$= (\Gamma^{-1}_\sigma \cup \Gamma^{-1}_\tau)(x).$$

∎

Let $G = (\sigma, \Gamma)$ be a fuzzy directed graph. If $x, y \in V$, we sometimes write $\Gamma(x, y)$ for $\Gamma_x(y)$. Then we can consider Γ as a fuzzy subset of $V \times V$, where (x, y) is considered to be a directed edge from x to y. Define the fuzzy subset Γ^{-1} of $v \times V$ by for all $(x, y) \in V \times V$, $\Gamma^{-1}(x, y) = \Gamma(y, x)$.

Definition 1.10.5 Let $G = (\sigma, \Gamma)$ be a fuzzy directed graph. Let Δ be the fuzzy subset of $V \times V$ defined by for all $(x, y) \in V \times V$, $\Delta(x, y) = \Gamma(x, y) \vee \Gamma^{-1}(x, y)$. Let $\widehat{\Gamma}, \widehat{\Gamma^{-1}}$, and $\widehat{\Delta}$ denote the transitive closure of Γ, Γ^{-1}, and Δ, respectively.

It follows that for all $x \in V$,

$$\widehat{\Gamma}_x = \Gamma^0_x \cup \Gamma^1_x \cup \Gamma^2_x \cup \cdots \cup \Gamma^{n-1}_x,$$
$$\widehat{\Gamma^{-1}}_x = \Gamma^0_x \cup \Gamma^{-1}_x \cup \Gamma^{-2}_x \cup \cdots \cup \Gamma^{-n+1}_x,$$

where for all $x, y \in V$, $\Gamma^0_x(y) = 0$ if $x \neq y$ and $\Gamma^0_x(y) = 1$, if $x = y$.

The numbers $\widehat{\Gamma}_x(y)$ and $\widehat{\Gamma^{-1}}_x(y)$ may be interpreted as the degree of a strongest directed path from x to y and that from y to x, respectively. Similarly, we have for all $x \in V$,

$$\widehat{\Delta}_x = \Delta^0_x \cup \Delta^1_x \cup \Delta^2_x \cup \cdots \cup \Delta^{n-1}_x.$$

The value $\widehat{\Delta}_x(y)$ may be interpreted as the degree of a semipath from x to y, whereby a semipath we mean an alternating sequence $v_0, e_0, v_1, \ldots, e_m, v_m$ of vertices v_i and directed edges e_i where each edge may be either (v_{i-1}, v_i) or (v_i, v_{i-1}).

Definition 1.10.6 The **grades of membership** of a fuzzy graph $G = (V, \Gamma)$ in U_3, U_2, U_1, and U_0 are defined by

$$\mu_{U_3}(G) = \ \wedge \{\widehat{\Gamma}_x(y) \mid x, y \in V\},$$
$$\mu_{U_2}(G) = \ \wedge \{\widehat{\Gamma}_x(y) \vee \widehat{\Gamma}_y(x) \mid x, y \in V\},$$
$$\mu_{U_1}(G) = \ \wedge \{\widehat{\Delta}_x(y) \mid x, y \in V\},$$
$$\mu_{U_0}(G) = 1 - \wedge \{\widehat{\Delta}_x(y) \mid x, y \in V\},$$

respectively.
It follows that

$$\mu_{U_3}(G) \le \mu_{U_2}(G) \le \mu_{U_1}(G) \tag{1.1}$$

for any $G = (X, \Gamma)$.

We can have that for any crisp digraph G in $C_i, \mu_{U_j}(G) = 0$ for $3 \ge j > i$; $\mu_{U_j}(G) = 1$ for $i \ge j \ge 1$.

1.11 Weakening and Strengthening Points of a Fuzzy Graph

In this section, we define weakening and strengthening points of a fuzzy directed graph.

Definition 1.11.1 Let $G = (V, \Gamma)$ be a fuzzy directed graph and let G_k be the fuzzy subgraph $(V \backslash \{x\}, \Gamma')$ obtained from G by the removal of a point x, where Γ' is defined in Definition 1.10.2. Then the point x is a **weakening point** for U_i (a W_i point, for short) if $\mu_{U_i}(G) < \mu_{U_i}(G_k)$; it is a **neutral point** for U_i (an N_i point) if $\mu_{U_i}(G) = \mu_{U_i}(G_k)$; and it is a **strengthening point** for U_i (an S_i point) if $\mu_{U_i}(G) > \mu_{U_i}(G_k)$, where $i = 1, 2, 3$.

Example 1.11.2 Let $G = (V, \Gamma)$ be the fuzzy directed graph with $V = \{x_1, x_2, x_3, x_4, x_5\}$ and with Γ defined as follows:

$$\Gamma(x_1, x_2) = \Gamma(x_2, x_1) = 1, \ \Gamma(x_2, x_5) = \Gamma(x_5, x_2) = 3/4,$$
$$\Gamma(x_4, x_2) = 1/2, \ \Gamma(x_5, x_4) = 3/4, \ \Gamma(x_4, x_3) = 1, \ \Gamma(x_3, x_5) = 3/4,$$

$\Gamma(w, w) = 1$ for all $w \in X$, and $\Gamma(u, v) = 0$ elsewhere. Then G is given by Fig. 1.6.

Fig. 1.6 Graph G

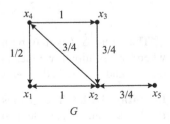

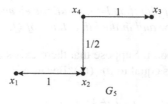

Fig. 1.7 Graph G_5

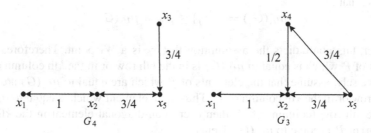

Fig. 1.8 Graphs G_4 and G_3

Now G_5 is given by Fig. 1.7,
and G_4 and G_3 are given by Fig. 1.8.
Thus,

$$\mu_{U_3}(G) = 3/4, \ \mu_{U_2}(G) = 3/4, \ \mu_{U_1}(G) = 3/4,$$
$$\mu_{U_3}(G_5) = 0, \ \mu_{U_2}(G_5) = 0, \ \mu_{U_1}(G_5) = 1/2,$$
$$\mu_{U_3}(G_4) = 0, \ \mu_{U_2}(G_4) = 3/4, \ \mu_{U_1}(G_4) = 3/4,$$
$$\mu_{U_3}(G_3) = 1/2, \ \mu_{U_2}(G_3) = 1/2, \ \mu_{U_1}(G_3) = 3/4.$$

Hence, x_5 is an (S_1, S_2, S_3) point and x_4 and x_3 are (S_1, N_2, N_3) and (N_1, S_2, S_3) points, respectively.

In the following, let $p_{ij} = \widehat{\Gamma}_{x_i}(x_j)$, $i, \ j = 1, 2, \ldots, n$, $q_{ij} = \widehat{\Delta}_{x_i}(x_j)$, $i, \ j = 1, 2, \ldots, n$, and $r_{ij} = \widehat{\Gamma}'_{x_i}(x_j)$, $i, \ j \neq k$; $i, \ j = 1, 2, \ldots, n$, where $\widehat{\Gamma}'$ is the transitive closure of Γ' for the fuzzy directed graph $G \backslash x_k$.

Let $P = [p_{ij}]$ and $Q = [q_{ij}]$ and let R be an $n \times n$ matrix, whose elements in the kth row and in the kth column are zeros and each $(i, \ j)$ element is r_{ij}, where $i, j \neq k$; $i, \ j = 1, 2, \ldots, n$.

The next result characterizes weakening points for each connectedness category.

Lemma 1.11.3 *(i) A point x_k is a W_3 point if and only if the elements of P which are equal to $\mu_{U_3}(G)$ are all in the kth row or in the kth column of P.*

(ii) A point x_k is a W_2 point if and only if any $(i, \ j)$ elements of P such that $p_{ij} \vee p_{ji} = \mu_{U_2}(G)$ are in the k -th row and in the kth column of P.

(iii) A point x_k is a W_1 point if and only if all the elements of Q which are equal to $\mu_{U_1}(G)$ are in the kth row and in the kth column of Q.

Proof (i) Let x_k be a W_3 point. Suppose that there exists an element, say an (l, m) element, $l, m \neq k$, which is equal to $\mu_{U_3}(G)$. Because

$$r_{ij} \leq p_{ij}, \quad i, \, j \neq k; \, i, \, j = 1, 2, \ldots, n,$$

it follows that

$$\mu_{U_3}(G_k) = \wedge\{r_{ij}\} \leq p_{lm} = \mu_{U_3}(G),$$

However, this contradicts the assumption that x_k is a W_3 point. Therefore, every element of P which is equal to $\mu_{U_3}(G)$ is in the kth row or in the kth column of P.

Conversely, assume that the elements of P which are equal to $\mu_{U_3}(G)$ are all in the kth row or in the kth column of P. Thus, if an element which is equal to $\mu_{U_3}(G)$ is in the kth row (column) of P, then every non-diagonal element in the kth row (column) of P is equal to $\mu_{U_3}(G)$. Hence,

$$p_{ik} \wedge p_{kj} = \mu_{U_3}(G) < p_{ij}, \quad i, \, j \neq k; i, \, j = 1, 2, \ldots, n.$$

Thus,

$$r_{ij} = p_{ij} > \mu_{U_3}(FG), \quad i, \, j \neq k; \, i, \, j = 1, 2, \ldots, n.$$

Therefore,

$$\mu_{U_3}(G_k) = \wedge\{r_{ij} \mid i, \, j \neq k\} > \mu_{U_3}(G).$$

Hence, x_k is a W_3 point. This completes the proof of (i).
    The proofs of (ii) and (iii) are similar to that of (i). ∎

The following theorem is an immediate consequence of Lemma 1.11.3.

Theorem 1.11.4 *There exist at most two W_i points in any fuzzy graph, where $i = 1, 2, 3$. Further, any fuzzy graph with $n (n \geq 3)$ points has at most one $W_1(W_3)$ point.*

We make use of the following concepts in the proof of the next result. An **oriented graph** is a digraph having no symmetric pair of directed edges and a **tournament** is an oriented complete graph. A **Hamiltonian path** in a directed graph D is a spanning path in D.

Lemma 1.11.5 *For any fuzzy graph $G = (V, \Gamma)$, there exists a path $\{x_{i_1}, x_{i_2}, \ldots, x_{i_s}\}(s \geq n)$ such that*
    (i) every point of X appears in the path;
    (ii) $\Gamma_{x_{l+1}}(x_{i_{l+1}}) \geq \mu_{U_2}(G), l = 1, 2, \ldots, s - 1.$

Proof The result is immediate if $\mu_{U_2}(G) = 0$. Suppose that $\mu_{U_2}(G) > 0$. We construct an ordinary digraph $G = (V, \Gamma'')$ from G as follows:

$$\Gamma''_{x_i}(x_j) = \begin{cases} 1 \text{ if } p_{ij} \geq \mu_{U_2}(G), \\ 0 \text{ if } p_{ij} < \mu_{U_2}(G), \end{cases}$$

$i, j = 1, 2, \ldots, n$. Because $p_{ij} \vee p_{ji} \geq \mu_{U_2}(G)$, G includes a tournament as a partial graph of G. Because every tournament has a Hamiltonian path, G has a Hamiltonian path. It follows from Definition 1.10.5 that if $p_{ij} \geq \mu_{U_2}(G)$, then there exists at least a path $\{x_i, x_u, \ldots, x_v, x_j\}$ such that

$$\Gamma_{x_i}(x_u) \geq \mu_{U_2}(G),$$

$$\vdots$$

$$\Gamma_{x_v}(x_j) \geq \mu_{U_2}(G).$$

Thus, we obtain the desired result. ■

The following theorem shows that in any fuzzy graph with $n(n \geq 2)$ points, it is impossible for all points to be strengthening ones for $U_2(U_1)$.

Theorem 1.11.6 *Let G be a fuzzy directed graph with $n(n \geq 2)$ points. Then there exist at least two points which are either weakening or neutral ones for $U_2(U_1)$.*

Proof Let $\{x_{i_1}, x_{i_2}, \ldots, x_{i_s}\}$ be a path satisfying (i) and (ii) of Lemma 1.11.5. Without loss of generality, we can assume that the initial and final points x_{i_1} and x_{i_s} appear exactly once in the path. For if the initial point (the final point) appears more than once in the path, we can delete the first point (the last point) of the path, so that the remaining path also meets the requirements (i) and (ii).

By the above assumption, a path $\{x_{i_2}, x_{i_3}, \ldots, x_{i_s}\}$ and a path $\{x_{i_1}, x_{i_2}, \ldots, x_{i_{s-1}}\}$ contain, respectively, all points in $X \backslash \{x_{i_1}\}$ and all points in $X \backslash \{x_{i_s}\}$. Thus,

$$\mu_{U_2}(G_{i_1}) \geq \mu_{U_2}(G),$$

and

$$\mu_{U_2}(G_{i_s}) \geq \mu_{U_2}(G).$$

Hence, each of x_{i_1} and x_{i_s} is either a W_2 or N_2 point. This completes the proof for U_2.

The proof for U_1 is similar. ■

Corollary 1.11.7 *Any fuzzy graph with n $(n \geq 3)$ points has at least one N_1 point.*

Proof The result holds by Theorems 1.11.4 and 1.11.6. ■

Theorem 1.11.8 *If a fuzzy graph G with n $(n \geq 3)$ points has two W_2 points, then*

$$\mu_{U_2}(G) < \mu_{U_2}(G).$$

Proof Let x_k and x_l be W_2 points. Then

$$\mu_{U_2}(G_k) > \mu_{U_2}(G), \tag{1.2}$$

$$\mu_{U_2}(G_l) > \mu_{U_2}(G). \tag{1.3}$$

Suppose that

$$\mu_{U_2}(G) = \mu_{U_1}(G). \tag{1.4}$$

From (1.1) and (1.2)–(1.4), we have that x_k and x_l must be W_1 points, which contradicts Theorem 1.11.4. Hence, $\mu_{U_2}(G) < \mu_{U_1}(G)$. ∎

Theorem 1.11.9 *Any W_3 point is either a W_2 one or an N_2 one.*

Proof Let x_k be a W_3 point. From the proof of Lemma 1.11.3, we have that

$$r_{ij} = p_{ij}, \quad i, \; j = 1, 2, \ldots, n.$$

Thus,

$$\mu_{U_2}(G_k) \geq \mu_{U_2}(G).$$

∎

The following theorem follows from Definition 1.11.1 and (1.1).

Theorem 1.11.10 *If $\mu_{U_i}(G) = \mu_{U_j}(G)$ for some $i < j$, then an S_i point is also an S_j point.*

Theorem 1.11.11 *If $\mu_{U_i}(G) = \mu_{U_j}(G)$ for some $i > j$, then a W_i point is also a W_l point, where $1 \leq l \leq i$.*

Proof Let $i = 3$ and $j = 1$. Then

$$\mu_{U_3}(G) = \mu_{U_1}(G). \tag{1.5}$$

Let x_k be a W_3 point. From (1.5) we have

$$\mu_{U_2}(G_k) > \mu_{U_2}(G),$$

and

$$\mu_{U_1}(G_k) > \mu_{U_1}(G).$$

Thus, x_k is a W_l point, where $1 \leq l \leq 3$.

Assume that x_k is a W_2 point and that $\mu_{U_2}(G) = \mu_{U_1}(G)$. It follows that

$$\mu_{U_1}(G_k) > \mu_{U_1}(G).$$

Hence, x_k is a W_l point, where $1 \leq l \leq 2$.

Finally, we shall prove that if x_k is a W_3 point and $\mu_{U_3}(G) = \mu_{U_2}(G)$, then it is a W_l point, where $1 \leq l \leq 3$. Because x_k is a W_2 point, it suffices to show that x_k is a W_l point. Using Lemma 1.11.3, it follows that in both the kth row and in the kth column of P there exists an element which is equal to $\mu_{U_3}(G)$. Hence, we get from the proof of Lemma 1.11.3 that

$$p_{kj} = p_{jk} = \mu_{U_3}(G), \quad j \neq k; \ j = 1, 2, \ldots, n.$$

Thus, we have

$$(\Gamma_{x_k} \cup \Gamma_{x_k}^{-1})(x_j) \leq \mu_{U_3}(G), \quad j \neq k; \ j = 1, 2, \ldots, n.$$

Hence,

$$q_{kj} = q_{jk} \leq \mu_{U_3}(G), \quad j \neq k; \ j = 1, 2, \ldots, n.$$

Therefore, we have

$$\mu_{U_1}(G) = \mu_{U_3}(G),$$

so that

$$\mu_{U_1}(G_k) > \mu_{U_1}(G).$$

∎

Theorem 1.11.12 *Let x_k be a W_i point. If $\mu_{U_i}(G_k) = \mu_{U_j}(G_k)$ for some $i < j$, then x_k is also a W_l point, where $1 \leq l \leq j$.*

Proof The proof of this theorem is similar to that of Theorem 1.11.11 . ∎

We now show how results for the crisp case can be obtained from the results of Sect. 1.7 as special cases. In the case of the ordinary digraph G, $\mu_{U_i}(G_k) > \mu_{U_i}(G)$ if and only if $\mu_{U_i}(G_k) = 1$ and $\mu_{U_i}(G) = 0$, that is, $G_k \in U_i$ and $G \notin U_i$. With the understanding that a weakening point for U_0 is one whose presence makes its fuzzy graph more highly disconnected than it would be without the point, the W_0 point is defined to be the W_1 point. We can see from Theorem 1.11.4 that any digraph has at most two weakening points and from Theorem 1.11.11 that there are no $(1, 3)$ points in any digraph.

1.12 Intuitionistic Fuzzy Graphs: Weakening and Strengthening Members of a Group

In this section, we consider intuitionistic fuzzy graphs. Counterintuitive outcomes frequently occur in many situations such as human interaction for example. These situations can be modelled using intuitionistic fuzzy sets, [5]. We wish to set the ground work to apply these concepts to the deplorable situation of human trafficking. We are also setting the ground work for combining the notions of intuitionistic fuzzy graphs and $(s, t]$-fuzzy graphs, [7]. The work here is based on [10]. Results on intuitionistic fuzzy graphs can be found in [1–6, 12, 24, 25].

Let V be a finite nonempty set such that $|V| = n$. Let Γ and Ψ denote functions from $V \times V$ into [0, 1]. Let $x \in V$. Define the functions Γ_x and Ψ_x of V into [0, 1] by for all $y \in V$, $\Gamma_x(y) = \Gamma(x, y)$ and $\Psi_x(y) = \Psi(x, y)$. Then Γ_x and Ψ_x can be thought of as fuzzy subsets of V.

The elements of V are called **points** or **vertices** or **nodes** . Let $x, y \in V$, $x \neq y$. Then the ordered pair (x, y) is a called **directed line** or a **directed edge** from x to y. Let $x_0, x_1, \ldots, x_k$ be distinct members of V. Then the set $\{(x_0, x_1), (x_1, x_2), \ldots, (x_{k-1}, x_k)\}$ is called a **directed path** from x_0 to x_k of length k.

Let X be a set. Let $\mathcal{I}(X) = \{\langle \sigma, \tau \rangle \mid \sigma, \tau \in \mathcal{FP}(X), \sigma(x) + \tau(x) \leq 1$ for all $x \in X\}$. Then $\langle \sigma, \tau \rangle$ is called an **intuitionistic fuzzy set** on X. Let $\langle \Gamma, \Psi \rangle \in \mathcal{I}(V \times V)$. Then for all $(x, y) \in V \times V$, $\Gamma(x, y) + \Psi(x, y) \leq 1$. The intuitionistic fuzzy set $\langle \Gamma, \Psi \rangle$ provides the strength of the edge between vertices $x, y \in V$. An **intuitionistic directed path** is defined to be a directed path $\{(x_0, x_1), (x_1, x_2), \ldots, (x_{k-1}, x_k)\}$ such that $\Gamma(x_i, x_{i+1}) > 0$ and $\Psi(x_i, x_{i+1}) < 1$, $i = 0, 1, \ldots, k - 1$. The **strength** of an intuitionistic directed path $\{(x_0, x_1), (x_1, x_2), \ldots, (x_{k-1}, x_k)\}$ is defined to be the pair $\langle \wedge \{\Gamma(x_i, x_{i+1}) \mid i = 0, 1, \ldots, k - 1\}, \vee \{\Psi(x_i, x_{i+1}) \mid i = 0, 1, \ldots, k - 1\} \rangle$. If

$$P = \{(x = x_0, x_1), (x_1, x_2), \ldots, (x_{k-1}, x_k = y)\} \text{ and}$$
$$Q = \{(x = y_0, y_1), (y_1, y_2), \ldots, (y_{m-1}, y_m = y)\}$$

are two intutionistic direct paths from x to y, then P is said to be **stronger** than Q if $\wedge \{\Gamma(x_i, x_{i+1}) \mid i = 0, 1, \ldots, k - 1\} \geq \wedge \{\Gamma(y_i, y_{i+1}) \mid i = 0, 1, \ldots, m - 1\}$ and $\vee \{\Psi(x_i, x_{i+1}) \mid i = 0, 1, \ldots, k - 1\} \leq \vee \{\Psi(y_i, y_{i+1}) \mid i = 0, 1, \ldots, m - 1\}$ with at least one of the inequalities being strict.

Let $\langle \sigma, \tau \rangle \in \mathcal{I}(V)$. We consider the intuitionistic fuzzy set $\langle \sigma, \tau \rangle$ as providing the strength of the vertices of V.

The results here and their proofs are very similar to those in the previous section as we now show. Consider an involutive fuzzy complement c. Then it is easily shown that for all $a, b \in [0, 1]$, $a \wedge b = c(c(a) \vee c(b))$ and $a \vee b = c(c(a) \wedge c(b))$. From this observation, it can be seen that many results in this section are obtainable from those in the previous section. We say that $\wedge$ and $\vee$ are dual with respect to c. Consider the intuitionistic fuzzy sets $\langle \sigma, \tau \rangle$ and $\langle \Gamma, \Psi \rangle$. It does not necessarily hold that $\tau = \sigma^c$, where $\sigma^c(x) = c(\sigma(x))$ for all $x \in V$ or that $\Psi = \Gamma^c$. However, a result R for $\langle \sigma, \Gamma \rangle$ has a complementary result for $\langle \sigma^c, \Gamma^c \rangle$. Assume a result R holds in general for any $\langle \sigma, \Gamma \rangle$. Now given $\langle \tau, \Psi \rangle$. Then result R hold for $\langle \tau^c, \Psi^c \rangle$. Hence, its dual holds for $\langle \tau, \Psi \rangle$ even though it may not be the case that $\tau = \sigma^c$ and $\Psi = \Gamma^c$. Nevertheless, we include the proofs here not only for the sake of completeness, but also to lay the foundation for future research which will unite the two concepts, intuitionistic fuzzy graphs and $(s, t]$-fuzzy graphs.

Definition 1.12.1 Let $\langle \sigma, \tau \rangle$ and $\langle \Gamma, \Psi \rangle$ be intuitionistic fuzzy sets. The triple $G = (V, \langle \sigma, \tau \rangle, \langle \Gamma, \Psi \rangle)$ is called an **intuitionistic fuzzy directed graph** if for all $x, y \in V$, $\Gamma(x, y) \leq \sigma(x) \wedge \sigma(y)$ and $\Psi(x, y) \geq \tau(x) \vee \tau(y)$.

If we do not wish to consider the strength of the vertices, then we let $\sigma(x) = 1$ and $\tau(x) = 0$ for all $x \in V$. In this case, we write $G = (V, \langle \Gamma, \Psi \rangle)$.

Proposition 1.12.2 *Let $\langle \sigma, \tau \rangle \in \mathcal{I}(V)$. Let Γ be a fuzzy subset of $V \times V$ such that for all $x, y \in V$, $\Gamma(x, y) \leq \sigma(x) \wedge \sigma(y)$. Then $\exists$ a fuzzy subset Ψ of $V \times V$ such that $G = (V, \langle \sigma, \tau \rangle, \langle \Gamma, \Psi \rangle)$ is an intuitionistic fuzzy directed graph.*

Proof Define Ψ by for all $x, y \in V$, $\Psi(x, y) = \tau(x) \vee \tau(y)$. Then

$$\begin{aligned} \Gamma(x, y) + \Psi(x, y) &\leq \sigma(x) \wedge \sigma(y) + \tau(x) \vee \tau(y) \\ &\leq (1 - \tau(x)) \wedge (1 - \tau(y)) + \tau(x) \vee \tau(y) \\ &= 1 - (\tau(x) \vee \tau(y)) + \tau(x) \vee \tau(y) = 1. \end{aligned}$$

∎

Definition 1.12.3 Let $G = (V, \langle \sigma, \tau \rangle, \langle \Gamma, \Psi \rangle)$ be an intuitionistic fuzzy directed graph. Then the triple $G' = (V', \langle \sigma', \tau' \rangle, \langle \Gamma', \Psi' \rangle)$ is called an **intuitionistic fuzzy directed subgraph** of G if G' is an intuitionistic fuzzy directed graph such that $V' \subseteq V$, $\sigma' = \sigma|_{V'}$, $\tau' = \tau|_{V'}$, $\Gamma' = \Gamma|_{V' \times V'}$, and $\Psi' = \Psi|_{V' \times V'}$.

Let $\mathcal{M}_n$ denote the set of all $n \times n$ fuzzy matrices with entries from $[0, 1]$. Define the binary relations $\circ, \diamond$ on $\mathcal{M}_n$ by for all $[b_{ik}], [c_{kj}] \in \mathcal{M}_n$,

$$[b_{ik}] \circ [c_{kj}] = [\vee \{b_{ik} \wedge c_{kj} \mid k = 1, .., n\}],$$
$$[b_{ik}] \diamond [c_{kj}] = [\wedge \{b_{ik} \vee c_{kj} \mid k = 1, .., n\}].$$

Let $\langle \Gamma, \Psi \rangle \in \mathcal{I}(V \times V)$. Let $\langle \Gamma, \Psi \rangle^2 = \langle \Gamma \circ \Gamma, \Psi \diamond \Psi \rangle$, where $\circ$ denotes max-min composition and $\diamond$ denotes min-max composition. Let Γ be denoted by $[a_{ij}]$ and Ψ by $[b_{ij}]$. Suppose for all $x, y \in V$ that $\Psi(x, y) = 1 - \Gamma(x, y)$. Then $b_{ij} = 1 - a_{ij}, i, j = 1, \ldots, n$. Let $\mathbf{1}$ denote the $n \times n$ matrix every entry of which equals 1. Then

$$\begin{aligned} [b_{ik}] \diamond [b_{kj}] &= [\wedge \{b_{ik} \vee b_{kj} \mid k = 1, \ldots, n\}] \\ &= [\wedge \{(1 - a_{ik}) \vee (1 - a_{kj}) \mid k = 1, \ldots, k\}] \\ &= [\wedge \{1 - (a_{ik} \wedge a_{kj}) \mid k = 1, \ldots, n\}] \\ &= [1 - \vee \{a_{ik} \wedge a_{kj} \mid k = 1, \ldots, n\}] \\ &= \mathbf{1} - [a_{ik}] \circ [a_{kj}]. \end{aligned}$$

We have shown that $\Psi \diamond \Psi = \mathbf{1} - \Gamma \circ \Gamma$.

Let m be an integer ≥ 2. For $\langle \Gamma, \Psi \rangle \in \mathcal{I}(V \times V)$, we define Ψ^m recursively as follows: for all $x, y \in V$, $\Psi^0(x, y) = 1$ if $x \neq y$ and $\Psi^0(x, y) = 0$ if $x = y$ and

$\Psi^1 = \Psi$. Assume Ψ^{m-1} has been defined. Define $\Psi^m = \Psi^{m-1} \diamond \Psi$. Define Γ^m in terms of $\diamond$ similarly.

Proposition 1.12.4 *Let* $\langle \Gamma, \Psi \rangle \in \mathcal{I}(V \times V)$. *Suppose* $\Gamma = 1 - \Psi$. *Then* $\Psi^m = 1 - \Gamma^m$ *for all positive integers* m.

Proof The result is given for $m = 1$. Assume the result is true for m. Let $\Gamma^m = [c_{ij}]$ and $\Gamma = [a_{ij}]$. Then $\Psi^{m+1} = \Psi^m \diamond \Psi = (1 - \Gamma^m) \diamond (1 - \Gamma)$. Now

$$[\wedge\{(1 - c_{ik}) \vee (1 - a_{kj}) \mid k = 1, \ldots, n\}]$$
$$= [\wedge\{1 - c_{ik} \wedge a_{kj} \mid k = 1, \ldots, n\}]$$
$$= [1 - \vee\{c_{ik} \wedge a_{kj} \mid k = 1, \ldots, n\}]$$
$$= 1 - [\vee\{c_{ik} \wedge a_{kj} \mid k = 1, \ldots, n\}]$$
$$= 1 - [c_{ik}] \circ [a_{kj}].$$

Thus, $\Psi^{m+1} = 1 - \Gamma^{m+1}$. ∎

Definition 1.12.5 Let $G = (V, \langle \Gamma, \Psi \rangle)$ be an intuitionistic fuzzy directed graph. Define the intuitionistic fuzzy subset $\langle \Gamma^{-1}, \Psi^{-1} \rangle$ of $V \times V$ as follows: for all $x, y \in V$, $\Gamma^{-1}(y, x) = \Gamma(x, y)$ and $\Psi^{-1}(y, x) = \Psi(x, y)$. Let Δ, Φ be the fuzzy subsets of $V \times V$ defined by $\Delta = \Gamma \cup \Gamma^{-1}$ and $\Phi = \Psi \cap \Psi^{-1}$. Let $\widehat{\Gamma}$, $\widehat{\Gamma^{-1}}$, $\widehat{\Delta}$, be the max-min transitive closures of Γ, Γ^{-1}, Δ, respectively, and $\widehat{\Psi}$, $\widehat{\Psi^{-1}}$, and $\widehat{\Phi}$ be the min-max transitive closures of Ψ, Ψ^{-1}, Φ, respectively.

Because M_n is $n \times n$, it follows that for all $x \in V$,

$$\langle \widehat{\Gamma}_x, \widehat{\Psi}_x \rangle = \langle \Gamma_x^0 \cup \Gamma_x \cup \Gamma_x^2 \cup \cdots \cup \Gamma_x^{n-1}, \Psi_x^0 \cap \Psi_x \cap \Psi_x^2 \cap \cdots \cap \Psi_x^{n-1} \rangle,$$

$$\langle \widehat{\Gamma_x^{-1}}, \widehat{\Psi_x^{-1}} \rangle = \langle \Gamma_x^0 \cup \Gamma_x^{-1} \cup \Gamma_x^{-2} \cup \cdots \cup \Gamma_x^{-n+1}, \Psi_x^0 \cap \Psi_x^{-1} \cap \Psi_x^{-2} \cap \cdots \cap \Psi_x^{-n+1} \rangle,$$

where $\Gamma_x^0(y) = 0$ if $x \neq y$ and $\Gamma_x^0(y) = 1$ if $x = y$ and $\Psi_x^0(y) = 1$ if $x \neq y$ and $\Psi_x^0(y) = 0$ if $x = y$. The grades of membership $\langle \widehat{\Gamma}_x(y), \widehat{\Psi}_x(y) \rangle$ and $\langle \widehat{\Gamma}_x^{-1}(y), \widehat{\Psi}_x^{-1}(y) \rangle$ are the degrees of a strongest intuitionistic directed path from x to y and that from y to x, respectively. Similarly, for all $x \in V$,

$$\langle \widehat{\Delta}_x, \widehat{\Phi}_x \rangle = \langle \Delta_x^0 \cup \Delta_x \cup \Delta_x^2 \cup \cdots \cup \Delta_x^{n-1}, \Phi_x^0 \cap \Phi_x \cap \Phi_x^2 \cap \cdots \cap \Phi_x^{n-1} \rangle.$$

The value $\langle \widehat{\Delta}_x(y), \widehat{\Phi}_x(y) \rangle$ is the degree of a strongest intuitionistic semipath between two points x and y.

Example 1.12.6 Let $V = \{x_1, x_2, x_3\}$. Define the fuzzy subsets Γ and Ψ of V as follows: $\Gamma(x_1, x_2) = 0.5$, $\Gamma(x_2, x_3) = 0.4$ and $\Psi(x_1, x_2) = 0.4$, $\Psi(x_2, x_3) = 0.3$. Then the strength of the path $\{x_1, x_2, x_3\}$ is $\langle 0.5, 0.4 \rangle$, but there is no edge of strength $\langle 0.5, 0.4 \rangle$.

Note that if $a_i, b_i \in [0, 1]$ and $a_i + b_i \leq 1$ for $i = 1, 2$, then $a_1 \vee a_2 + b_1 \wedge b_2 \leq 1$: Suppose $a_1 \vee a_2 = a_2$. Then $1 \geq a_2 + b_2 \geq a_1 \vee a_2 + b_1 \wedge b_2$.

Let G be an intuitionistic fuzzy directed graph. Then G is said to be **strongly connected** if for every two (distinct) points x and y, there exists a directed path from x to y and one from y to x. G is said to be **unilaterally connected** if for every two points x and y, there is a directed path from x to y or one from y to x. G is called **disconnected** if the points can be divided into two sets with no line joining any point in one set with a point in the other set. G is called **weakly connected** if it is not disconnected.

Let U_3 be the collection of all strongly connected intuitionistic fuzzy directed graphs, U_2 the collection of all unilaterally connected intuitionistic fuzzy directed graphs, U_1 the set of all weakly connected intuitionistic fuzzy directed graphs, and U_0 the set of all disconnected intuitionistic fuzzy directed graphs. Then $U_3 \subseteq U_2 \subseteq U_1$ and $U_1 \cap U_0 = \emptyset$.

Definition 1.12.7 The grades of membership of a intuitionistic fuzzy directed graph $G = (V, \langle \Gamma, \Psi \rangle)$ in U_3, U_2, U_1, and U_0 are defined as follows:

$$\mu_{U_3}(G) = \wedge \{ \widehat{\Gamma}_{x_i}(x_j) \mid i, j = 1, \dots, n \},$$
$$\nu_{U_3}(G) = \vee \{ \widehat{\Psi}_{x_i}(x_j) \mid i, j = 1, \dots, n \},$$
$$\mu_{U_2}(G) = \wedge \{ \widehat{\Gamma}_{x_i}(x_j) \vee \widehat{\Gamma}_{x_j}(x_i) \mid i, j = 1, \dots, n \},$$
$$\nu_{U_2}(G) = \vee \{ \widehat{\Psi}_{x_i}(x_j) \wedge \widehat{\Psi}_{x_j}(x_i) \mid i, j = 1, \dots, n \},$$
$$\mu_{U_1}(G) = \wedge \{ \widehat{\Delta}_{x_i}(x_j) \mid i, j = 1, \dots, n \},$$
$$\nu_{U_1}(G) = \vee \{ \widehat{\Phi}_{x_i}(x_j) \mid i, j = 1, \dots, n \},$$
$$\mu_{U_0}(G) = 1 - \wedge \{ \widehat{\Delta}_{x_i}(x_j) \mid i, j = 1, \dots, n \},$$
$$\nu_{U_0}(G) = 1 - \vee \{ \widehat{\Phi}_{x_i}(x_j) \mid i, j = 1, \dots, n \},$$

respectively.

Definition 1.12.8 Let $G = (V, \langle \Gamma, \Psi \rangle)$ be an intuitionistic fuzzy directed graph. Let G_k be the intuitionistic fuzzy directed graph obtained by the removal of a point x_k. Then the point x_k is a μ_k-**weakening point** for U_i (a W_i point) if $\mu_{U_i}(G) < \mu_{U_i}(G_k)$ and a ν_k-**weakening point** for U_i (a W_i point) if $\nu_{U_i}(G) > \nu_{U_i}(G_k)$; x_k is a μ_k-**neutral point** for U_i (an N_i point) if $\mu_{U_i}(G) = \mu_{U_i}(G_k)$ and a ν_k-**neutral point** for U_i (an N_i point) if $\nu_{U_i}(G) = \nu_{U_i}(G_k)$; x_k is a μ_k- **strengthening point** for U_i (an S_i point) if $\mu_{U_i}(G) > \mu_{U_i}(G_k)$ and ν_k-**strengthening point** for U_i (an S_i point) if $\nu_{U_i}(G) < \nu_{U_i}(G_k)$, $i = 1, 2, 3$. The point x_k is a **weakening point** for U_i if it is a μ_k-weakening point and a ν_k-weakening point; a **neutral point** for U_i if it is a μ_k-neutral point and a ν_k-neutral point; a **strengthening point** for U_i if it is a μ_k-strengthening point and a ν_k-strengthening point.

Example 1.12.9 Let $V = \{x_1, x_2, x_3, x_4\}$. Let

$$\langle \Gamma, \Psi \rangle (x_1, x_2) = \langle 0.7, 0.2 \rangle, \ \langle \Gamma, \Psi \rangle (x_1, x_3) = \langle 0.5, 0.1 \rangle,$$
$$\langle \Gamma, \Psi \rangle (x_1, x_4) = \langle 0.6, 0.3 \rangle,$$
and $\langle \Gamma, \Psi \rangle (x_i, x_j) = \langle 0, 1 \rangle$ for all other pairs $(x_i, x_j), i \neq j$.

Then

$$\widehat{\Gamma}_{x_1}(x_1) = 1, \ \widehat{\Gamma}_{x_1}(x_2) = 0.7, \ \widehat{\Gamma}_{x_1}(x_3) = 0.5, \ \widehat{\Gamma}_{x_1}(x_4) = 0.6,$$
$$\widehat{\Psi}_{x_1}(x_1) = 0, \ \widehat{\Psi}_{x_1}(x_2) = 0.2, \ \widehat{\Psi}_{x_1}(x_3) = 0.1, \ \widehat{\Psi}_{x_1}(x_4) = 0.3.$$

Also, $\widehat{\Gamma}_{x_i}(x_j) = 0$ and $\widehat{\Psi}_{xi}(x_j) = 1$ for $i = 2, 3, 4$ and $j \neq i$. From this it follows that $\mu_{U_i}(G) = 0$ and $\nu_{U_i}(G) = 1$ for $i = 3, 2$. It follows that

$$\mu_{U_1}(G) = 0.5 \text{ and } \nu_{U_1}(G) = 0.3.$$

Example 1.12.10 Let $V = \{x_1, x_2, x_3, x_4\}$. Let

$$\langle \Gamma, \Psi \rangle (x_1, x_2) = \langle 0.7, 0.2 \rangle, \quad \langle \Gamma, \Psi \rangle (x_2, x_4) = \langle 0.4, 0.3 \rangle,$$
$$\langle \Gamma, \Psi \rangle (x_1, x_3) = \langle 0.5, 0.4 \rangle, \quad \langle \Gamma, \Psi \rangle (x_3, x_4) = \langle 0.6, 0.1 \rangle$$
and $\langle \Gamma, \Psi \rangle (x_i, x_j) = \langle 0, 1 \rangle$ for all other pairs $(x_i, x_j), \ i \neq j$.

Then

$$\Gamma = \begin{bmatrix} 1 & 0.7 & 0.5 & 0 \\ 0 & 1 & 0 & 0.4 \\ 0 & 0 & 1 & 0.6 \\ 0 & 0 & 0 & 1 \end{bmatrix}, \ \Gamma^2 = \begin{bmatrix} 1 & 0.7 & 0.5 & 0.5 \\ 0 & 1 & 0 & 0.4 \\ 0 & 0 & 1 & 0.6 \\ 0 & 0 & 0 & 1 \end{bmatrix},$$

$$\Psi = \begin{bmatrix} 0 & 0.2 & 0.4 & 1 \\ 1 & 0 & 1 & 0.3 \\ 1 & 1 & 0 & 0.1 \\ 1 & 1 & 1 & 0 \end{bmatrix}, \ \Psi^2 = \begin{bmatrix} 0 & 0.2 & 0.4 & 0.3 \\ 1 & 0 & 1 & 0.3 \\ 1 & 1 & 0 & 0.1 \\ 1 & 1 & 1 & 0 \end{bmatrix}$$

and

$$\Gamma \cup \Gamma^2 = \begin{bmatrix} 1 & 0.7 & 0.5 & 0.5 \\ 0 & 1 & 0 & 0.4 \\ 0 & 0 & 1 & 0.6 \\ 0 & 0 & 0 & 1 \end{bmatrix}, \ \Psi \cap \Psi^2 = \begin{bmatrix} 0 & 0.2 & 0.4 & 0.3 \\ 1 & 0 & 1 & 0.3 \\ 1 & 1 & 0 & 0.1 \\ 1 & 1 & 1 & 0 \end{bmatrix}.$$

Note that $\widehat{\Gamma}_{x_1}(x_4) = 0.5$ and $\widehat{\Psi}_{x_1}(x_4) = 0.3$. Note also that there is no edge of strength $\langle 0.5, 0.3 \rangle$. In fact, there is no strongest intuitionistic directed path from x_1 to x_4.

It follows that $\mu_{U_i}(G) = 0$ and $\nu_{U_i}(G) = 1$ for $i = 3, 2$ and that

$$\mu_{U_1}(G) = 0.5 \text{ and } \nu_{U_1}(G) = 0.3.$$

$i =$	3	2	1	0
$\mu_{U_i}(G)$	0	0	0.5	0.5
$\nu_{U_i}(G)$	1	1	0.3	0.7

$i =$	3	2	1	0
$\mu_{U_i}(G_1)$	$0, N_3$	$0, N_2$	$0.4, S_1$	0.6
$\nu_{U_i}(G_1)$	$1, N_3$	$1, N_2$	$0.3, N_1$	0.7

$i =$	3	2	1	0
$\mu_{U_i}(G_2)$	$0, N_3$	$0.5, W_2$	$0.5, N_1$	0.5
$\nu_{U_i}(G_2)$	$1, N_3$	$0.4, W_2$	$0.4, S_1$	0.6

$i =$	3	2	1	0
$\mu_{U_i}(G_3)$	$0, N_3$	$0.4, W_2$	$0.4, S_1$	0.6
$\nu_{U_i}(G_3)$	$1, N_3$	$0.3, W_2$	$0.3, N_1$	0.7

$i =$	3	2	1	0
$\mu_{U_i}(G_4)$	$0, N_3$	$0, N_2$	$0.5, N_1$	0.5
$\nu_{U_i}(G_4)$	$1, N_3$	$1, N_2$	$0.4, S_1$	0.6

In the following, let

$$b_{ij} = \widehat{\Psi}_{x_i}(x_j), \quad i, \ j = 1, 2, \ldots, n,$$
$$c_{ij} = \widehat{\Delta}_{x_i}(x_j), \quad i, \ j = 1, 2, \ldots, n,$$
$$d_{ij}^k = \widehat{\Psi'}_{x_i}(x_j), \quad i, \ j = 1, 2, \ldots, n; \ i \neq k \neq j,$$

where Ψ' is Ψ restricted to $(V\backslash\{x_k\}) \times (V\backslash\{x_k\})$ and $\widehat{\Psi'}$ is the min-max transitive closure of Ψ'. Let $B = [b_{ij}]$, $C = [c_{ij}]$, and $D = [d_{ij}^k]$. Let $s_{ij}^k = \widehat{\Delta'}_{x_i}(x_j), i, j = 1, 2, \ldots, n; i \neq k \neq j$. Let $S = [s_{ij}^k]$. Recall that we do not allow for reciprocated edges.

Lemma 1.12.11 *Let* $G = (V, \langle\sigma, \tau\rangle, \langle\Gamma, \Psi\rangle)$ *be an intuitionistic fuzzy directed graph.*

(i) A point x_k *is a* W_3 *point with respect to* Ψ *if and only if the elements of* B *which are equal to* $\nu_{U_3}(G)$ *are all on the kth row or in the kth column of* B.

(ii) A point x_k *is a* W_2 *point with respect to* Ψ *if and only if any entry of* B *such that* $b_{ij} \wedge b_{ji} = \nu_{U_2}(G)$ *is in the kth row and the kth column of* B.

(iii) A point x_k is a W_1 point with respect to Ψ if and only if all the entries of C, which are equal to $\nu_{U_1}(G)$ are in the kth row and kth column of C.

Proof (i) Let x_k be W_3 point with respect to Ψ. Suppose that there exists an element b_{rs} with $r \neq k \neq s$ such that $b_{rs} = \nu_{U_3}(G)$. Because $d_{ij}^k \geq b_{ij}$ for $i, j = 1, 2, \ldots, n; i \neq k \neq j$, we have that

$$\nu_{U_3}(G_k) = \vee\{d_{ij}^k \mid i, j = 1, 2, \ldots, n; \ i \neq k \neq j\} \geq b_{rs} = \nu_{U_3}(G).$$

However, this contradicts the assumption that x_k is a W_3 point with respect to Ψ. Thus, every element of B which equals $\nu_{U_3}(G)$ is in the kth row or the kth column of B.

Conversely, assume that the elements of B which equal $\nu_{U_3}(G)$ are in the kth row or kth column of B. Then there exists an element b on the kth row or kth column of B such that $b > b_{ij}, i, j = 1, 2, \ldots, n; i \neq k \neq j$. Now $d_{ij}^k = b_{ij}, i, j = 1, 2, \ldots, n, i \neq k \neq j$. Hence, $\nu_{U_3}(G) > \nu_{U_3}(G_k)$. Thus, x_k is a W_3 point with respect to Ψ.

(ii) Let x_k be a W_2 point with respect to Ψ. Suppose there exist entries b_{rs}, b_{sr} of B such that $b_{rs} \wedge b_{sr} = \nu_{U_2}(G)$ with $r \neq k \neq s$. Then $\nu_{U_2}(G) \leq \nu_{U_2}(G_k)$, contrary to the assumption that x_k is a W_2 point with respect to Ψ. Thus, every b_{rs}, b_{sr} such that $b_{rs} \wedge b_{sr} = \nu_{U_2}(G)$ is such that $r = k$ or $s = k$.

Conversely, suppose that any entry of B such that $b_{ij} \wedge b_{ji} = \nu_{U_2}(G)$ is in the kth row or the kth column of B. Then there exists an element b_{rs} with $r = k$ or $s = k$ such that $b_{rs} \wedge b_{sr} > b_{ij} \wedge b_{ji}, i, j = 1, 2, \ldots, n; i \neq k \neq j$. Thus, $\nu_{U_3}(G) > \nu_{U_3}(G_k)$. Hence, x_k is a W_2 point with respect to Ψ.

(iii) Suppose x_k is a W_1 point with respect to Ψ. Suppose that there exists an entry c_{rs} of C with $r \neq k \neq s$ such that $c_{rs} = \nu_{U_1}(G)$. Because $s_{ij}^k \geq c_{ij}$ for $i, j = 1, 2, \ldots, n; i \neq k \neq j$, we have that

$$\nu_{U_3}(G_k) = \vee\{s_{ij}^k \mid i, j = 1, 2, \ldots, n; i \neq k \neq j\} \geq c_{rs} = \nu_{U_3}(G).$$

However, this contradicts the assumption that x_k is a W_1 point with respect to Ψ. Thus, every element of C which equals $\nu_{U_3}(G)$ is in the kth row or the kth column of C.

Conversely, assume that the elements of C which equal $\nu_{U_1}(G)$ are in the kth row and kth column of C. Then there exists an element c on the kth row or kth column of C such that $c > c_{ij}, i, j = 1, 2, \ldots, n; i \neq k \neq j$. Now $s_{ij}^k = c_{ij}, i, j = 1, 2, \ldots, n, i \neq k \neq j$. Hence, $\nu_{U_1}(G) > \nu_{U_1}(G_k)$. Thus, x_k is a W_1 point with respect to Ψ. ∎

Example 1.12.12 Let $X = \{x_1, x_2, x_3\}$ and let Γ be given by the following matrix

$$
\begin{array}{c}
\begin{array}{ccc} x_1 & x_2 & x_3 \end{array} \\
\begin{array}{c} x_1 \\ x_2 \\ x_3 \end{array}
\left[\begin{array}{ccc}
1 & 1 & 1 \\
0 & 1 & 0 \\
0 & 0 & 1
\end{array} \right]
\end{array}
$$

Then it is easily seen that $\Gamma = \widehat{\Gamma}$. Thus, $\widehat{\Gamma}_{x_2}(x_3) \vee \widehat{\Gamma}_{x_3}(x_2) = 0$. Hence, $\mu_{U_2}(G) = 0$. Consider Γ' given by the deletion of x_3. Then Γ' is given by

$$
\begin{array}{c}
\begin{array}{cc} x_1 & x_2 \end{array} \\
\begin{array}{c} x_1 \\ x_2 \end{array}
\left[\begin{array}{cc}
1 & 1 \\
0 & 1
\end{array} \right]
\end{array}
$$

It is easily seen that $\widehat{\Gamma'}$ is given by

$$
\begin{array}{c}
\begin{array}{cc} x_1 & x_2 \end{array} \\
\begin{array}{c} x_1 \\ x_2 \end{array}
\left[\begin{array}{cc}
1 & 1 \\
1 & 1
\end{array} \right]
\end{array}
$$

Thus, $\mu_{U_2}(G_3) = 1$. Therefore, x_3 is a W_2 point. Similarly, x_2 is a W_2 point. That is, G has two W_2 points.

Theorem 1.12.13 *There exists at most two W_i points in any intuitionistic fuzzy directed graph, $i = 1, 2, 3$. Any intuitionistic directed graph with $n \geq 3$ points has at most one W_1 point and at most one W_3 point.*

Proof The proof follows from (i) and (iii) of Lemma 1.12.11 and [21, Lemma 1, p. 223]. ∎

Lemma 1.12.14 *Let $G = (V, \langle \Gamma, \Psi \rangle)$ be an intuitionistic fuzzy directed graph. Suppose $\Gamma_{x_i}(x_j) > 0$ if and only if $\Psi_{x_i}(x_j) < 1$ for $i, j = 1, \ldots, n$. Then there exists a path $\{x_{i_1}, x_{i_2}, \ldots, x_{i_s}\}, s \geq n$, such that the following properties hold:*

(i) every point of V appears in the path;
(ii) $\Gamma_{x_{i_j}}(x_{i_{j+1}}) \geq \mu_{U_2}(G), j = 1, 2, \ldots, s - 1$;
(iii) $\Psi_{i_j}(x_{i_{j+1}}) \leq \nu_{U_2}(G), j = 1, 2, \ldots, s - 1$.

Proof Because $\Gamma_{x_i}(x_j) > 0$ if and only if $\Psi_{x_i}(x_j) < 1$, it follows that $\mu_{U_2}(G) = 0$ if and only if $\nu_{U_2}(G) = 1$. If it is the case that $\mu_{U_2}(G) = 0$ and $\nu_{U_2}(G) = 1$, then the desired result is immediate. Suppose that $\mu_{U_2}(G) > 0$ if and only if $\mu_{U_2}(G) < 1$. We construct an ordinary intuitionistic fuzzy graph from G as follows: for all $i, j \in V$,

$$
\Gamma''_{x_i}(x_j) = \begin{cases} 1 \text{ if } \widehat{\Gamma}_{x_i}(x_j) \geq \mu_{U_2}(G), \\ 0 \text{ if } \widehat{\Gamma}_{x_i}(x_j) < \mu_{U_2}(G). \end{cases}
$$

and

$$
\Psi''_{x_i}(x_j) = \begin{cases} 0 \text{ if } \widehat{\Psi}_{x_i}(x_j) \leq \nu_{U_2}(G), \\ 1 \text{ if } \widehat{\Psi}_{x_i}(x_j) > \nu_{U_2}(G). \end{cases}
$$

Now $\widehat{\Gamma}_{x_i}(x_j) \vee \widehat{\Gamma}_{x_j}(x_i) \geq \mu_{U_2}(G)$ and $\widehat{\Psi}_{x_i}(x_j) \wedge \widehat{\Psi}_{x_j}(x_i) \leq \nu_{U_2}(G)$. We have that the $\Gamma''_{x_i}(x_j)$ induce a Hamiltonian graph on V. Also, $\widehat{\Gamma}_{x_i}(x_j) \geq \mu_{U_2}(G)$ if and only if $\widehat{\Psi}_{x_i}(x_j) \leq \nu_{U_2}(G)$. Thus, if $\widehat{\Gamma}_{x_i}(x_j) \geq \mu_{U_2}(G)$, then $\widehat{\Psi}_{x_i}(x_j) \leq \nu_{U_2}(G)$ and there exists a path $\{x_i, x_u, \ldots, x_v, x_j\}$ such that

$$\Gamma_{x_i}(x_u) \geq \mu_{U_2}(G), \ldots, \Gamma_{x_v}(x_j) \geq \mu_{U_2}(G),$$

and by a dual argument

$$\Psi_{x_i}(x_u) \leq \nu_{U_2}(G), \ldots, \Psi_{x_v}(x_j) \leq \nu_{U_2}(G).$$

∎

Theorem 1.12.15 *Let $G = (V, \langle \Gamma, \Psi \rangle)$ be an intuitionistic fuzzy directed graph. Suppose $\Gamma_{x_i}(x_j) > 0$ if and only if $\Psi_{x_i}(x_j) < 1$ for $i, j = 1, \ldots, n$. If G has two or more points, then there exist at least two points which are either weakening or neutral points for $U_2(U_1)$.*

Proof Let $\{x_{i_1}, x_{i_2}, \ldots, x_{i_s}\}, s \geq n$, be a path satisfying $(i), (ii)$, and (iii) of Lemma 1.12.14. We can assume that the initial point (final point) appears exactly once in the path. For, if the initial point (the final point) appears more than once in the path, we can delete the first point (the final point) of the path so that the remaining path also meets the requirements $(i), (ii)$, and (iii).

Hence, consider a path $\{x_{i_2}, x_{i_3}, \ldots, x_{i_s}\}$ and a path $\{x_{i_1}, x_{i_2}, \ldots, x_{i_{s-1}}\}$ containing all points in $V \setminus \{x_{i_1}\}$ and all points in $V \setminus \{x_{i_s}\}$, respectively. Thus,

$$\mu_{U_2}(G_{i_1}) \geq \mu_{U_2}(G), \nu_{U_2}(G_{i_1}) \leq \nu_{U_2}(G)$$

and

$$\mu_{U_2}(G_{i_s}) \geq \mu_{U_2}(G), \nu_{U_2}(G_{i_s}) \leq \nu_{U_2}(G).$$

Hence, each of x_{i_1} and x_{i_s} is either a W_2 or a N_2 point. The proof for U_1 is similar. ∎

Corollary 1.12.16 *Any intuitionistic fuzzy graph with $n \geq 3$ points has at least one N_1 point.*

Proof The desired result follows from Theorems 1.12.13 and 1.12.15. ∎

Let $G = (V, \langle \Gamma, \Psi \rangle)$ be an intuitionistic fuzzy directed graph. Suppose $\Psi = 1 - \Gamma$. Then $\mu_{U_2}(G) = \mu_{U_1}(G)$ if and only if $\nu_{U_2}(G) = \nu_{U_1}(G)$.

Lemma 1.12.17 *Let $G = (V, \langle \Gamma, \Psi \rangle)$ be an intuitionistic fuzzy directed graph. Suppose $\Gamma = 1 - \Psi$. Then $\mu_{U_2}(G) = 1 - \nu_{U_2}(G)$.*

Proof By Proposition 1.12.4, $\Psi^m = 1 - \Gamma^m$ for all positive integers m. For all $x \in V$, $\widehat{\Gamma}_x = 1 - (\Gamma_x^0 \vee \Gamma_x^1 \vee \cdots \vee \Gamma_x^{n-1}) = (1 - \Gamma_x^0) \wedge (1 - \Gamma_x^1) \wedge \ldots \wedge (1 - \Gamma_x^{n-1}) =$

$\Psi_x^0 \wedge \Psi_x^1 \wedge \ldots \wedge \Psi_x^{n-1} = \widehat{\Psi_x}$. Thus, $\nu_{U_2}(G) = 1 - \mu_{U_2}(G)$. The desired result follows because $\mu_{U_2}(G) = \wedge\{\widehat{\Gamma}_{x_i}\{(x_j) \vee \widehat{\Gamma}_{x_j}\{(x_i) \mid i, j = 1, \ldots, n\}$ and $\nu_{U_2}(G) = \vee\{\widehat{\Psi}_{x_i}\{(x_j) \wedge \widehat{\Psi}_{x_j}\{(x_i) \mid i, j = 1, \ldots, n\}$. ∎

Theorem 1.12.18 *Let $G = (V, \langle \Gamma, \Psi \rangle)$ be an intuitionistic fuzzy directed graph with $n \geq 3$ points. Suppose that $\mu_{U_2}(G) = \mu_{U_1}(G)$ if and only if $\nu_{U_2}(G) = \nu_{U_1}(G)$. If G has two W_2 points, then*

$$\mu_{U_2}(G) < \mu_{U_1}(G) \text{ and } \nu_{U_2}(G) > \nu_{U_1}(G).$$

Proof Let w_k and w_m be W_2 points. Then

$$\mu_{U_2}(G_k) > \mu_{U_2}(G), \nu_{U_2}(G_k) < \mu_{U_2}(G)$$

and

$$\mu_{U_2}(G_m) > \mu_{U_2}(G), \nu_{U_2}(G_m) < \mu_{U_2}(G).$$

Suppose that $\mu_{U_2}(G) = \mu_{U_1}(G)$ and $\nu_{U_2}(G) = \nu_{U_1}(G)$. Then

$$\mu_{U_1}(G) < \mu_{U_2}(G_k) \leq \mu_{U_1}(G_k) \text{ and } \mu_{U_1}(G) < \mu_{U_2}(G_m) \leq \mu_{U_1}(G_m).$$

Also,

$$\nu_{U_1}(G) > \nu_{U_2}(G_k) \geq \nu_{U_1}(G_k) \text{ and } \nu_{U_1}(G) > \nu_{U_2}(G_m) \geq \nu_{U_1}(G_m).$$

Thus, x_k and x_m are W_1 points. However, this contradicts Theorem 1.12.13. Hence, $\mu_{U_2}(G) < \mu_{U_1}(G)$ and $\nu_{U_2}(G) > \nu_{U_1}(G)$. ∎

Theorem 1.12.19 *Let $G = (V, \langle \Gamma, \Psi \rangle)$ be an intuitionistic fuzzy directed graph. Then any W_3 point is either a W_2 point or an N_2 point.*

Proof Let x_k be a W_3 point. Then from the proof of Lemma 1.12.11 , we have $\widehat{\Gamma}_{x_i}(x_j) = \widehat{\Gamma'}_{x_i}(x_j)$ and $\widehat{\Psi}_{x_i}(x_j) = \widehat{\Psi'}_{x_i}(x_j), i, j = 1, \ldots, n, i \neq k \neq j$. Thus, $\mu_{U_2}(G_k) \geq \mu_{U_2}(G)$ and $\nu_{U_2}(G_k) \leq \nu_{U_2}(G)$. ∎

Theorem 1.12.20 *Let $G = (V, \langle \Gamma, \Psi \rangle)$ be an intuitionistic fuzzy directed graph. If $\mu_{U_i}(G) = \mu_{U_j}(G)$ and $\nu_{U_i}(G) = \nu_{U_j}(G)$ for some $i < j$, then an S_i point is also an S_j point.*

Proof The proof follows from the inequalities $\mu_{U_3}(G) \leq \mu_{U_2}(G) \leq \mu_{U_3}(G)$ and $\nu_{U_1}(G) \geq \nu_{U_2}(G) \geq \nu_{U_3}(G)$, and Definition 1.12.7. ∎

Theorem 1.12.21 *Let $G = (V, \langle \Gamma, \Psi \rangle)$ be an intuitionistic fuzzy directed graph. If $\mu_{U_i}(G) = \mu_{U_j}(G)$ and $\nu_{U_i}(G) = \nu_{U_j}(G)$ for some $i > j$, then a W_i point is also a W_m point, where $1 \leq m \leq i$.*

Proof Let $i = 3$ and $j = 1$. Then $\mu_{U_3}(G) = \mu_{U_1}(G)$ and $\nu_{U_3}(G) = \nu_{U_1}(G)$. Let x_k be a W_3 point. Because $\mu_{U_3}(G) \leq \mu_{U_2}(G) \leq \mu_{U_1}(G)$ and $\nu_{U_3}(G) \geq \nu_{U_2}(G) \geq \nu_{U_1}(G)$, we have $\mu_{U_3}(G_k) \leq \mu_{U_2}(G_k)$ and $\mu_{U_1}(G) = \mu_{U_2}(G) = \mu_{U_3}(G_k)$ and also $\nu_{U_3}(G_k) \geq \nu_{U_2}(G_k)$ and $\nu_{U_1}(G) = \nu_{U_2}(G) = \nu_{U_3}(G_k)$. Thus,

$$\mu_{U_2}(G_k) > \mu_{U_2}(G), \mu_{U_1}(G_k) > \mu_{U_1}(G)$$

and

$$\nu_{U_2}(G_k) < \nu_{U_2}(G), \nu_{U_1}(G_k) < \nu_{U_1}(G).$$

Hence, x_k is a W_m point for $1 \leq m \leq 3$.

Now assume that x_k is a W_2 point and that $\mu_{U_2}(G) = \mu_{U_1}(G)$ and $\nu_{U_2}(G) = \nu_{U_1}(G)$. It follows that $\mu_{U_1}(G_k) > \mu_{U_1}(G)$ and $\nu_{U_1}(G_k) < \nu_{U_1}(G)$. Thus, x_k is a W_m point for $1 \leq m \leq 2$.

We now show that if x_k is a W_3 point and $\mu_{U_3}(G) = \mu_{U_2}(G)$ and $\nu_{U_3}(G) = \nu_{U_2}(G)$, then x_k is a W_m point for $1 \leq m \leq 3$. Clearly, x_k is a W_2 point. We show x_k is a W_1 point. By Lemma 1.12.11, it follows that both in the kth row and in the kth column of B there is an element equal to $\nu_{U_3}(G)$. Hence, from the proof of Lemma 1.12.11, we have $b_{kj} = b_{jk} = \nu_{U_3}(G), j \neq k; j = 1, \ldots, n$. Thus, $(\Psi_{x_k} \cap \Psi_{x_k}^{-1})(x_j) \geq \nu_{U_3}(G)$, $j \neq k; j = 1, \ldots, n$, and so $c_{kj} = c_{jk} \geq \nu_{U_3}(G), j \neq k; j = 1, \ldots, n$. Therefore, $\nu_{U_1}(G) = \nu_{U_3}(C)$ and so $\nu_{U_1}(G_k) < \nu_{U_1}(G)$. (The proof for Γ is similar and can be found in [25].) ∎

Theorem 1.12.22 *Let x_k be a W_i point. If $\mu_{Ui}(G_k) = \mu_{U_j}(G_k)$ and $\nu_{Ui}(G_k) = \nu_{U_j}(G_k)$ for some $i < j$, then x_k is also a W_m point for $1 \leq m \leq j$.*

Proof The proof is similar to that of Theorem 1.12.21. ∎

1.13 Fuzzy Graphs and Complementary Fuzzy Graphs

We consider the work in [23] in this section. The following is stated in [23].

"As stated by the Central Intelligence Agency, trafficking in persons is modern day slavery involving victims who are forced, defrauded, or coerced into labor or sexual exploitation. Human trafficking deprives people of their human rights and freedoms, risks global health, promotes social breakdown, inhibits development of countries by depriving countries of their human capital, and helps to fuel the growth of organized crime. We take an approach based on the findings of the U. S. Department of State, Office to Monitor and Combat Trafficking in Persons. Each country is assigned a tier and each country is designated as a source, transit, or destination country, [13]. We also take a similar approach based on the CIA World Factbook, [11]."

Let σ be a fuzzy subset of S. Then $\sigma^* = \{x \in S \mid \sigma(x) > 0\}$ is called the **support** of σ. The set $\sigma^{\#} = \{x \in S \mid \sigma(x) < 1\}$ is called the **cosupport** of σ. In [21], the idea of a fuzzy point and its membership to and quasi-coincidence with a fuzzy subset with respect to the standard complement were used to define and study certain kinds of fuzzy topological spaces. In [7], an arbitrary complement was used. In this section, we introduce these ideas to fuzzy graph theory. Many of the results are easily extended to fuzzy directed graphs.

Recall that a **graph** G is a pair (V, E), where V is a finite set and E is a subset of the set of all subsets of V such that for all $X \in E$, $|X| = 2$. We use the notation $\{x, y\}$ to denote an element of E. The set V is called the set of **vertices** of G and E is called the set of **edges** of G. A **fuzzy graph** $G = (\sigma, \mu)$ is a pair such that σ is a fuzzy subset of V and μ is a fuzzy subset of E such that for all $x, y \in V$, $\mu(x, y) \leq \sigma(x) \wedge \sigma(y)$, where we write $\mu(x, y)$ for $\mu(\{x, y\})$. Let $G^* = (\sigma^*, \mu^*)$. Clearly G^* is a graph.

Let $c : [0, 1] \to [0, 1]$ be such that $(i)\, c(0) = 1$ and $c(1) = 0$ and (ii) for all $a, b \in [0, 1]$, $a \leq b \Rightarrow c(a) \geq c(b)$. Let λ be a fuzzy subset of V. Define the fuzzy subset $c\lambda$ of V by for all $x \in V$, $(c\lambda)(x) = c(\lambda(x))$. Then $c\lambda$ is called the **complement** of λ with respect to c. If there is a point $e_c \in [0, 1]$ such that $c(e_c) = e_c$, then e_c is called an **equilibrium point** for c. If e_c exists, it is unique. If $c(c(a)) = a$ for all $a \in [0, 1]$, then c is called **involutive** . If c is involutive, then c has an equilibrium point, [20, p. 48].

A **complementary fuzzy graph** $\widehat{G} = (\tau, \nu)$ is a pair such that τ is a fuzzy subset of V and ν is a fuzzy subset of E such that for all $x, y \in V$, $\nu(x, y) \geq \tau(x) \vee \tau(y)$. Let $\tau^{\#}$ denote the cosupport of τ and $\nu^{\#}$ the cosupport of ν. Let $\widehat{G}^{\#} = (\tau^{\#}, \nu^{\#})$.

Let $x \in V$ and let $t \in [0, 1]$. Define the fuzzy subset x_t of V by for all $y \in V$, $x_t(y) = t$ if $y = x$ and $x_t(y) = 0$ if $y \neq x$. Then x_t is called a **fuzzy singleton** . Let c be an involutive complement of $[0, 1]$. Define the fuzzy subset x_t^c of V as follows: for all $y \in V$,

$$x_t^c(y) = \begin{cases} t & \text{if } y \neq x, \\ c(t) & \text{if } y = x. \end{cases}$$

Then x_t^c is called a **complementary fuzzy singleton** .

The notions of a fuzzy graph and a complementary fuzzy graph can be combined in a natural way to form an intuitionistic fuzzy graph. Results on intuitionistic fuzzy graphs can be found in [1–6, 12, 24, 25]. In [10, 30, 33] and the previous sections, strengthening and weakening members of a group were considered.

We next provide the basic results needed for the remainder of the chapter. The proofs of parts (i) and (ii) of the following results are duals of each other as has been shown in Sect. 1.12 by using an involutive fuzzy complement c. Consequently, we prove only part (ii) in the following results because the case for part (i) has been considered in [7].

Definition 1.13.1 Let c and $\widehat{c}$ be involutive fuzzy complements. Let $G = (\sigma, \mu)$ and $\widehat{G} = (\tau, \nu)$ be a fuzzy graph and a complementary fuzzy graph, respectively. Then

(i) G is called a **quasi-fuzzy graph** with respect to c if for all $x, y \in V$ and for all $t \in [0, 1]$, $\mu(x, y) \geq t \Rightarrow t \leq \sigma(x) \wedge \sigma(y)$ or $t > c(\sigma(x) \wedge \sigma(y))$ and

(ii) $\widehat{G}$ is called a **complementary quasi-fuzzy graph** with respect to $\widehat{c}$ if for all $x, y \in V$ and for all $t \in [0, 1]$, $\nu(x, y) \leq t \Rightarrow t \geq \tau(x) \vee \tau(y)$ or $t < \widehat{c}(\tau(x) \vee$

$\tau(y))$. Then $(G, \widehat{G})$ is called a **quasi-fuzzy graph** with respect to $(c, \widehat{c})$ if G is a quasi-fuzzy graph with respect to c and $\widehat{G}$ is a complementary quasi-fuzzy graph with respect to $\widehat{c}$.

Theorem 1.13.2 (*i*) $G = (\sigma, \mu)$ *is a quasi-fuzzy graph with respect to* c *if and only if for all* $x, y \in V$, $\mu(x, y) \wedge e_{\underline{c}} \le \sigma(x) \wedge \sigma(y)$.

(*ii*) *Let* $\widehat{G} = (\tau, \nu)$. *Then* $\widehat{G}$ *is a complementary quasi-fuzzy graph with respect to* $\widehat{c}$ *if and only if for all* $x, y \in V$, $\nu(x, y) \vee e_{\widehat{c}} \ge \tau(x) \vee \tau(y)$.

Proof (*ii*) Let $x, y \in V$. Suppose $\nu(x, y) \le t$. Suppose $t < \tau(x) \vee \tau(y)$. Then by hypothesis, $t < \widehat{c}(\tau(x) \vee \tau(y))$. Thus, $t < e_{\widehat{c}}$. Suppose $\nu(x, y) \vee e_{\widehat{c}} < \tau(x) \vee \tau(y)$. We obtain a contradiction. Then $e_{\widehat{c}} < \tau(x) \vee \tau(y)$. Hence, $\tau(x) \vee \tau(y) > e_{\widehat{c}} \ge c(\tau(x) \vee \tau(y)) > t$ and $e_{\widehat{c}} = c(e_{\widehat{c}}) \ge \widehat{c}(\tau(x) \vee \tau(y))$. Let t' be such that $\tau(x) \vee \tau(y) > t' > e_{\widehat{c}}$. Then $t' > t$ and so $\nu(x, y) \le t'$. Because $t' < \tau(x) \vee \tau(y)$, we have by an argument as above that $t' < \widehat{c}(\tau(x) \vee \tau(y))$ and $t' < e_{\widehat{c}}$, a contradiction. Thus, $\nu(x, y) \vee e_{\widehat{c}} \ge \tau(x) \vee \tau(y)$.

Conversely, suppose for all $x, y \in V$, $\nu(x, y) \vee e_{\widehat{c}} \ge \tau(x) \vee \tau(y)$. Suppose $\nu(x, y) \le t$. If $t \ge \tau(x) \vee \tau(y)$, we have the desired result. Suppose $t < \tau(x) \vee \tau(y)$. Then $\nu(x, y) < \tau(x) \vee \tau(y)$. Because also $\nu(x, y) \vee e_{\widehat{c}} \ge \tau(x) \vee \tau(y)$, $\tau(x) \vee \tau(y) \le e_{\widehat{c}}$. Hence, $\widehat{c}(\tau(x) \vee \tau(y)) \ge c(e_{\widehat{c}}) = e_{\widehat{c}}$. Thus, $t < \tau(x) \vee \tau(y) \le e_{\widehat{c}} \le \widehat{c} (\tau(x) \vee \tau(y))$. ∎

Definition 1.13.3 (*i*) $G = (\sigma, \mu)$ is called a **quasi-fuzzy graph** with respect to $t \in (0, 1]$ if for all $\{x, y\} \in E$, $\mu(x, y) \wedge t \le \sigma(x) \wedge \sigma(y)$.

(*ii*) $\widehat{G} = (\tau, \nu)$ is called a **complementary quasi-fuzzy graph** with respect to $\widehat{t} \in [0, 1)$ if for all $\{x, y\} \in E$, $\nu(x, y) \vee \widehat{t} \ge \tau(x) \vee \tau(y)$.

In fact, we consider the following definition.

Definition 1.13.4 (*i*) Let $s, t \in [0, 1]$ with $s < t$. Then $G = (\sigma, \mu)$ is called an $(s, t]$-**fuzzy graph** if G^* is a graph and for all $x, y \in V$, $\mu(x, y) \wedge t \le (\sigma(x) \wedge \sigma(y)) \vee s$.

(*ii*) Let $\widehat{s}, \widehat{t} \in [0, 1)$ with $\widehat{s} > \widehat{t}$. Then $\widehat{G} = (\tau, \nu)$ is called a **complementary** $[\widehat{t}, \widehat{s})$-**fuzzy graph** if $\widehat{G}^{\#}$ is a graph and for all $x, y \in V$, $\nu(x, y) \vee \widehat{t} \ge (\tau(x) \vee \tau(y)) \wedge \widehat{s}$.

Let $\widehat{r}, \widehat{s}, \widehat{t} \in [0, 1]$. Suppose $\widehat{t} < \widehat{s}$. Let X be a set and λ be a fuzzy subset of X. Let $\lambda_{\widehat{ts}}^{\widehat{r}} = \{x \in X \mid \lambda(x) \wedge \widehat{s} \le \widehat{r} \vee \widehat{t}\}$. Then $\lambda_{\widehat{ts}}^{\widehat{r}}$ is called an $\widehat{r}_{\widehat{ts}}$-**level set** of λ. Let $G_{\widehat{ts}}^{\widehat{r}} = \left(\tau_{\widehat{ts}}^{\widehat{r}}, \nu_{\widehat{ts}}^{\widehat{r}}\right)$.

Theorem 1.13.5 (*i*) G *is an* $(s, t]$-*fuzzy graph if and only if for all* $r \in (s, t]$, G^r *is a subgraph of* (V, E).

(*ii*) $\widehat{G}$ *is a* $[\widehat{t}, \widehat{s})$-*fuzzy complementary graph if and only if for all* $\widehat{r} \in [0, 1]$, $\widehat{G}^{\widehat{r}}$ *is a complementary subgraph of* (V, E).

Proof (*ii*) Suppose $\widehat{G}$ is an $[\widehat{t}, \widehat{s}]$-fuzzy complementary graph. Let $\widehat{r} \in [0, 1]$. Suppose $\{x, y\} \in \nu_{\widehat{ts}}^{\widehat{r}}$. Then $\nu(x, y) \wedge \widehat{s} \leq \widehat{r} \vee \widehat{t}$. Now $(\tau(x) \vee \tau(y)) \wedge \widehat{s} \leq \nu(x, y) \vee \widehat{t}$ and so $(\tau(x) \vee \tau(y)) \wedge \widehat{s} \leq (\nu(x, y) \vee \widehat{t}) \wedge \widehat{s} = (\nu(x, y) \wedge \widehat{s}) \vee (\widehat{t} \wedge \widehat{s}) \leq (\widehat{r} \vee \widehat{t}) \vee \widehat{t} = (\widehat{r} \vee \widehat{t})$. Thus, $\tau(x) \wedge \widehat{s} \leq (\widehat{r} \vee \widehat{t})$ and $\tau(y) \wedge \widehat{s} \leq (\widehat{r} \vee \widehat{t})$. Hence, $x, y \in \tau_{\widehat{ts}}^{\widehat{r}}$. Thus, $\widehat{G}_{\widehat{ts}}^{\widehat{r}}$ is a subgraph of (V, E).

Conversely, suppose for all $\widehat{r} \in [0, 1]$, $\widehat{G}_{\widehat{ts}}^{\widehat{r}}$ is a subgraph of (V, E). Let $\{x, y\} \in E$ and let $\nu(x, y) = \widehat{r_0}$. Then $\nu(x, y) \wedge \widehat{s} \leq \widehat{r_0} \vee \widehat{t}$. Hence, $\{x, y\} \in \nu_{\widehat{ts}}^{\widehat{r_0}}$. Thus, $x, y \in \tau_{\widehat{ts}}^{\widehat{r_0}}$. Hence, $\tau(x) \wedge \widehat{s} \leq \widehat{r_0} \vee \widehat{t}$ and $\tau(y) \wedge \widehat{s} \leq \widehat{r_0} \vee \widehat{t}$. Thus, $(\tau(x) \vee \tau(y)) \wedge \widehat{s} \leq \widehat{r_0} \vee \widehat{t} = \nu(x, y) \vee \widehat{t}$. Hence, $\widehat{G}$ is a $[\widehat{t}, \widehat{s}]$-fuzzy complementary graph. ∎

Theorem 1.13.6 (*i*) G *is an* $(s, t]$-*fuzzy graph if and only if for all* $r \in [0, 1]$, G_{st}^{r} *is a subgraph of* (V, E).

(*ii*) $\widehat{G}$ *is a* $[\widehat{t}, \widehat{s}]$-*fuzzy complementary graph if and only if for all* $\widehat{r} \in [0, 1]$, $\widehat{G}_{\widehat{ts}}^{\widehat{r}}$ *is a subgraph of* (V, E).

Proof (ii) Suppose $\widehat{G}$ is a $[\widehat{t}, \widehat{s}]$-fuzzy complementary graph. Let $\widehat{r} \in [0, 1]$. Suppose $\{x, y\} \in \nu_{\widehat{ts}}^{\widehat{r}}$. Then $\nu(x, y) \wedge \widehat{s} \leq \widehat{r} \vee \widehat{t}$. Now $(\tau(x) \vee \tau(y)) \wedge \widehat{s} \leq \nu(x, y) \vee \widehat{t}$ and so $(\tau(x) \vee \tau(y)) \wedge \widehat{s} \leq (\tau(x, y) \vee \widehat{t}) \wedge \widehat{s} = (\nu(x, y) \wedge \widehat{s}) \vee (\widehat{t} \wedge \widehat{s}) \leq (\widehat{r} \vee \widehat{t}) \vee \widehat{t} = \widehat{r} \vee \widehat{t}$. Thus, $\tau(x) \wedge \widehat{s} \leq \widehat{r} \vee \widehat{t}$ and $\tau(y) \wedge \widehat{s} \leq \widehat{r} \vee \widehat{t}$. Hence, $x, y \in \tau_{\widehat{ts}}^{\widehat{r}}$. Thus, $(\tau_{\widehat{ts}}^{\widehat{r}}, \nu_{\widehat{ts}}^{\widehat{r}})$ is a subgraph of (V, E).

Conversely, suppose $(\tau_{\widehat{ts}}^{\widehat{r}}, \nu_{\widehat{ts}}^{\widehat{r}})$ is a subgraph of (V, E) for all $\widehat{r} \in [0, 1]$. Let $\{x, y\} \in E$ and let $\nu(x, y) = \widehat{r_0}$. Then $\nu(x, y) \wedge \widehat{s} \leq \widehat{r_0} \vee t$. Hence, $\{x, y\} \in \nu_{\widehat{ts}}^{\widehat{r}}$. Thus, $\tau(x) \wedge \widehat{s} \leq \widehat{r_0} \vee t$. Hence, $(\tau(x) \vee \tau(y)) \wedge \widehat{s} \leq \widehat{r_0} \vee t = \nu(x, y) \vee \widehat{t}$. Thus, $\widehat{G}$ is $[\widehat{t}, \widehat{s}]$-fuzzy complementary graph. ∎

Let $G = (\sigma, \mu)$ be a fuzzy graph. Let $x, y \in V$. A path from x to y in G is a sequence $\pi : x = x_0, x_1, \ldots, x_n = y$ of points x_i in V such that $\sigma(x_i) > 0$ for $i = 0, 1, \ldots, n$ and $\mu(x_{i-1}, x_i) > 0$ for $i = 1, \ldots, n$. The **strength** of the path π is $\wedge \{\mu(x_{i-1}, x_i) \mid i = 1, \ldots, n\}$. Define $\text{CONN}_G(x, y)$ to be the maximum of the strengths of all paths from x to y. If $x \in V$, we use the notation $G \backslash x$ to denote the fuzzy graph determined by setting $\sigma(x) = 0$ and $\mu(x, y) = 0$ for all $y \in V$ such that $\{x, y\} \in E$. Properties of fuzzy cut vertices and fuzzy end vertices can be found in [5, 6], respectively.

We next present the main results concerning fuzzy cut vertices and fuzzy end vertices. We provide an interpretation of a fuzzy graph, where vertices can be considered to be countries and edges connections between countries that allow trafficking.

Definition 1.13.7 Let $G = (\sigma, \mu)$ be a fuzzy graph and let $x \in V$. Then x is called a **fuzzy cut vertex** if there exists $y, z \in V \backslash \{x\}$ such that $\text{CONN}_G(y, z) > \text{CONN}_{G \backslash x}(y, z)$.

Definition 1.13.8 Let $G = (\sigma, \mu)$ be a fuzzy graph and let $x \in V$. Then x is called a **fuzzy end vertex** if there exists $y \in V \backslash \{x\}$ such that $\mu(x, y) > \vee \{\mu(x, z) \mid z \in V, z \neq y\}$.

Definition 1.13.9 Let $s, t \in [0, 1]$ be such that $s < t$. Let $G = (\sigma, \mu)$ be a fuzzy graph and let $x \in V$. Then x is called a $(s, t]$-**fuzzy cut vertex** if there exists $y, z \in V \setminus \{x\}$ such that $\text{CONN}_G(y, z) \wedge t > \text{CONN}_{G \setminus x}(y, z) \vee s$.

Definition 1.13.10 Let $s, t \in [0, 1]$ be such that $s < t$. Let $G = (\sigma, \mu)$ be a fuzzy graph and let $x \in V$. Then x is called an $(s, t]$-**fuzzy end vertex** if there exists $y \in V \setminus \{x\}$ such that $\mu(x, y) \wedge t > \vee\{\mu(x, z) \mid z \in V, z \neq y\} \vee s$.

If $s = 0$ and $t = 1$, then $(0, 1]$-fuzzy end vertices and $(0, 1]$-fuzzy cut vertices are simply fuzzy end vertices and fuzzy cut vertices of G, respectively.

Suppose that $x \in V$ is an $(s, t]$-fuzzy end vertex. Then the $y \in V$ such that $\mu(x, y) \wedge t > \vee\{\mu(x, z) \mid z \in V, z \neq y\} \vee s$ is unique. This follows because for all $u \in V \setminus \{y\}$, $\mu(x, y) \geq \mu(x, y) \wedge t > \mu(x, u)$. Also, if x is an $(s, t]$-fuzzy end vertex, then x is also an $(r, w]$-fuzzy vertex for all $r \in [0, s]$ and $w \in [t, 1]$.

The following example shows that a fuzzy end vertex may not be a an $(s, t]$-fuzzy end vertex for certain s and t. An interpretation of this may be as follows. The values s and t may represent a benefit of supporting the fuzzy end node x. The support of a fuzzy end node is not of sufficient value because if edge $\{x, y\}$ were deleted, $\mu(x, z) \not\leq t = 0.95$.

Example 1.13.11 Let $V = \{x, y, z\}$ and $E = \{\{x, y\}, \{x, z\}, \{y, z\}\}$. Let $s = 0.05$ and $t = 0.95$. Define the fuzzy subset σ of V and the fuzzy subset μ of E as follows: $\sigma(x) = \sigma(y) = \sigma(z) = 1$ and $\mu(x, y) = \mu(y, z) = 1$ and $\mu(x, z) = 0.99$. Then x is a fuzzy end vertex, but not an $(s, t]$-fuzzy end vertex. We also have that y is a fuzzy cut vertex, but not an $(s, t]$-fuzzy cut vertex. This can be interpreted as saying that attacking or deleting the fuzzy cut vertex is not of sufficient value because the connectedness of the graph is not sufficiently reduced, where sufficiently is defined by $t = 0.95$.

Example 1.13.12 Let $V = \{x, y, z\}$ and $E = \{\{x, y\}, \{x, z\}, \{y, z\}\}$. Let $s = 0.05$ and $t = 0.95$. Define the fuzzy subset σ of V and the fuzzy subset μ of E as follows: $\sigma(x) = \sigma(y) = \sigma(z) = 1$ and $\mu(x, y) = \mu(y, z) = 0.04$ and $\mu(x, z) = 0.01$. Then x is a fuzzy end vertex, but not an $(s, t]$-fuzzy end vertex. We also have that y is a fuzzy cut vertex, but not an $(s, t]$-fuzzy cut vertex.

Remark 1.13.13 Let $x \in V$. If x is an end vertex in (V, E), then x is a fuzzy end vertex and an $(s, t]$-fuzzy end vertex for all $t \in (0, 1]$ and for all $s < t \leq \mu(x, y)$, where y is the unique neighbor of x in G^*.

Definition 1.13.14 Let $x, y \in V$. Then y is called a **neighbor** of x if $\mu(x, y) > 0$. y is called a t-**strong neighbor** of x if y is a neighbor of x and $\mu(x, y) \wedge t \geq \text{CONN}_G(x, y) \wedge t$.

Definition 1.13.15 Let $\widehat{G} = (\tau, \nu)$ be a complementary fuzzy graph with respect to $\widehat{c}$. Let $x, y \in V$. A **complementary path** from x to y is a sequence P : $x_0 = x, x_1, \ldots, x_n = y$ of points x_i in V such that $\tau(x_i) < 1$ for $i = 0, 1, \ldots, n$ and $\nu(x_{i-1}, x_i) < 1$ for $i = 1, \ldots, n$. Define the **strength** of the P to be $s(P) = \vee\{\nu(x_{i-1}, x_i) \mid i = 1, .., n\}$. Define $\text{CONN}_{\widehat{G}}^c(x, y)$ to be the minimum of the strengths of all complementary paths from x to y.

Definition 1.13.16 Let $\widehat{G} = (\tau, \nu)$ be a complementary fuzzy graph and let $x \in V$. Then x is called a **complementary fuzzy cut vertex** if there exists $y, z \in V \setminus \{x\}$ such that $\mathrm{CONN}^c_{\widehat{G}}(y, z) < \mathrm{CONN}^c_{\widehat{G} \setminus x}(y, z)$.

Definition 1.13.17 Let $\widehat{G} = (\tau, \nu)$ be a complementary fuzzy graph and let $x \in V$. Then x is called a **complementary fuzzy end vertex** if there exists $y \in V \setminus \{x\}$ such that $\nu(x, y) < \wedge \{\nu(x, z) \mid z \in V, z \neq y\}$.

Definition 1.13.18 Let $\widehat{t}, \widehat{s} \in [0, 1)$ be such that $\widehat{t} < \widehat{s}$. Let $\widehat{G} = (\tau, \nu)$ be a complementary fuzzy graph and let $x \in V$. Then x is called a **complementary $[\widehat{t}, \widehat{s})$-fuzzy cut vertex** if there exists $y, z \in V \setminus \{x\}$ such that $\mathrm{CONN}^c_{\widehat{G}}(y, z) \vee \widehat{t} < \mathrm{CONN}^c_{\widehat{G} \setminus x}(y, z) \wedge \widehat{s}$.

Definition 1.13.19 Let $\widehat{t}, \widehat{s} \in [0, 1]$ be such that $\widehat{t} < \widehat{s}$. Let $\widehat{G} = (\tau, \nu)$ be a complementary fuzzy graph and let $x \in V$. Then x is called a **complementary $[\widehat{t}, \widehat{s})$-fuzzy end vertex** if there exists $y \in V \setminus \{x\}$ such that $\nu(x, y) \vee \widehat{t} < \wedge \{\nu(x, z) \mid z \in V, z \neq y\} \wedge \widehat{s}$.

If $\widehat{t} = 0$ and $\widehat{s} = 1$, then complementary $[0, 1)$-fuzzy end vertices and complementary $[1, 0)$-fuzzy cut vertices are simply fuzzy end vertices and fuzzy cut vertices of G, respectively.

Suppose that $x \in V$ is a complementary $[\widehat{t}, s)$-fuzzy end vertex. Then the $y \in V$ such that $\nu(x, y) \vee \widehat{t} < \wedge \{\nu(x, z) \mid z \in V, z \neq y\} \wedge \widehat{s}$ is unique. This follows because for all $u \in V \setminus \{y\}$, $\nu(x, y) \leq \nu(x, y) \vee \widehat{t} < \nu(x, u)$. Also, if x is a complementary $[\widehat{t}, \widehat{s})$-fuzzy end vertex, then x is also an $[w, r)$-fuzzy vertex for all $r \in [\widehat{s}, 1]$ and $w \in [0, \widehat{t}]$.

Example 1.13.20 Let $V = \{x, y, z\}$ and $E = \{\{x, y\}, \{x, z\}, \{y, z\}\}$. Let $\widehat{s} = 0.95$ and $\widehat{t} = 0.05$. Define the fuzzy subset τ of V and the fuzzy subset ν of E as follows: $\tau(x) = \tau(y) = \tau(z) = 0$ and $\nu(x, y) = \nu(y, z) = 0$ and $\nu(x, z) = 0.01$. Then x is a complementary fuzzy end vertex, but not a complementary $[\widehat{t}, s)$-fuzzy end vertex. We also have that y is a complementary fuzzy cut vertex, but not a complementary $[\widehat{t}, \widehat{s})$-fuzzy cut vertex.

Example 1.13.21 Let $V = \{x, y, z\}$ and $E = \{\{x, y\}, \{x, z\}, \{y, z\}\}$. Let $\widehat{s} = 0.95$ and $\widehat{t} = 0.05$. Define the fuzzy subset τ of V and the fuzzy subset ν of E as follows: $\tau(x) = \tau(y) = \tau(z) = 0$ and $\nu(x, y) = \nu(y, z) = 0.96$ and $\eta(x, z) = 0.99$. Then x is a complementary fuzzy end vertex, but not an $[\widehat{t}, s)$-fuzzy end vertex. We also have that y is a complementary fuzzy cut vertex, but not a complementary $[\widehat{t}, s)$-fuzzy cut vertex.

Remark 1.13.22 Let $x \in V$. If x is a complementary end vertex in (V, E), then x is a complementary fuzzy end vertex and an $[\widehat{t}, s)$-fuzzy end vertex for all $\widehat{t} \in [0, 1)$ and for all $\widehat{s} > \widehat{t} \geq \nu(x, y)$, where y is the unique complementary neighbor of x in $G^{\#}$.

Definition 1.13.23 Let $x, y \in V$. Then y is called a **complementary neighbor** of x if $\nu(x, y) < 1$. y is called a $\widehat{t}$-**strong complementary neighbor** of x if y is a complementary neighbor of x and $\nu(x, y) \vee \widehat{t} \leq \text{CONN}_{\widehat{G}}^c(x, y) \vee \widehat{t}$.

Theorem 1.13.24 (i) *Let $z \in V$ be an $(s, t]$-fuzzy cut vertex of G. Then z has at least two t-strong neighbors.*

(ii) *Let $z \in V$ be a complementary $[\widehat{t}, s)$-fuzzy cut vertex of $\widehat{G} = (\tau, \nu)$. Then z has at least two complementary $\widehat{t}$-strong neighbors.*

Proof (ii) Because z is a complementary $[\widehat{t}, s)$-fuzzy cut vertex of $\widehat{G}$, there exists $x, y \in V \setminus \{z\}$ such that $\text{CONN}_{\widehat{G}}^c(x, y) \vee \widehat{t} < \text{CONN}_{\widehat{G} \setminus z}^c(x, y) \wedge \widehat{s}$. Hence, there exists a complementary path P from x to y and passing through z such that $s(P) \vee \widehat{t} = \text{CONN}_{\widehat{G}}^c(x, y) \vee \widehat{t}$. Because $x \neq z \neq y$, z has two complementary neighbors in P, say u and v. Because $\{u, z\}$ and $\{z, v\}$ are on P, $\nu(u, z) \vee \widehat{t} \leq s(P) \vee \widehat{t}$ and $\nu(z, v) \vee \widehat{t} \leq s(P) \vee \widehat{t}$. Suppose $\nu(u, z) \vee \widehat{t} > \text{CONN}_{\widehat{G}}^c(u, z) \vee \widehat{t}$. Then there exists a complementary path Q from u to z such that $s(Q) \vee \widehat{t} < \nu(u, z) \vee \widehat{t}$. Let u' in Q be a complementary neighbor of z. We show $u' \neq v$. Suppose $u' = v$. Let Q' be the path $x \ldots u \ldots u' \ldots v \ldots y$. Then $s(Q') \vee \widehat{t} \leq s(P) \vee \widehat{t}$, but Q' does not pass through z. However, this is impossible because $\text{CONN}_{\widehat{G} \setminus z}^c(x, y) \vee \widehat{t} > \text{CONN}_{\widehat{G}}^c(x, y) \vee \widehat{t}$. Let P' denote the path $x \ldots u \ldots u' z v \ldots y$. Then $s(P') \vee \widehat{t} \leq s(P) \vee \widehat{t} = \text{CONN}_{\widehat{G}}^c(x, y) \vee \widehat{t}$. Suppose $\nu(u', z) \vee \widehat{t} > \text{CONN}_{\widehat{G}}^c(u, z) \vee \widehat{t}$. Then we can apply the above process again. Now length$(P') >$ length(P). Hence, this process must end in a finite number of steps yielding a complementary neighbor u^* of z such that $\nu(u^*, z) \vee \widehat{t} \leq \text{CONN}_{\widehat{G}}^c(u, z) \vee \widehat{t}$. As above $u^* \neq v$. Hence, a similar argument can be used to yield a complementary neighbor v^* of z such that $u^* \neq v^*$ and $\mu(v^*, z) \vee \widehat{t} \leq \text{CONN}_{\widehat{G}}^c(u, z) \vee \widehat{t}$. Thus, z has two complementary $\widehat{t}$-strong neighbors, namely u^* and v^*. ∎

Example 1.13.25 Let $V = \{x, y, z\}$ and $E = \{\{x, y\}, \{x, z\}, \{y, z\}\}$. Define $\mu : E \to [0, 1]$ as follows: $\mu(x, y) = 0.7$, $\mu(x, z) = 0.9$, and $\mu(y, z) = 0.8$. Then x is not a fuzzy cut vertex. Now $\text{CONN}_G(x, y) = 0.8$ and $\text{CONN}_G(x, z) = 0.9$. Now z is a 1-strong neighbor of x because $\mu(x, z) \wedge 1 \geq \text{CONN}_G(x, z) \wedge 1$, but y is not a 1-strong neighbor of x because $\mu(x, y) \not\geq \text{CONN}_G(x, y)$. However, z is a fuzzy cut vertex of G because $\text{CONN}_G(x, y) = 0.8 > 0.7 = \text{CONN}_{G \setminus z}(x, y)$. Now z has two 1-strong neighbors, namely x and y, because $\mu(z, x) = \text{CONN}_G(z, x)$ and $\mu(z, y) = \text{CONN}_G(z, y)$.

Definition 1.13.26 Let $x \in V$. Then x is called a **complementary weak fuzzy end node** if there exists $\widehat{t} \in [0, h^c(\nu))$ such that x is an end node in $\widehat{G^t}$, where $h^c(\nu)$ is the complementary height of ν, i.e., $h^c(\nu) = \wedge \{\nu(x, y) \mid \{x, y\} \in E\}$.

Definition 1.13.27 Let $x \in V$. Then x is called a **complementary partial fuzzy end node** if x is a complementary end node in $\widehat{G^t}$ for all $\widehat{t} \in [h^c(\nu), d^c(\nu)) \cup \{d^c(\nu)\}$, where $d^c(\mu)$ is the complementary depth of ν, i.e., $d^c(\mu) = \vee \{\nu(x, y) \mid \{x, y\} \in E\}$.

Definition 1.13.28 Let $x \in V$. Then x is called a **weak fuzzy end node** if there exists $t \in (0, h(\mu)]$ such that x is an end node in G^t, where $h(\mu)$ is the **height** of μ, i.e., $h(\mu) = \vee \{\mu(x, y) \mid \{x, y\} \in E\}$.

Definition 1.13.29 Let $x \in V$. Then x is called a **partial fuzzy end node** if x is an end node in G^t for all $t \in (d(\mu), h(\mu)] \cup \{h(\mu)\}$, where $d(\mu)$ is the depth of μ, i.e., $d(\mu) = \wedge \{\mu(x, y) \mid \{x, y\} \in E\}$.

Proposition 1.13.30 (*i*) *Let $x \in V$. Then x is an $(0, 1]$-fuzzy end vertex if and only if x is a weak fuzzy end vertex.*

(*ii*) *Let $x \in V$. Then x is a complementary $[0, 1)$-fuzzy end vertex if and only if x is a complementary weak fuzzy end vertex.*

Proof (ii) Suppose x is a complementary $[0, 1)$-fuzzy end vertex. Then there exists $y \in V$ such that $\nu(x, y) = \nu(x, y) \vee 0 < \wedge \{\nu(x, z) \mid z \in V, z \neq y\}$. Thus, x is a complementary end vertex in $\widehat{G}^a$, where $a = \nu(x, y)$. Hence, x is a complementary weak fuzzy end vertex. Conversely, suppose x is a complementary weak fuzzy end vertex. Then x is a complementary end node in $\widehat{G}^b$ for some $b \in [0, h^c(\nu))$. Hence, there exists a unique $y \in V$ such that $\nu(x, y) < b$ and for all other $z \in V$, $\nu(x, z) > b$. Thus, x is a complementary $[0, 1)$-fuzzy end vertex. ∎

Example 1.13.31 Let $V = \{x, y, z\}$. Define the fuzzy relation ν on V as follows:

$$\nu(x, y) = 0.7, \ \nu(y, z) = 0.8, \ \text{and} \ \nu(x, z) = 0.9.$$

Then $0.7 = \nu(x, y) = \nu(x, y) \vee 0 < \wedge \{\nu(x, u) \mid u \in V, u \neq y\} = \nu(x, x) \wedge \nu(x, z) = 0.9$. Also, x is a complementary weak fuzzy end vertex for $\widehat{G}^a = (\tau^a, \nu^a)$, where $a = \nu(x, y) = 0.7$ because $\nu^{0.7} = \{(x, y)\}$. Now $h^c(\nu) = 0.7$. Let $b = 0.7$. Then there exists an unique $y \in V$ such that $\nu(x, y) \leq b$ and for all other $z \in V, \nu(x, z) > b$. Thus, x is a complementary $[0, 1)$-fuzzy end vertex.

Corollary 1.13.32 (*i*) *Let $x \in V$. Then x is a partial fuzzy end vertex if and only if x is a $(d(\mu), 1]$-fuzzy end vertex.*

(*ii*) *Let $x \in V$. Then x is a complementary partial fuzzy end vertex if and only if x is a $[d^c(\nu), 1)$-fuzzy end vertex.*

Corollary 1.13.33 (*i*) *Let $x \in V$. Then x is an end vertex of G^* if and only if x is a $(0, d(\mu)]$-fuzzy end vertex.*

(*ii*) *Let $x \in V$. Then x is a complementary end vertex of $\widehat{G}^{\#}$ if and only if x is a complementary $[0, d^c(\nu))$-fuzzy end vertex.*

Application.
We apply our results to social network analysis.

Definition 1.13.34 The pair $(G, \widehat{G})$ is called an **intuitionistic** $(s, t]$, $[\widehat{t}, \widehat{s})$-**fuzzy graph** if $G = (\sigma, \mu)$ is an $(s, t]$-fuzzy graph and $\widehat{G} = (\tau, \nu)$ is a complementary $[\widehat{t}, \widehat{s})$-fuzzy graph such that $0 \leq s + \widehat{s} \leq 1, 0 \leq t + \widehat{t} \leq 1$ and $0 \leq \sigma(x) + \tau(x) \leq 1, 0 \leq \mu(x, y) + \nu(x, y) \leq 1$ for all $x, y \in V$.

Let $x, y \in V$. Suppose $\mu(x, y) > \sigma(x) \wedge \sigma(y)$. Let

$$s_{x,y} = \wedge\{s \in (0, 1] \mid \mu(x, y) \leq (\sigma(x) \wedge \sigma(y)) \vee s\},$$
$$t_{x,y} = \vee\{t \in (0, 1] \mid \mu(x, y) \wedge t \leq \sigma(x) \wedge \sigma(y)\}.$$

Then $s_{x,y} = \mu(x, y)$ and $t_{x,y} = \sigma(x) \wedge \sigma(y)$. Let

$$s^* = \vee\{s_{x,y} \mid x, y \in X, \mu(x, y) > \sigma(x) \wedge \sigma(y), \mu(x, y) \leq (\sigma(x) \wedge \sigma(y)) \vee s_{x,y}\},$$
$$t^* = \wedge\{t_{x,y} \mid x, y \in X, \mu(x, y) > \sigma(x) \wedge \sigma(y), \mu(x, y) \wedge t_{x,y} \leq (\sigma(x) \wedge \sigma(y))\}.$$

Then $s^* = \vee\{\mu(x, y) \mid x, y \in X, \mu(x, y) > \sigma(x) \wedge \sigma(y)\}$ and $t^* = \wedge\{\sigma(x) \wedge \sigma(y) \mid x, y \in X, \mu(x, y) > \sigma(x) \wedge \sigma(y)\}$.

Proposition 1.13.35 *(i) Let $a, b, s, t \in [0, 1]$ be such that $a > b$ and $0 < s < t \leq 1$. Then $a \wedge t \leq b \vee s \Leftrightarrow t \leq b$ or $a \leq s$.*

(ii) Let $a, b, \widehat{s}, \widehat{t} \in [0, 1]$ be such that $b > a$ and $0 \leq \widehat{t} < \widehat{s} < 1$. Then $a \vee \widehat{t} \geq b \wedge \widehat{s} \Leftrightarrow \widehat{t} \geq b$ or $a \geq \widehat{s}$.

Proof (i) Suppose $a \wedge t \leq b \vee s$. Suppose $t > b$. Then because $a > b$ and $a \wedge t \leq b \vee s$, we have $s \geq a \wedge t$ because $t > s$ and $a \leq s$. Suppose $a > b$. Because $a > b$ and $a \wedge t \leq b \vee s$, we have $t \leq b \vee s$. But $s < t$ and so $t \leq b$.

Conversely, suppose $t \leq b$ or $a \leq s$. If $t \leq b$, then clearly $a \wedge t \leq b \vee s$. If $a \leq s$, then clearly $a \wedge t \leq b \vee s$. ∎

For a, b fixed with $a > b$, there exists two solutions for $a \wedge t \leq b \vee s$, namely

$$t = b, \ s = 0 \text{ and } t = 1, \ s = a$$

for which t is as large as possible and s is as small as possible (keeping $0 \leq s < t \leq 1$). These two solutions are candidates for $\wedge\{1 - t + s \mid a \wedge t \leq b \vee s\}$, where $a > b$ and a, b are fixed.

For all pairs (a, b) (finite in number) with $a > b$,

$$a \wedge 1 \leq b \vee \max a,$$
$$a \wedge \min b \leq b \vee 0$$

with $t = 1 > \max a$ and $\min b > 0$ yield $t = 1, s = \max a$ and $t = \min b, s = 0$ as solutions to $a \wedge t \leq b \vee s$ for all a, b with $a > b$. Thus, there exists $s, t, 0 \leq s < t \leq 1$, which is a solution to $a \wedge t \leq b \vee s$ for all a, b. Hence, there exists a solution for which $\wedge\{1 - t + s \mid a \wedge t \leq b \vee s\}$ exists because we are assuming we have a given finite set of pairs (a, b) with $a > b$.

In the above discussion, we can substitute $\mu(x, y)$ for a and $\sigma(x) \wedge \sigma(y)$ for b. Let $\widehat{s}_{x,y}$ and $\widehat{t}_{x,y}$ be defined in a complementary manner such as $s_{x,y}$ and $t_{x,y}$, respectively.

Corollary 1.13.36 (*i*) *Suppose* $\sigma : V \to [0, 1]$ *and* $\mu : E \to [0, 1]$. *Let* $s, t \in [0, 1]$ *be such that* $0 \leq s < t \leq 1$. *Let* $x, y \in X$. *Then* $\mu(x, y) \wedge t_{x,y} \leq (\sigma(x) \wedge \sigma(y)) \vee s_{x,y} \Leftrightarrow t_{x,y} \leq \sigma(x) \wedge \sigma(y)$ *or* $\mu(x, y) \leq s_{x,y}$.

(*ii*) *Suppose* $\tau : V \to [0, 1]$ *and* $\nu : E \to [0, 1]$. *Let* $\widehat{s}, \widehat{t} \in [0, 1]$ *be such that* $0 \leq \widehat{t} < \widehat{s} < 1$. *Let* $x, y \in X$. *Then* $\nu(x, y) \vee \widehat{t}_{x,y} \geq ((\tau(x) \vee \tau(y)) \wedge \widehat{s}_{x,y} \Leftrightarrow \widehat{t}_{x,y} \geq \tau(x) \vee \tau(y)$ *or* $\nu(x, y) \geq \widehat{s}_{x,y}$.

In attacking or influencing a particular vertex (country) or edge, there may be a cost such as time and/or money. In the next example, we provide an example of a cost function.

Example 1.13.37 Let $V = \{x, y, z, w\}$. Define $\mu : E \to [0, 1]$ and $\sigma : X \to [0, 1]$ as follows:

$$\mu(x, y) = 7/16, \ \sigma(x) = \sigma(y) = 3/8, \tag{1.6}$$

$$\mu(z, w) = 15/16, \ \sigma(z) = \sigma(w) = 7/8, \tag{1.7}$$

and so that $\mu(u, v) \leq \sigma(u) \wedge \sigma(v)$ for all other pairs (u, v). Consider the cost function $\text{Cost}(s, t) = 1 - t + s$. From (1.6), we have $s = 7/16, t = 1$ and $s = 0, t = 3/8$. From (1.7), we have $s = 15/16, t = 1$ and $s = 0, t = 7/8$. Thus, $\text{Cost}(15/16, 1) = 1 - 1 + 15/16 = 15/16$ and $\text{Cost}(0, 3/8) = 1 - 3/8 + 0 = 5/8$. Now

$$\mu(x, y) \wedge 7/8 = 7/16 \wedge 7/8 = 7/16 \leq 3/8 \vee 7/16 = (\sigma(x) \wedge \sigma(y)) \vee 7/16,$$

and

$$\mu(z, w) \wedge 7/8 = 15/16 \wedge 7/8 = 7/8 \leq 7/8 \vee 7/16 = (\sigma(z) \wedge \sigma(w)) \vee 7/16.$$

However, $\text{Cost}(7/16, 7/8) = 1 - 7/8 + 7/16 = 9/16$, which is smaller than the previous two costs.

Let $\sigma : V \to [0, 1], \mu : E \to [0, 1], \tau : V \to [0, 1]$, and $\nu : E \to [0, 1]$. We interpret $\sigma(x)$ as the degree of ease in influencing (attacking) x. We interpret $\mu(x, y)$ as the degree of failure in influencing (attacking) edge $\{x, y\}$. We interpret $\tau(x)$ as the degree of difficulty influencing x. We interpret $\nu(x, y)$ as the degree of success in influencing edge $\{x, y\}$.

Let $x, y \in X$. Suppose $\mu(x, y) > \sigma(x) \wedge \sigma(y)$. We wish to find $s_{x,y}, t_{x,y} \in [0, 1]$, $0 < s_{x,y} \leq t_{x,y} \leq 1$, so that $\mu(x, y) \wedge t_{x,y} \leq (\sigma(x) \wedge \sigma(y)) \vee s_{x,y}$. In general, we wish to find "suitable" $s, t \in [0, 1]$ such that for all $x, y \in X, \mu(x, y) \wedge t \leq (\sigma(x) \wedge \sigma(y)) \vee s, 0 < s \leq t \leq 1$.

Let $x, y \in X$. Suppose $\nu(x, y) < \tau(x) \wedge \tau(y)$. We wish to find $\widehat{s}_{x,y}, \widehat{t}_{x,y} \in [0, 1]$, $0 \leq \widehat{t}_{x,y} < \widehat{s}_{x,y} < 1$, so that $\nu(x, y) \vee \widehat{t}_{x,y} \geq (\tau(x) \vee \tau(y)) \wedge \widehat{s}_{x,y}$. In general, we wish to find "suitable" $\widehat{s}, \widehat{t} \in [0, 1]$ such that for all $x, y \in X, \nu(x, y) \vee \widehat{t} \geq (\tau(x) \vee \tau(y)) \wedge \widehat{s}, 0 \leq \widehat{t} < \widehat{s} < 1$.

For all $x, y \in X, \mu(x, y) \wedge 1 = \mu(x, y)$ and $\sigma(x) \wedge \sigma(y)) \vee 0 = \sigma(x) \wedge \sigma(y)$. Thus, a "cost" or "effort" in obtaining $\mu(x, y) \wedge t \leq (\sigma(x) \wedge \sigma(y)) \vee s$ is $1 - t + s$.

A suitable (s, t) is one which minimizes $1 - t + s$. Similarly, a suitable $(\widehat{t}, \widehat{s})$ is one which minimizes $\widehat{t} + 1 - \widehat{s}$.

The number $\mu(x, y)$ might also denote the degree of flow or the degree of capacity in edge $\{x, y\}$. The number $\sigma(x)$ might denote the degree of inefficient attack or failure of attack on node x.

Proposition 1.13.38 (i) (σ, μ) is a fuzzy graph if and only if (σ, μ) is an $(s, t]$-fuzzy graph for all $s, t \in (0, 1]$ such that $0 < s < t \leq 1$.

(ii) (τ, ν) is a complementary fuzzy graph if and only if (τ, ν) is a complementary $[\widehat{t}, \widehat{s})$-fuzzy graph for all $\widehat{s}, \widehat{t} \in [0, 1]$ such that $0 \leq \widehat{t} < \widehat{s} < 1$.

Proof (i) Suppose (σ, μ) is a fuzzy graph. Let $s, t \in [0, 1]$. Because $\mu(x, y) \leq \sigma(x) \wedge \sigma(y)$, $\mu(x, y) \wedge t \leq (\sigma(x) \wedge \sigma(y)) \vee s$. Conversely, suppose (σ, μ) is an $(s, t]$-fuzzy graph for all $s, t \in (0, 1]$ such that $0 < s < t \leq 1$. Let $t = 1$. Then $\mu(x, y) \leq (\sigma(x) \wedge \sigma(y)) \vee s$ for all $s \in (0, 1)$. Because s can be taken arbitrarily close to 0 and X is finite, it follows that $\mu(x, y) \leq \sigma(x) \wedge \sigma(y)$.

It follows easily that $\exists s, t \in (0, 1]$ such that $0 < s < t \leq 1$ and (σ, μ) is an $(s, t]$-fuzzy graph $\not\Rightarrow$ (σ, μ) is a fuzzy graph: Take $\mu(x, y) > \sigma(x) \wedge \sigma(y)$ and $s > \mu(x, y)$. Then $\mu(x, y) \wedge t \leq (\sigma(x) \wedge \sigma(y)) \vee s$ for any $t \geq s$. Take $t \leq \sigma(x) \wedge \sigma(y)$. Then $\mu(x, y) \wedge t \leq (\sigma(x) \wedge \sigma(y)) \vee s$ for any $s \leq t$. ∎

In the following example we show that a cost function may have solutions other than the extreme ones, where $t = 1$ or $s = 0$.

Example 1.13.39 Let $V = \{x, y, z, w, u, v\}$. Define $\sigma : V \to [0, 1]$ and $\mu : E \to [0, 1]$ as follows:

$$\mu(x, y) = 1/2, \ \sigma(x) = \sigma(y) = 3/8,$$
$$\mu(z, w) = 3/4, \ \sigma(z) = \sigma(w) = 5/8,$$
$$\mu(u, v) = 19/32, \ \sigma(u) = \sigma(v) = 17/32$$

and for all other pairs $\{a, b\} \in E$, define σ and μ so that $\mu(a, b) \leq \sigma(a) \wedge \sigma(b)$. We have that $\mu(x, y) > \sigma(x) \wedge \sigma(y)$, $\mu(z, w) > \sigma(z) \wedge \sigma(w)$, and $\mu(u, v) > \sigma(u) \wedge \sigma(v)$. Now

$$\mu(x, y) \wedge t_{x,y} \leq (\sigma(x) \wedge \sigma(y)) \vee s_{x,y}$$
$$1/2 \wedge t_{x,y} \leq 3/8 \vee s_{x,y}$$
$$t_{x,y} = 1, s_{x,y} = 1/2, \ \text{Cost} = 1 - 1 + 1/2 = 1/2$$
$$\text{or}$$
$$t_{x,y} = 3/8, s_{x,y} = 0, \ \text{Cost} = 1 - 3/8 + 0 = 5/8$$

$$\mu(z, w) \wedge t_{z,w} \leq (\sigma(z) \wedge \sigma(w)) \vee s_{z.w}$$

$$3/4 \wedge t_{z,w} \leq 5/8 \vee s_{z,w}$$

$$t_{z,w} = 1, \ s_{z,w} = 3/4, \ \text{Cost} = 1 - 1 + 3/4 = 3/4$$

or

$$t_{z,w} = 5/8, \ s_{z,w} = 0, \ \text{Cost} = 1 - 5/8 + 0 = 3/8$$

$$\mu(u, v) \wedge t_{u,v} \leq (\sigma(u) \wedge \sigma(v)) \vee s_{u.v}$$

$$19/32 \wedge t_{u,v} \leq 17/32 \vee s_{u,v}$$

$$t_{u,v} = 1, \ s_{x,y} = 19/32, \ \text{Cost} = 1 - 1 + 19/32 = 19/32$$

or

$$t_{u,v} = 17/32, \ s_{u,yv} = 0, \ \text{Cost} = 1 - 17/32 + 0 = 15/32.$$

Now $t = t_{z,w} = 5/8, s = s_{u,v} = 19/32$ and $t = t_{u,v} = 17/32, s = s_{x,y} = 1/2$ are such that $\mu(a, b) \wedge t \leq (\sigma(a) \wedge \sigma(b)) \vee s$ for all $a, b \in X$. We have that

$$\text{Cost}(19/32, 5/8) = 1 - 5/8 + 19/32 = 31/32,$$

$$\text{Cost}(1/2, 17/32) = 1 - 17/32 + 1/2 = 31/32,$$

$$\text{Cost}\left(\frac{1}{2} \vee \frac{3}{4} \vee \frac{19}{32}, \ 1\right) = 1 - 1 + 3/4 = 3/4,$$

$$\text{Cost}\left(0, \ \frac{3}{8} \wedge \frac{5}{8} \wedge \frac{17}{32}\right) = 1 - 3/8 + 0 = 5/8.$$

Let V be a set and σ a fuzzy subset of V and μ a fuzzy subset of $V \times V$. Let $s, t \in (0, 1]$ with $s < t$. Then $G = (\sigma, \mu)$ is called an (s, t)-**fuzzy graph** if $G^* = (\text{Supp}(\sigma),$ Supp $(\mu))$ is a graph and for all $x, y \in V, \mu(x, y) \wedge t \leq (\sigma(x) \wedge \sigma(y)) \vee s$. For a possible interpretation with respect to trafficking, $\mu(x, y)$ might represent the degree of capacity (or flow) for an edge (x, y) and $\sigma(x)$ the degree of country x allowing flow. Then t might represent the degree of possible reduction of the capacity or flow while s might represent the degree increasing the attack on countries x and y.

"We take an approach based on the U. S. Department of State, Office to Monitor and Combat Trafficking in Persons, [13]. Each country is assigned a tier. Tier 1 consists of countries whose governments fully comply with the Trafficking Victims Protection Act's (TVPA) minimum standards. Tier 2 consists of countries whose governments do not fully comply with the TVPA's minimum standards, but are making significant efforts to bring themselves into compliance with those standards. Tier 2 Watch List (WL) is made up of countries whose governments do not fully comply with the TVPA's minimum standards, but are making efforts to bring themselves into compliance and have severe forms of trafficking. Tier 3 consists of countries whose governments do not fully comply with the minimum standards and are not making significant efforts to do so [23]."

Let V denote the set of countries. Define $\sigma : V \rightarrow [0, 1]$ as follows: for all $x \in V$,

$$\sigma(x) = \begin{cases} 0.8 & \text{if } x \text{ is Tier 1,} \\ 0.6 & \text{if } x \text{ is Tier 2,} \\ 0.4 & \text{if } x \text{ is Tier 2 WL,} \\ 0.2 & \text{if } x \text{ is Tier 3,} \end{cases}$$

Let $\mu : E \rightarrow [0, 1]$. Let $x, y \in X$. Suppose $\mu(x, y) > \sigma(x) \wedge \sigma(y)$. Determine suitable $s, t \in [0, 1]$ such that $0 < s < t \leq 1$ and $\mu(x, y) \wedge t > (\sigma(x) \wedge \sigma(y)) \vee s$, where $\mu(x, y)$ is the degree of a weak attack of edge $\{x, y\}$. One can interpret t as lessening the weakness of attack on $\{x, y\}$, i.e., increasing the attack and s can be interpreted as increasing the compliance of the countries.

Define $\tau : V \rightarrow [0, 1]$ as follows: for all $x \in V$,

$$\tau(x) = \begin{cases} 0.2 & \text{if } x \text{ is Tier 1,} \\ 0.4 & \text{if } x \text{ is Tier 2,} \\ 0.6 & \text{if } x \text{ is Tier 2 WL,} \\ 0.8 & \text{if } x \text{ is Tier 3,} \end{cases}$$

Let $\nu : E \rightarrow [0, 1]$. Let $x, y \in X$. Suppose $\nu(x, y) < \tau(x) \vee \tau(y)$. Determine suitable $\widehat{s}, \widehat{t} \in [0, 1]$ such that $0 \leq \widehat{t} < \widehat{s} < 1$ and $\nu(x, y) \wedge \widehat{t} \geq (\tau(x) \wedge \tau(y)) \vee \widehat{s}$, where $\nu(x, y)$ is the degree of a strong attack of edge $\{x, y\}$. One can interpret $\widehat{t}$ as increasing the strength of attack on $\{x, y\}$, i.e., increasing the attack and $\widehat{s}$ can be interpreted as decreasing the noncompliance of the countries.

We take a different approach based on the CIA World Factbook, [11]. Each country is assigned a tier and each country is designated as a source, transit, or destination country. Let x denote a country and S_x denote the subset of $\{s, t, d\}$ that contains x's designation, where s denotes source, t transit, and d destination. Let σ denote the fuzzy subset of V, the set countries, defined as follows: for all $x \in V$,

$$\sigma(x) = \begin{cases} 0.2 & \text{if } x \text{ is Tier 1,} \\ 0.4 & \text{if } x \text{ is Tier 2,} \\ 0.6 & \text{if } x \text{ is Tier 2 WL,} \\ 0.8 & \text{if } x \text{ is Tier 3.} \end{cases}$$

Let $\veebar$ denote exclusive or. Let μ be the fuzzy subset of E defined as follows: for all $\{x, y\} \in E$,

$$\mu(x, y) = \begin{cases} 0.8 & \text{if } s, t \in S_x \text{ and } t, d \in S_y, \\ 0.6 & \text{if } s \veebar t \in S_x \text{ and } t, d \in S_y, \\ 0.6 & \text{if } s, t \in S_x \text{ and } t \veebar d \in S_y, \\ 0.4 & \text{if } s \veebar t \in S_x \text{ and } t \veebar d \in S_y, \\ 0.2 & \text{if } s \veebar t \in S_x \text{ and } t \notin S_y, d \notin S_y, \\ 0.2 & \text{if } s \notin S_x, t \notin S_x \text{ and } t \veebar d \in S_y, \\ 0 & \text{otherwise.} \end{cases}$$

Then (σ, μ) is not a fuzzy graph because for example if x and y are tier 2 WL, then $\sigma(x) = 0.6 = \sigma(y)$ and it may be the case that $\mu(x, y) = 0.8$.

It may make sense to restrict the use of this approach to regions because if countries x and y are in different regions, y being a destination country may not necessarily mean x will traffic humans to y.

Let τ be a fuzzy subset of V such that for all $x \in V, \sigma(x) + \tau(x) \le 1$ and let ν a fuzzy subset of E such that for all $\{x, y\} \in E, \mu(x, y) + \nu(x, y) \le 1$. We can think of $\tau(x)$ as the degree of country x rejecting trafficking and $\nu(x, y)$ as the degree of the edge $\{x, y\}$ of restricting trafficking.

Define the fuzzy subset γ of E as follows: for all $x, y \in V, \gamma(x, y) = (\sigma(x) \lor \tau(x)) \land (\tau(y) \lor \delta(y))$. The fuzzy subset γ is thought of as the capacity of flow between x and y. Let μ be the actual flow between x and y. Then $\mu(x, y) \le \gamma(x, y)$ for all $x, y \in V$.

We close the chapter with some ideas that could be pursued in the future.

Let V be a set and σ a fuzzy subset of V and μ a fuzzy subset of $V \times V$. Let $s, t \in (0, 1]$ with $s < t$. Then $G = (\sigma, \mu)$ is called an $(s, t]$-**fuzzy graph** if $G^* = (\text{Supp}(\sigma),$ Supp $(\mu))$ is a graph and for all $x, y \in V, \mu(x, y) \land t \le (\sigma(x) \land \sigma(y)) \lor s$. For a possible interpretation with respect to trafficking, $\mu(x, y)$ might represent the degree of capacity (or flow) for an edge (x, y) and $\sigma(x)$ the degree of country x allowing flow. Then t might represent the degree of possible reduction of the capacity or flow while s might represent the degree increasing the attack on countries x and y. Of course other interpretations are possible. A complementary fuzzy graph $\widehat{G} = (\tau, \nu)$ is a pair such that τ is a fuzzy subset of V and ν is a fuzzy subset of $V \times V$ such that for all $x, y \in V, \nu(x, y) \ge \tau(x) \lor \tau(y)$. Let $\tau^\#$ denote the cosupport of τ and $\nu^\#$ the cosupport of ν. Let $\widehat{G}^\# = (\tau^\#, \nu^\#)$. Let $\widehat{s}, \widehat{t} \in [0, 1)$ with $\widehat{s} > \widehat{t}$. Then $\widehat{G} = (\tau, \nu)$ is called a **complementary** $[\widehat{s}, \widehat{t})$-**fuzzy graph** if $\widehat{G}^\#$ is a graph and for all $x, y \in V, \nu(x, y) \lor \widehat{t} \ge (\tau(x) \lor \tau(y)) \land \widehat{s}$. The pair $(G, \widehat{G})$ is called an **intuitionistic** $(s, t], [\widehat{s}, \widehat{t})$-**fuzzy graph** if $G = (\sigma, \mu)$ is an $(s, t]$-fuzzy graph and $\widehat{G} = (\tau, \nu)$ is a complementary $[\widehat{s}, \widehat{t})$-fuzzy graph such that $0 \le s + \widehat{s} \le 1, 0 \le t + \widehat{t} \le 1$ and $0 \le \sigma(x) + \tau(x) \le 1, 0 \le \mu(x, y) + \nu(x, y) \le 1$ for all $x, y \in V$.

We take an approach based on the U. S. department of state, Office to Monitor and Combat Trafficking in Persons. Each country is assigned a tier and each country is designated as a source, transit, or destination country. Let x denote a country and S_x denote the subset of $\{s, t, d\}$ that contains $x's$ designation, where s denotes source, t transit, and d destination. Let σ denote the fuzzy subset of $\mathcal{C}$, the set of countries, defined as follows: for all $x \in \mathcal{C}$,

$$\sigma(x) = \begin{cases} 0.2 & \text{if } x \text{ is Tier 1,} \\ 0.4 & \text{if } x \text{ is Tier 2,} \\ 0.6 & \text{if } x \text{ is Tier 2 WL,} \\ 0.8 & \text{if } x \text{ is Tier 3.} \end{cases}$$

Let $\underline{\lor}$ denote exclusive or. Let μ be the fuzzy subset of $\mathcal{C} \times \mathcal{C}$ defied as follows: for all $(x, y) \in \mathcal{C} \times \mathcal{C}$,

$$\mu(x, y) = \begin{cases} 0.8 & \text{if } s, t \in S_x \text{ and } t, d \in S_y, \\ 0.6 & \text{if } s \veebar t \in S_x \text{ and } t, d \in S_y, \\ 0.6 & \text{if } s, t \in S_x \text{ and } t \veebar d \in S_y, \\ 0.4 & \text{if } s \veebar t \in S_x \text{ and } t \veebar d \in S_y, \\ 0.2 & \text{if } s \veebar t \in S_x \text{ and } t \notin S_y, d \notin S_y, \\ 0.2 & \text{if } s \notin S_x, t \notin S_x \text{ and } t \veebar d \in S_y, \\ 0 & \text{otherwise.} \end{cases}$$

Then (σ, μ) is not necessarily a fuzzy graph because for example if x and y are tier 2WL, then $\sigma(x) = 0.6 = \sigma(y)$ and it may be the case that $\mu(x, y) = 0.8$.

It may make sense to restrict the use of this approach to regions because if countries x and y are in different regions, y being a destination country may not necessarily mean x will traffic humans to y.

Let τ be a fuzzy subset of $\mathcal{C}$ such that for all $x \in \mathcal{C}$, $\sigma(x) + \tau(x) \leq 1$ and let ν a fuzzy subset of $\mathcal{C} \times \mathcal{C}$ such that for all $(x, y) \in \mathcal{C} \times \mathcal{C}$, $\mu(x, y) + \nu(x, y) \leq 1$. We can think of $\tau(x)$ as the degree of country x rejecting trafficking and $\nu(x, y)$ as the degree of the edge (x, y) of restricting trafficking.

References

1. M. Akram, N.O. Alshehri, Intuitionistic fuzzy cycles and intuitionistic fuzzy trees. Sci. World J. ID 305836, 1–11 (2014). (Hindawi Publishing Corporation)
2. M. Akram, B. Davvaz, Strong intuitionistic fuzzy graphs. Filomat **26**, 177–195 (2012)
3. M. Akram, M.G. Karunambigai, O.K. Kalaivani, Some metric aspects of intuitionistic fuzzy graphs. World Appl. Ser. J. **17**, 1789–1801 (2012)
4. K.T. Atanassov, On intuitionistic fuzzy graphs and intuitionistic fuzzy relations, in *Proceedings of the 6th IFSA World Congress* vol. 1 (San Paulo, Brazil, 1995), pp. 551–554
5. K.T. Atanassov, Intuitionistic fuzzy sets. Fuzzy Sets Syst. **20**, 87–96 (1986)
6. K.T. Atanassov, A. Shannon, On a generalization of intuitionistic fuzzy graphs. Notes Intuit. Fuzzy Sets **12**, 24–29 (2006)
7. K.R. Bhutani, J.N. Mordeson, P.K. Saha, $(s, t]$-fuzzy graphs, in *JCIS Proceedings* (2005), pp. 37–40
8. K.R. Bhutani, A. Rosenfeld, Fuzzy end nodes in fuzzy graphs. Inf. Sci. **152**, 323–326 (2003)
9. K.R. Bhutani, J.N. Mordeson, A. Rosenfeld, On degrees of end nodes and cut nodes in fuzzy graphs. Iranian J. Fuzzy Syst. **1**, 57–64 (2004)
10. C. Caniglia, B. Cousino, S.-C. Cheng, D.S. Malik, J.N. Mordeson, Intuitionistic fuzzy graphs: weakening and strengthening members of a group. J. Fuzzy Math. **24**, 87–102 (2016)
11. Central Intelligence Agency, Field Listing: Trafficking in Persons (2014). http://www.cia.gov/librasry/publications/the-world-factbook/fields/2196.html
12. P. Chountas, A. Shannon, P. Rangasamy, K. Atassov, On intuitionistic fuzzy trees and their index matrix interpretation. Notes Intuit. Fuzzy Sets **15**, 52–56 (2009)
13. U.S. Department of State, Diplomacy in Action, Trafficking in Persons Report 2013: Tier Placements. http://www.state.gov/j/rls/tiprpt/2013/210548.htm
14. F. Harary, R.Z. Norman, *Graph Theory as a Mathematical Model in Social Science*, Ann Arbor, Mich.: Institute for Social Research, 1953)
15. F. Harary, R.Z. Norman, D. Cartwright, Introduction to digraph theory for social scientists, in *Process of Publication*
16. F. Harary, Graph theoretic methods in the management sciences. Manag. Sci. **5**, 387–403 (1959)

17. F. Harary, I.C. Ross, The number of complete cycles in a communication network. J. Soc. Psychol. **40**, 329–332 (1953)
18. F. Harary, I.C. Ross, A procedure for clique detention using the group matrix. Sociometry **20**, 205–215 (1957)
19. A. Kaufmann, *Introduction to the Theory of Fuzzy Subsets*, vol. 1 (Academic Press, New York, 1975)
20. G.J. Klir, B. Yuan, *Fuzzy Sets and Fuzzy Logic: Theory and Applications* (Prentice Hall PTR, Upper Saddle River, 1995)
21. P.P. Ming, L.Y. Ming, Fuzzy topology I: neighborhood structure of a fuzzy point and Moore-Smith convergence. J. Math. Anal. Appl. **76**, 571–599 (1980)
22. J.N. Mordeson, P.S. Nair, *Fuzzy Graphs and Fuzzy Hypergraphs*. Studies in Fuzziness and Soft Computing, vol. 46 (Physica-Verlag, Heidelberg, 2000)
23. J.N. Mordeson, D.S. Malik, C.D. Richards, J.A. Trebbian, M.A. Boyce, M.P. Byrne, B.J. Cousino, Fuzzy graphs and complementary fuzzy graphs. J. Fuzzy Math. **24**, 271–288 (2016)
24. R. Parvathi, M.G. Karunambigigai, Intuitionistic fuzzy graphs. Adv. Soft Comput. **38**, 139–159 (2006)
25. G. Pasi, R. Yager, K. Atanassov, Intuitionistic fuzzy graph interpretations of multi-person multi-critera decision making: generalized net approach, in *Proceedings of the 2nd International IEEE Conference in Intelligent Systems* (2004), pp. 434–439
26. D. Rosenblatt, On linear models and the graphs of Minkowski -Leontief matrices. Econometrica **25**, 325–338 (1975)
27. A. Rosenfeld, Fuzzy graphs, in *Fuzzy Sets and Their Applications,* ed. by L.A. Zadeh, K.S. Fu, M. Shimura (Academic Press, 1975), pp. 77–95
28. I.C. Ross, F. Harary, On the determination of redundancies in sociometric chains. Psychometrika **17**, 195–208 (1952)
29. I.C. Ross, F. Harary, Identification of the liaison persons of an organization using the structure matrix. Manag. Sci. **1**, 251–258 (1955)
30. I.C. Ross, F. Harary, A description of strengthening and weakening members of a group. Sociometry **22**, 139–147 (1959)
31. M.S. Sunitha, A. Vijayakumar, Blocks in fuzzy graphs. J. Fuzzy Math. **13**(1) 13–23 (2005)
32. M.S. Sunitha, A. Vijayakumar, A characterization of fuzzy trees. Inf. Sci. **113**, 293–300 (1999)
33. E. Takeda, T. Nishida, An application of fuzzy graph to the problem concerning group structure. J. Oper. Res. Soc. Jpn. **10**, 217–227 (1976)
34. L.A. Zadeh, Fuzzy sets. Inf. Control **8**, 338–353 (1965)
35. L.A. Zadeh, Similarity relations and fuzzy orderings. Inf. Sci. **3**, 177–200 (1971)

Chapter 2
Domination in Fuzzy Graphs

2.1 Preliminaries

The results of the next two sections are based on [29]. The formal mathematical definition of domination was given by Ore, [22]. Cockayne and Hedetnieme, [3], published a survey paper on this topic in 1977 and since then hundreds of papers have been published on this subject. According to [29], the rapid growth of research in this area is due to the following three factors.

(1) The diversity of applications of domination theory to both real world and mathematical coverings or location problems.

(2) The wide variety of domination parameters that can be defined.

(3) The NP-completeness of the basic domination problem, its close and natural relationship to other NP-complete problems and the subsequent interest in finding polynomial time solutions to domination problems in special classes of graphs.

In some fuzzy networks, flow is allowed in both the directions. Internet and power grids are examples. But there are networks that undergo one dimensional flows with imprecise capacities and flows. Human and drug trafficking and illegal migration are examples [23, 31, 32]. To model such phenomena, directed fuzzy networks (DFN) are studied here.

The crime of human trafficking affects nearly every country in every region in the world. Human trafficking flows are mainly intraregional. That is, the origin and the destination of those trafficked is within the same region. Victims tend to be trafficked from poor countries to more affluent ones within the region. We apply techniques of digraph theory to examine the regional flow of trafficking in persons. We take a regional study. The regions are the ones defined in [32], namely, Western and Central Europe, Eastern and Central Europe, North America, Central America and the Caribbean, South America, Middle East, Sub-Saharan Africa, Central Europe and the Balkans, East Asia and the Pacific, South Asia. We determine destination dominating sets, minimal destination dominating sets, and maximal independent sets.

© Springer International Publishing AG 2018
J. N. Mordeson et al., *Fuzzy Graph Theory with Applications to Human Trafficking*, Studies in Fuzziness and Soft Computing 365,
https://doi.org/10.1007/978-3-319-76454-2_2

Let $G = (\sigma, \mu)$ be a fuzzy graph on V and $V_1 \subseteq V$. Define σ_1 on V_1 by $\sigma_1(x) = \sigma(x)$ for all $x \in V$ and μ_1 on the collection E_1 of two element subsets of V_1 by $\mu_1(xy) = \mu(xy)$ for all $x, y \in V_1$. Then (σ_1, μ_1) is called the **fuzzy subgraph** of G induced by V_1 and is denoted by $\langle V_1 \rangle$.

The **order** p and **size** q of a fuzzy graph $G = (\sigma, \mu)$ are defined by $p = \sum_{x \in V} \sigma(x)$ and $q = \sum_{xy \in E} \mu(xy)$.

Let $\sigma : V \to [0,1]$ be a fuzzy subset of V. Then the **complete fuzzy graph** on σ is defined to be (σ, μ), where $\mu(xy) = \sigma(x) \wedge \sigma(y)$ for all $xy \in E$ and is denoted by K_σ.

Let $G = (\sigma, \mu)$ be a fuzzy graph on V and $S \subseteq V$. Then the **fuzzy cardinality** of S is defined to be $\sum_{v \in S} \sigma(v)$.

The **complement** of a fuzzy graph G denoted by $\overline{G}$ is defined to be $\overline{G} = (\sigma, \overline{\mu})$, where $\overline{\mu}(xy) = \sigma(x) \wedge \sigma(y) - \mu(xy)$.

Definition 2.1.1 Let $G = (\sigma, \mu)$ be a fuzzy graph on V. Let P be a subset of E. Let $x, y \in V$. Then x is said to **dominate** y in G with respect to P if $xy \in P$. A subset S of V is called a **dominating set** in G if for all $v \in V \backslash S$, $\exists u \in S$ such that u dominates v. A dominating set S is called a **minimal dominating set** if no proper subset of S is a dominating set.

Let $x, y \in V$. If $xy \in E$, then x and y are called **adjacent**. If $xy \in P$, then x and y are called **adjacent with respect to** P.

Definition 2.1.2 Let $G = (\sigma, \mu)$ be a fuzzy graph on V. Let P be a subset of E. Let $x \in V$ and $N(x) = \{y \in V \mid xy \in P\}$. Then $N(x)$ is called the **neighborhood** of x with respect to P and $N(x) \cup \{x\}$ is called the **closed neighborhood** of x with respect to P.

Theorem 2.1.3 *Let D be a dominating set of G with respect to P. Then D is a minimal dominating set if and only if for all $d \in D$, one of the following two conditions hold:*

(i) $N(d) \cap D = \emptyset$;
(ii) $\exists c \in V \backslash D$ such that $N(c) \cap D = \{d\}$.

Proof Suppose D is minimal. Let $d \in D$. Then $D_d = D \backslash \{d\}$ is not a dominating set. Hence, $\exists x \in V \backslash D_d$ such that x is not dominated by any element of D_d. If $x = d$, then (i) holds. If $x \neq d$, then (ii) holds with $c = x$.

Conversely, suppose for all $d \in D$ that (i) or (ii) holds. Let $d \in D$. If (i) holds for d, then $D \backslash \{d\}$ is not a dominating set because there does not exist $d' \in D$ such that $d'd \in P$. If (ii) holds for d, there does not exist $d' \in D \backslash \{d\}$ such that $cd' \in P$. Hence, $D \backslash \{d\}$ is not a dominating set. Thus, D is minimal. ∎

Definition 2.1.4 A vertex u of a fuzzy graph is said to be an **isolated vertex** with respect to P if for all $v \in V$, $uv \notin P$, i.e., $N(u) = \emptyset$.

Theorem 2.1.5 *Let G be a fuzzy graph without isolated vertices. Let D be a minimal dominating set. Then $V \backslash D$ is a dominating set of G.*

Proof Let $d \in D$. Because G has no isolated vertices with respect to P, there exists $c \in V$ such that $c \in N(d)$. If $c \notin D$, then d is dominated by an element of $V \setminus D$. If $c \in D$, then (i) of Theorem 2.1.3 doesn't hold. Hence, because D is minimal, (ii) of Theorem 2.1.3 holds and so there exists $c' \in V \setminus D$ such that $c' \in N(d)$. Thus, every element of D is dominated by some element of $V \setminus D$. ∎

Definition 2.1.6 A set S of vertices of a fuzzy graph is said to be **independent** with respect to P if for all $x, y \in S$, $xy \notin P$.

Theorem 2.1.7 *If D is an independent dominating set of a fuzzy graph with respect to P, then D is both a minimal dominating set and a maximal independent set. Conversely, any maximal independent set D is a dominating set.*

Proof Suppose D is independent and dominating. Because D is independent, $D \setminus \{d\}$ is not dominating for all $d \in D$ because there does not exist $d' \in D$ such that $d'd \in P$. Thus, D is a minimal dominating set. Also, for all $x \in V \setminus D$, $D \cup \{x\}$ is not independent because there exists $d \in D$ such that $dx \in P$. Hence, D is a maximal independent set.

Conversely, suppose D is a maximal independent set with respect to P. Then for all $x \in V \setminus D$, $D \cup \{x\}$ is not independent so there exists $d \in D$ such that $xd \in P$. Hence, D is a dominating set. ∎

Corollary 2.1.8 *An independent set is maximal independent if and only if it is independent and dominating.*

Corollary 2.1.9 *Every maximal independent set is a minimal dominating set.*

Definition 2.1.10 Let $S \subseteq V$. Define $N(S) = \cup_{s \in S} N(s)$ and $N[S] = N(S) \cup S$.

Definition 2.1.11 Let $S \subseteq V$. Let $v \in V$. Then v is called a **private neighbor** of $u \in S$ with respect to S or an S **-private neighbor** of u if $N[v] \cap S = \{u\}$. The **private neighborhood** of $u \in S$ with respect to S is defined to be $PN[u, S] = \{v \in V \mid N[v] \cap S = \{u\}\}$.

It follows in Definition 2.1.11 that $PN[u, S] = N[u] \setminus N[S \setminus \{u\}]$.

Definition 2.1.12 $S \subseteq V$ is said to be **irredundant** with respect to P if $PN[u, S] \neq \emptyset$ for all $u \in S$. The **irredundant number** $ir(G)$ is the minimum cardinality over all maximal irredundant sets of nodes of G. An irredundant set with cardinality $ir(G)$ is called an ir-**set**.

Theorem 2.1.13 *D is a minimal dominating set of nodes of a graph G with respect to P if and only if D is irredundant and dominating.*

Proof Suppose D is a minimal dominating set. Suppose D is not irredundant. Then $\exists d \in D$ such that $PN[d, D] = \{v \in V \mid N[v] \cap D = \{d\}\} = \emptyset$. Thus, $\nexists v \in V$ such that $N[v] \cap D = \{d\}$. Hence, for all $v \in V$, either $N[v] \cap D \supset \{d\}$ or $d \notin N[v] \cap D$ so $d \notin N[v]$. If $v \in V$ is such that $N[v] \cap D \supset \{d\}$, then $\exists d' \in D \setminus \{d\}$ such that

$vd' \in P$ and if v is such that $d \notin N[v]$, then $\exists d'' \in D \backslash \{d\}$ such that $vd'' \in P$. Thus, $D \backslash \{d\}$ is a dominating, contradicting the minimality of D.

Conversely, suppose D is irredundant and dominating. Suppose D is not a minimal dominating set. Then $\exists d \in D$ such that $D \backslash \{d\}$ is dominating. Hence, $\exists d' \in D$ such that $d'd \in P$. Thus, $N[d'] \supseteq dd'$. Hence, $N[d'] \cap D \supseteq \{d, d'\} \neq \{d\}$. Thus, it is not the case that for all $d'' \in D$, $N[d''] \cap D = \{d\}$. Hence, D is not irredudant, a contradiction. ■

Theorem 2.1.14 *If D is a minimal dominating set, then D is a maximal irredundant set.*

Proof Suppose D is a minimal dominating set. Then D is irredundant and dominating by Theorem 2.1.13. Suppose D is not maximal irredundant. Then there exists $S \subseteq V$ such that S is irredundant and $D \subset S$. Because S must be dominating, we have by Theorem 2.1.13 that S is a minimal dominating set. However, this is impossible because D is dominating. ■

A fuzzy graph $G = (\sigma, \mu)$ is said to be **bipartite** if the vertex set V can be partitioned into two nonempty sets V_1 and V_2 such that $\mu(v_1 v_2) = 0$ if $v_1, v_2 \in V_1$ or $v_1, v_2 \in V_2$. Further, if $\mu(uv) = \sigma(u) \wedge \sigma(v)$ for all $u \in V_1$ and $v \in V_2$, then G is called a **complete bipartite graph** and is denoted by K_{σ_1, σ_2}, where σ_1 and σ_2 are, respectively, the restrictions of σ to V_1 and V_2.

2.2   Domination in Fuzzy Graphs

In this section, we introduce the concepts of domination and total domination in fuzzy graphs. We determine the domination number γ and the total domination number γ_t for several classes of fuzzy graphs and obtain bounds for the same. We also obtain Nordhaus–Gaddum type results for these parameters.

An edge $e = xy$ of a fuzzy graph is called an **effective edge** if $\mu(xy) = \sigma(x) \wedge \sigma(y)$. $N(x) = \{y \in V \mid \mu(xy) = \sigma(x) \wedge \sigma(y)\}$ is called the **neighborhood** of x and $N[x] = N(x) \cup \{x\}$ is the **closed neighborhood** of x.

The degree of a vertex can be generalized in different ways for a fuzzy graph.

The **effective degree** of a vertex u is defined to be the sum of the weights of the effective edges incident at u and is denoted by $dE(u)$. $\sum_{v \in N(u)} \sigma(v)$ is called the **neighborhood degree** of u and is denoted by $dN(u)$. The **minimum effective degree** $\delta_E(G) = \wedge \{dE(u) \mid u \in V\}$ and the **maximum effective degree** $\Delta_E(G) = \vee \{dE(u) \mid u \in V\}$.

The **neighborhood degree** of a vertex v is defined as the sum of the membership value of the neighborhood vertices of v, and is denoted by $d_N(v)$. The **minimum** and **maximum neighborhood degrees** are defined by $\delta_N(G) = \wedge \{d_N(v) \mid v \in V\}$ and $\Delta_N(G) = \vee \{d_N(v) \mid v \in V\}$, respectively.

We observe that these concepts reduce to the usual degree of a vertex in crisp case.

Let $G = (V, E)$ be a graph. A subset S of V is called a **dominating set** in G if every vertex in $V \backslash S$ is adjacent to some vertex in S. The **domination number** of G is the minimum cardinality taken over all dominating sets in G and is denoted by $\gamma(G)$, or simply γ.

Let G be a graph without isolated vertices. A dominating set S of G is called a **total dominating set** if the subgraph $\langle S \rangle$ induced by S has no isolated vertices. The minimum cardinality taken over all total dominating sets of G is called the **total domination number** of G and is denoted by γ_t.

A dominating set S of a graph G is called an **independent dominating set** of G if no two vertices in S are adjacent.

We now proceed to extend these concepts to fuzzy graphs.

Let $G = (\sigma, \mu)$ be a fuzzy graph on V. Let $P = \{xy \mid \mu(xy) = \sigma(x) \wedge \sigma(y)\}$.

Let $G = (\sigma, \mu)$ be a fuzzy graph on V. Let $x, y \in V$. From Definition 2.1.1, we have that x dominates y in G if $\mu(xy) = \sigma(x) \wedge \sigma(y)$. The minimum fuzzy cardinality of a dominating set in G is called the **domination number** of G and is denoted by $\gamma(G)$ or γ.

We note that the results of the previous section hold if we let $P = \{\{x, y\} \mid \mu(yx) = \sigma(x) \wedge \sigma(y)\}$.

Remark 2.2.1 (*i*) Note that for any $x, y \in V$, if x dominates y then y dominates x and hence, domination is a symmetric relation on V.

(*ii*) For any $x \in V$, $N(x)$ is precisely the set of all $y \in V$ which are dominated by x.

(*iii*) If $xy \notin P$ and $(\mu(xy) < \sigma(x) \wedge \sigma(y))$ for all $x, y \in V$, then the only dominating set in G is V.

The above definition of domination in fuzzy graph is motivated by the following situation.

Let G be a graph which represents the road network of a city. Let the vertices denote the junctions and the edges denote the roads connecting the junctions. From the statistical data that represents the number of vehicles passing through various junctions and the number of vehicles passing through various roads during a peak hour, the membership functions σ and μ on the vertex set and edge set of G can be constructed by using the standard techniques given in Bobrowicz et al., [1]; Reha Civanlar and Joel Trussel, [24]. In this fuzzy graph, a dominating set S can be interpreted as a set of junctions which are busy in the sense that every junction not in S is connected to a member in S by a road in which the traffic flow is full.

Example 2.2.2 Let $V = \{x, y, u, v\}, \sigma(x) = \sigma(y) = 3/4, \sigma(u) = \sigma(v) = 1/2$, and $\mu(xu) = \mu(xv) = \mu(yu) = \mu(yv) = 1/2$. Then

$$dN(x) = \sigma(u) + \sigma(v) = 1/2 + 1/2 = 1,$$
$$dN(y) = \sigma(u) + \sigma(v) = 1/2 + 1/2 = 1,$$
$$dN(u) = \sigma(x) + \sigma(y) = 3/4 + 3/4 = 3/2,$$
$$dN(v) = \sigma(x) + \sigma(y) = 3/4 + 3/4 = 3/2.$$

Thus, $\Delta_N = dN(v) = dN(u) = 3/2$. Let $V_1 = \{x, y\}$ and $V_2 = \{u, v\}$. Then G is the complete fuzzy bipartite graph K_{σ_1,σ_2}, where $\sigma_1 = \sigma|_{V_1}$ and $\sigma_2 = \sigma|_{V_2}$. Now $\gamma(K_{\sigma_1,\sigma_2}) = 1$ because $\{u, v\}$ is a dominating set with minimal fuzzy cardinality $1/2 + 1/2 = 1 \neq 3/4 + 1/2 = \min_{z \in V_1} \sigma_1(z) + \min_{w \in V_2} \sigma_2(w)$.

Example 2.2.3 (*i*) Because $\{v\}$ is a dominating set of K_σ for all $v \in V$, we have $\gamma(K_\sigma) = \min_{x \in V} \sigma(x)$.

(*ii*) $\gamma(G) = p$ if and only if $\mu(xy) < \sigma(x) \wedge \sigma(y)$ for all $x, y \in V$ ($P = \emptyset$). In particular, $\gamma(\overline{K_\sigma}) = p$.

For the domination number γ, the following theorem gives a Nordhaus–Gaddum type result.

Theorem 2.2.4 *For any fuzzy graph* G, $\gamma + \overline{\gamma} \leq 2p$, *where* $\overline{\gamma}$ *is the domination number of* $\overline{G}$ *and equality holds if and only if* $0 < \mu(xy) < \sigma(x) \wedge \sigma(y)$ *for all* $x, y \in V$.

Proof The inequality is trivial. Further $\gamma = p$ if and only if $\mu(xy) < \sigma(x) \wedge \sigma(y)$ for all $x, y \in V$ and $\overline{\gamma} = p$ if and only if $\sigma(x) \wedge \sigma(y) - \mu(xy) < \sigma(x) \wedge \sigma(y)$ for all $x, y \in V$, which is equivalent to $\mu(xy) > 0$. Hence, $\gamma + \overline{\gamma} = 2p$ if and only if $0 < \mu(xy) < \sigma(x) \wedge \sigma(y)$. ∎

For the general case, the condition, $0 < \mu(xy) < \sigma(x) \wedge \sigma(y)$ in Theorem 2.2.4 is equivalent to $P = \{\{x, y\} \mid \mu(xy) = \sigma(x) \wedge \sigma(y)\} = \emptyset$ and $\overline{P} = \{\{x, y\} \mid \overline{\mu}(xy) = \sigma(x) \wedge \sigma(y)\} = \emptyset$. Clearly, $\overline{P} = \emptyset$ if and only if $0 < \mu(xy)$ for all $x, y \in V$.

The following theorem gives a characterization of minimal dominating sets, which is analogous to the result of Ore [22], 1962, in the crisp case. See Theorem 2.1.3.

Theorem 2.2.5 *A dominating set* D *of* G *is a minimal dominating set if and only if for each* $d \in D$ *one of the following two conditions holds.*
(*i*) $N(d) \cap D = \emptyset$
(*ii*) *There is a vertex* $c \in V \backslash D$ *such that* $N(c) \cap D = \{d\}$.

In this section, a vertex u of a fuzzy graph is an isolated vertex if $\mu(uv) < \sigma(u) \wedge \sigma(v)$ for all $v \in V \backslash \{u\}$, i.e., $N(u) = \emptyset$.

Thus, an isolated vertex does not dominate any other vertex in G.

The following result follows from Theorem 2.1.5.

Theorem 2.2.6 *Let* G *be a fuzzy graph without isolated vertices. Let* D *be a minimal dominating set of* G. *Then* $V \backslash D$ *is a dominating set of* G.

Corollary 2.2.7 *For any graph without isolated vertices* $\gamma \leq p/2$.

Proof Any graph without isolated vertices has two disjoint dominating sets and hence the result follows. ∎

Fig. 2.1 A fuzzy graph

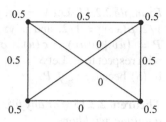

If $\overline{G}$ is the complement of a fuzzy graph G, we let $\overline{\gamma}$ or $\gamma(\overline{G})$denote the domination number of $\overline{G}$.

Corollary 2.2.8 *Let G be a fuzzy graph such that both G and $\overline{G}$ have no isolated vertices. Then $\gamma + \overline{\gamma} \leq p$. Furthermore, equality holds if and only if $\gamma = \overline{\gamma} = p/2$.*

Example 2.2.9 For the fuzzy graph given in Fig. 2.1, $\gamma = \overline{\gamma} = 1$, $p = 2$, so that $\gamma + \overline{\gamma} = 2$.

In this section, a set S of vertices of a fuzzy graph is independent if $\mu(xy) < \sigma(x) \wedge \sigma(y)$ for all $x, y \in S$.

The following theorem gives a characterization of independent dominating sets. It follows from Theorem 2.1.7.

Theorem 2.2.10 *If D is an independent dominating set of a fuzzy graph G, then D is both a minimal dominating set and a maximal independent set. Conversely, any maximal independent set D in G is an independent dominating set of G.*

The following theorem gives an upper bound for the domination number of a fuzzy graph.

We note that $V \setminus N(v)$ is a dominating set for any $v \in V$: Let $u \in N(v)$. Then there exists $w \in V \setminus N(v)$ such that $uw \in P$, namely, $w = v$ because $v \notin N(v)$.

Theorem 2.2.11 $\gamma \leq p - \Delta_N$.

Proof Let v be a vertex such that $dN(v) = \Delta_N$. Now $V \setminus N(v)$ is a dominating set of G so that $\gamma \leq \sum_{u \in V \setminus N(v)} \sigma(s) = \sum_{u \in V} \sigma(u) - \sum_{u \in N(v)} \sigma(u) = p - \Delta_N$. ∎

(i) The inequality in Theorem 2.2.11, cannot be improved further. For example, for the complete graph K_σ, $\gamma = p - \Delta_N$.

(ii) Clearly $\Delta_E \leq \Delta_N$ and hence $\gamma \leq p - \Delta_E$.

Definition 2.2.12 Let G be a fuzzy graph without isolated vertices. A subset D of V is said to be a **total dominating set** if every vertex in V is dominated by a vertex in D.

Definition 2.2.13 The minimum fuzzy cardinality of a total dominating set is called the **total domination number** of G and is denoted by γ_t.

Example 2.2.14 Let $V = \{x, y, z\}$. Define $\sigma(x) = \sigma(y) = \sigma(z) = 1$ and $\mu(xy) = 1/2, \mu(xz) = 1/2$, and $\mu(yz) = 1$. Then x is an isolated vertex with respect to $P = \{uv \mid \mu(uv) = \sigma(u) \wedge \sigma(v)\}$. Let $\mu(xy) = 1$. Then G has no isolated vertex with respect to P. Let $S = \{x, y\}$. Then S is a dominating set and x and z are isolated in $\langle S \rangle$ because $xz \notin P$.

Theorem 2.2.15 *For any fuzzy graph G, $\gamma_t = p$ if and only if every vertex of G has a unique neighbor.*

Proof If every vertex of G has a unique neighbor, then V is the only total dominating set of G so that $\gamma_t = p$.

Conversely suppose $\gamma_t = p$. If there exists a vertex v with two neighbors x and y, then $V \backslash \{x\}$ is a total dominating set of G so that $\gamma_t < p$, which is a contradiction. ∎

Corollary 2.2.16 *If $\gamma_t = p$, then the number of vertices in G is even.*

Recall that in (iv) of the next result, $0 < \mu(xy) < \sigma(x) \wedge \sigma(y)$ is equivalent to $P = \{xy \mid \mu(xy) = \sigma(x) \wedge \sigma(y)\} = \emptyset$ and $\overline{P} = \{xy \mid \overline{\mu}(xy) = \sigma(x) \wedge \sigma(y)\} = \emptyset$. We say two edges are **mutually disjoint** if they share no vertex.

Theorem 2.2.17 *Let G be a fuzzy graph without isolated vertices. Then $\gamma_t + \overline{\gamma}_t \leq 2p$ and equality holds if and only if the following conditions hold:*
(i) the number of vertices in G is even, say $2n$,
(ii) there is a set S_1 of n mutually disjoint effective edges in G,
(iii) there is a set S_2 of n mutually disjoint effective edges in $\overline{G}$, and
(iv) for any edge $xy \notin S_1 \cup S_2$, $0 < \mu(xy) < \sigma(x) \wedge \sigma(y)$.

Proof Because $\gamma_t \leq p$ and $\overline{\gamma}_t \leq p$, the inequality follows. Further, $\gamma_t + \overline{\gamma}_t = 2p$ if and only if $\gamma_t = \overline{\gamma}_t = p$ and hence by Corollary 2.2.16, the number of vertices in G is even, say $2n$. Because $\gamma_t = p$, there is a set S_1 of n mutually disjoint effective edges in G. Similarly, there is a set S_2 of n mutually disjoint effective edges in $\overline{G}$. Further if $xy \notin S_1 \cup S_2$, $0 < \mu(xy) < \sigma(x) \wedge \sigma(y)$. The converse is obvious. ∎

The concept of domination in fuzzy graph theory is very important both in theoretical development and applications. More than thirty domination parameters have been investigated by various authors. In this section, we introduced the concept of domination, total domination and independent domination for fuzzy graphs.

2.3 Strong and Weak Domination

We introduce the concept of strong and weak domination in fuzzy graphs, and provide some examples to explain various notions introduced. The results in this and the next section are based on [21].

Somasundaram and Somasundaram [29] introduced the concept of domination in fuzzy graphs. In this view, we obtain the analog of strong and weak domination in fuzzy graphs.

Let $P = \{xy \in E \mid \sigma(x, y) = \mu(x) \wedge \mu(y)\}$. Then this definition is a special case of Definition 2.1.1.

Example 2.3.1 Consider a fuzzy graph $G = (V, \sigma, \mu)$, where $V = \{v_1, v_2, v_3, v_4, v_5\}$,

	v_1	v_2	v_3	v_4	v_5
σ	0.3	0.6	0.8	0.9	0.8

	$v_1 v_2$	$v_2 v_3$	$v_2 v_4$	$v_3 v_4$	$v_3 v_5$	$v_4 v_5$	$v_5 v_1$
μ	0.3	0.5	0.4	0.8	0.8	0.6	0.2

Here v_3 dominates v_4 and v_5, and v_2 dominates v_1. Clearly $S = \{v_1, v_3\} \subset V$, is the minimum dominating set of G, and therefore $\gamma(G) = 1.1$.

Example 2.3.2 In a complete fuzzy graph G, $\{v\}$ is a dominating set for all $v \in V$, and we have $\gamma(G) = \wedge\sigma(v)$.

Definition 2.3.3 ([29]) Let $G = (\sigma, \mu)$ be a fuzzy graph on V. A subset S of V is called a **vertex cover** of G if for every effective edge $e = vw$, at least one of $v, w \in S$. The minimum fuzzy cardinality of a vertex cover is called the **covering number** of G, and is denoted by $\alpha_0(G)$ or simply by α_0.

Definition 2.3.4 ([29]) Let $G = (\sigma, \mu)$ be a fuzzy graph on V. The set S is said to be **maximal independent set** if $S \cup \{y\}$ is not an independent set for any $y \in V \setminus S$. The minimum fuzzy cardinality of an independent set in G is called the **independence number** of G, and is denoted by $\beta_0(G)$ or simply β_0.

Example 2.3.5 Consider a fuzzy graph $G = (V, \sigma, \mu)$, where $V = \{v_1, v_2, v_3, v_4, v_5, v_6\}$,

	v_1	v_2	v_3	v_4	v_5	v_6
σ	0.4	0.6	0.8	0.1	0.2	0.3

	$v_1 v_2$	$v_2 v_3$	$v_2 v_6$	$v_3 v_4$	$v_3 v_5$	$v_4 v_5$	$v_5 v_6$	$v_3 v_6$	$v_6 v_1$
μ	0.4	0.6	0.3	0.1	0.2	0.1	0.2	0.3	0.3

Here $\{v_1, v_3\}$ is an independent set of maximum fuzzy cardinality and therefore $\beta_0(G) = 1.2$. $\{v_2, v_4, v_5, v_6\}$ is a vertex cover of minimum fuzzy cardinality and therefore $\alpha_0(G) = 1.2$.

Finally, we recall the idea of strong and weak domination in graph (crisp) theory.

Let $G = (V, E)$ be an undirected connected loop-free graph. Given two adjacent vertices u and v, we say that u **strongly dominates** v, if $d(u) \geq d(v)$. Similarly, we say that v **weakly dominates** u, if $d(u) \geq d(v)$. A set $D \subseteq V(G)$ is a **strong-dominating set** if every vertex in $V \setminus D$ is strongly dominated by at least one vertex in D. Similarly, D is a **weak-dominating set** if every vertex in $V \setminus D$ is weakly dominated by at least one vertex in D. The strong domination number $\gamma_S(G)$ is the minimum number $\gamma_W(G)$ is the minimum cardinality of a weak dominating set of G.

We now introduce the concept of strong and weak domination in fuzzy graphs.

Definition 2.3.6 Let $G = (V, \sigma, \mu)$ be a fuzzy graph. For any $u, v \in V$, u **strongly dominates** v if $uv \in P$ $(\sigma(u, v) = \mu(u) \wedge \mu(v))$ and $d(u) \geq d(v)$. Similarly, u **weakly dominates** v if $uv \in P$ $(\sigma(u, v) = \mu(u) \wedge \mu(v))$ and $d(v) \geq d(u)$.

Definition 2.3.7 The minimum fuzzy cardinality of a strong dominating set is called the **strong dominaton number**, and is denoted by $\gamma_S(G)$. Similarly, the minimum fuzzy cardinality of a weak dominating set is called the **weak domination number**, and is denoted by $\gamma_W(G)$.

Example 2.3.8 Consider the fuzzy graph $G = (V, \sigma, \mu)$, where $V = \{v_1, v_2, v_3\}$,

	v_1	v_2	v_3
σ	1.0	0.7	0.6

	$v_1 v_2$	$v_2 v_3$	$v_3 v_1$
μ	0.7	0.5	0.6

Suppose $D = \{v_1\}$. We have $V \setminus D = \{v_2, v_3\}$. Here v_1 dominates v_2 and v_3 also $d(v_1) > d(v_2)$ and $d(v_1) > d(v_3)$. Therefore, v_1 strongly dominates both v_2 and v_3. There is no other strong-dominating set. Thus, $D = \{v_1\}$ is the strong dominating set. Therefore, we have $\gamma_S(G) = \sigma(v_1) = 1.0$. Suppose, $D' = \{v_2, v_3\}$ and $V \setminus D' = \{v_1\}$. D' is weak dominating set and $\gamma_W(G) = 1.3$.

Example 2.3.9 Consider the fuzzy graph $G = (V, \sigma, \mu)$, where $V = \{v_1, v_2, v_3, v_4\}$,

	v_1	v_2	v_3	v_4
σ	0.9	0.7	0.5	1.0

	$v_1 v_2$	$v_2 v_3$	$v_3 v_1$	$v_1 v_4$	$v_2 v_4$	$v_3 v_4$
μ	0.7	0.5	0.5	0.9	0.7	0.5

Suppose $D = \{v_1\}$. We have $V \setminus D = \{v_2, v_3, v_4\}$. We have every vertex of $V \setminus D$ is strongly dominating by the set D. Therefore, D is a strong dominating set. Suppose we have $D' = \{v_4\}$, $V \setminus D = \{v_1, v_2, v_3\}$, which is also strongly dominating set. Thus, $\gamma_S(G) = \sigma(v_1) = 0.9$ and $\gamma_W(G) = \sigma(v_1) = 0.5$.

Theorem 2.3.10 *In any complete fuzzy graph $G = (V, \sigma, \mu)$, the following inequality holds*

$$\gamma_W(G) \leq \gamma_S(G).$$

Proof Let $G = (V, \sigma, \mu)$ be a complete fuzzy graph.

Suppose for all $v_i \in V$, $\sigma(v_i)$ are equal. Because G is a complete fuzzy graph, $\mu(v_iv_j) = \sigma(v_i) \wedge \sigma(v_j)$ for all $v_i, v_j \in V$. We have $\mu(v_iv_i) = \sigma(v_i)$ for all $v_i \in V$. Thus,

$$\gamma_W(G) = \gamma_S(G) = \wedge\sigma(v_i) = \sigma(v_i). \tag{2.1}$$

Suppose for all $v_i \in V$, the $\sigma(v_i)$ are not equal. In a complete fuzzy graph, any one of the vertices dominates all other vertices; if it is least among them, then the dominating set with that vertex is called weak domination number. That is, $\gamma_W(G) = \wedge\sigma(v_i)$. Obviously, the strong dominating set has a vertex other than the least value of the vertex set. Therefore, the strong domination numbers strictly greater than weak domination number.

$$\gamma_W(G) < \gamma_S(G). \tag{2.2}$$

From the Eqs. (2.1) and (2.2), we get $\gamma_W(G) \leq \gamma_S(G)$. ∎

Example 2.3.11 In Example 2.3.9, $\gamma_W(G) = 0.5$; $\gamma_S(G) = 0.9$. Therefore, $\gamma_W(G) < \gamma_S(G)$.

Theorem 2.3.12 *In any fuzzy graph $G = (V, \sigma, \mu)$, the following inequalities hold:*
  (i) $\gamma(G) \leq \gamma_S(G) \leq p - \Delta(G)$;
  (ii) $\gamma(G) \leq \gamma_W(G) \leq p - \delta(G)$.

Proof (i) By Definitions 2.3.6 and 2.3.7, we have

$$\gamma(G) \leq \gamma_S(G). \tag{2.3}$$

Recall that $p = \sum \sigma(v)$ and $q = \sum \mu(vw)$. Therefore,

$$p - q$$
$$= \text{the sum of the degrees of } G \text{ excluding the maximum}$$
$$\text{degree of a vertex} \tag{2.4}$$

From (2.3) and (2.4), we get,

$$\gamma(G) \leq \gamma_S(G) \leq p - \Delta(G).$$

(ii) By Definitions 2.3.6 and 2.3.7, the weak domination number of G has greater or equal weight of a domination number of G, because the vertices of weakly dominating set D, it weakly dominates any one of the vertices of $V \setminus D$. But the dominating set has no limitations, it should be minimum fuzzy cardinality of a dominating set. Therefore, it be lesser or equal to the weakly domination number. That is,

$$\gamma(G) \leq \gamma_W(G). \tag{2.5}$$

Also, we gave

$$p - \delta(G)$$
$$= \sum \sigma(v) - \wedge\{d(v_i)\}$$
$$= \text{the sum of the degrees of } G \text{ excluding the}$$
$$\text{minimum degree of a vertex.} \tag{2.6}$$

From Eqs. (2.5) and (2.6), we have $\gamma(G) \leq \gamma_W(G) \leq p - \delta(G)$. ∎

Example 2.3.13 In Example 2.3.9, $\gamma(G) = 0.5$; $\gamma_W(G) = 0.5$, $\gamma_S(G) = 0.9$, $p = 3.1$; $\delta(G) = 1.5$; $\Delta(G) = 2.1$. Hence, Theorem 2.3.12 can be verified.

2.4   Independent Domination and Irredundance

Recall that the domination number $\gamma(G)$ of the fuzzy graph G is the minimum cardinality taken over all minimal dominating sets of G. The independent domination number $\iota(G)$ is the minimum cardinality taken over all maximal irredundant sets of G. The irredundant number $ir(G)$ is the minimum cardinality taken over all maximal irredundant sets of G. In this section, we prove a result that relates the numbers $ir(G)$, $\gamma(G)$, and $\iota(G)$. The results are from [12].

Nagoor Gani and Chandrasekaran [11] discussed domination in fuzzy graphs using strong arcs. We also discuss domination, independent domination, and irredundance in fuzzy graphs using strong arcs.

Recall that the **order** p and **size** q of the fuzzy graph $G = (\sigma, \mu)$ are defined by $p = \sum_{u \in V} \sigma(v)$ and $q = \sum_{(u,v) \in E} \mu(u, v)$. Further let G be a fuzzy graph on V and $S \subseteq V$. Then the fuzzy cardinality of S is defined by $\sum_{u \in S} \sigma(v)$. The **strength of connectedness** between two nodes u, v in the fuzzy graph G is $\mu^{\infty}(u, v) = \vee\{\mu^k(u, v) \mid k = 1, 2, 3, \ldots\}$, where $\mu^k(u, v) = \vee\{\mu(u, u_1) \wedge \mu(u_1, u_2) \wedge \cdots \wedge \mu(u_{k-1}, v) \mid u_i \in V, i = 1, \ldots, k\}\}$. An arc (u, v) is said to be a **strong arc** or **strong edge** if $\mu(u, v) \geq \mu^{\infty}(u, v)$ and the node v is said to be a **strong neighbor** of u.

In this section, we use a different definition of an isolated edge. A node u is said to be **isolated** if $\mu(u, v) = 0$ for all $u \neq v$. In this section, $P = \{\{x, y\} \mid \{x, y\}$ is a strong edge$\}$.

In a fuzzy graph, if every arc is strong arc, then the graph is called a **strong arc fuzzy graph**. A strong arc fuzzy graph is given in Fig. 2.2.

A path in which every arc is a strong arc, then the path is called a **strong path** and the path contains n strong arcs is denoted by P_n. Let u be a node in the fuzzy graph G, then $N(u) = \{v \mid \{u, v\}$ is a strong arc$\}$ is called **neighborhood** of u and $N[u] = N(u) \cup \{u\}$ is called **closed neighborhood** of u.

Two nodes in a fuzzy graph G are said to be fuzzy independent if there is no strong arc between them. A subset S of V is said to be a **fuzzy independent set** of G if

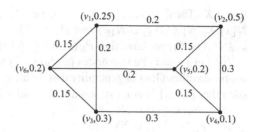

Fig. 2.2 A strong arc fuzzy graph

every two nodes of S are fuzzy independent. A fuzzy independent set S of G is said to be **maximal fuzzy independent**, if for every node $v \in V \setminus S$, the set $S \cup \{v\}$ is not fuzzy independent. The **independent domination number** $i(G)$ is the maximum cardinalities taken over all maximal independent sets of nodes of G.

A set of nodes S is said to be a **fuzzy irredundant set** if $P_N[u, S] \neq \emptyset$ for every node $u \in S$. A fuzzy irredundant set S is a **maximal irredundant set**, if for every node $u \in V \setminus S$, the set $S \cup \{u\}$ is not a fuzzy irredundant set. The **irredundant number** $ir(G)$ is the minimum cardinality taken over all maximal irredundant set of nodes of G. An irredundant set with cardinality $ir(G)$ is called an ir-set.

Definition 2.4.1 Let $G = (\sigma, \mu)$ and $G' = (\sigma', \mu')$ be two fuzzy graphs. A bijective map $h : V \to V'$ which satisfies $\sigma(x) = \sigma'(h(x))$ for all $x \in V$ and $\mu(x, y) = \mu'(h(x), h(y))$ for all $x, y \in V$ is called a **fuzzy isomorphism** from G into G'.

Definition 2.4.2 A set of edges E' in a fuzzy graph is said to be a **matching** or **independent set** if all vertices are non adjacent in E'. In a matching, if every arc is a strong arc, then the matching is called a **strong matching**.

Definition 2.4.3 A set of edges E' is said to be the **maximal strong matching** if $E' \cup \{e\}$ is not a maximal strong matching for every $e \in E \setminus E'$.

Example 2.4.4 The edges set $\{v_1 v_2, v_3 v_4, v_5 v_6\}$ is a strong matching of the fuzzy graph in Fig. 2.2.

Theorem 2.4.5 ([15]) *For every fuzzy graph G, $ir(G) \leq \gamma(G) \leq i(G)$.*

Theorem 2.4.6 ([10]) *If a fuzzy graph $G = (\sigma, \mu)$ has no induced fuzzy subgraph isomorphic to a strong fuzzy graph with underlying graph $K_{1,3}$, then $\gamma(G) = i(G)$.*

Theorem 2.4.7 *Let $G = (\sigma, \mu)$ be a fuzzy graph and u be a node which is not dominated by a maximal fuzzy irredundant set X. Then for some $x \in X$,*
  (i) $P_N(x, X) \subseteq N(u)$;
  (ii) for $x_1, x_2 \in P_N(x, X)$ such that $x_1 \neq x_2$ either (x_1, x_2) is a strong arc or for $i = 1, 2$ there exists $y_i \in X \setminus \{x\}$ such that x_i is adjacent to each node in $P_N(y_i, X)$.

Proof (i) Because X is a maximal fuzzy irredundant set, some node of $\{u\} \cup X$ is fuzzy redundant in $\{u\} \cup X$. Because u is not fuzzy dominated by X, we have that $u \in P_N(u, \{u\} \cup X) = N[u] \setminus N[X]$. Hence, some $x \in X$ is redundant in

$\{u\} \cup X$. Thus, $P[x, \{u\} \cup X] = \emptyset$, i.e., $N[x] \backslash N[\{u\} \cup X \backslash \{x\}] = \emptyset$. Therefore, $N[x] \subseteq N[X \backslash \{x\}] \cup N[u]$ and $P_N(x, X) = N[x] \backslash N[X \backslash \{x\}] \subseteq N[u]$. Because $u \notin P_N(x, X)$, we have that $P_N(x, X) \subseteq N(u)$.

(ii) Let $x_1 \neq x_2$ be two nodes of $P_N(x, X)$ such that $x_1 x_2$ is a not a strong arc and suppose without loss of generality that for all $y_i \in X \backslash \{x\}$, there exists $z_i \in P_N(y_i, X)$ such that $\{x_i, z_i\}$ is not a strong arc. Consider the set $\{x_i\} \cup X$. Then

$x_2 \in P_N(x, \{x_1\} \cup X$,

$u \in P_N(x_1, \{x_1\} \cup X)$, and

$z_i \in P_N(y_i, \{x_1\} \cup X)$ for each $y_i \in X \backslash \{x\}$.

It follows that $\{x_1\} \cup X$ is a fuzzy irredundant set, which contradicts the maximality of X. ∎

Corollary 2.4.8 *If u is not dominated by the maximal fuzzy irredundant set X and x is a node which satisfy the conditions in Theorem 2.4.7, then there exists a node $y \in X \backslash \{x\}$ such that for all $x' \in P_N(x, X)$, $G[u, x', x, y]$ is isomorphic to a strong path P_4.*

Proof From part (i) of Theorem 2.4.7, $x \notin N(u)$ implies $x \notin P_N(x, X)$. Therefore, for some $y \in X \backslash \{x\}$, xy is a strong arc. By the definition private neighbor of x with respect to X, xx' is a strong arc and $x'y$ is not a strong arc. By Theorem 2.4.7, $x'u$ is a strong arc. Because X does not dominate u, ux and uy are not strong arcs. It follows that $G[u, x', x, y]$ is a fuzzy path which is isomorphic to a strong path P_4. ∎

Theorem 2.4.9 *Let U be the set of nodes of a fuzzy graph $G = (\sigma, \mu)$ that are not dominated by X, a smallest maximal fuzzy irredundant set. For each $u \in U$, define $X_u = \{x \in X \mid P_N(x, X) \subseteq N(u)\}$. Let $M = \{x_1, x_2, x_3, \dots, x_r\}$ be a subset of X or smallest fuzzy cardinality m such that $X_u \cap M \neq \emptyset$ for each $u \in U$ and $M' = \{x_1', x_2', x_3', \dots, x_r'\}$ be a subset of $V \backslash X$ such that cardinality of M' is also m. Then $\gamma(G) < ir(G) + m$.*

Proof Let $X = \{x_1, x_2, x_3, \dots, x_n\}$ be a smallest maximal irredundant set such that $ir(G) = \sum_{v \in X} \sigma(x)$ and let $M = \{x_1, x_2, x_3, \dots, x_r\}$ be a subset of X with fuzzy cardinality of m, $0 < m \leq r$. For each $i = 1, 2, \dots, r$, $x_i \in X_u$ for some $u \in U$, therefore $P_N(x_i, X) \subseteq N(u)$. Because u is not dominated by X, $x_i \notin N(u)$ and hence $x_i \notin P_N(x_i, X)$. We deduce that there exists $x_i' \in P_N(x_i, X)$ such that $x_i' \neq x_i$. Let $D = X \cup M'$, where $M' = \{x_1', x_2', x_3', \dots, x_r'\}$. For each $u \in U$, by the definition of M, there exists i, $1 \leq i \leq r$ such that $x_i \in X_u$, hence x_i' is a strong neighbor of u. Fuzzy cardinality of M' is m and hence D is a dominating set of G of fuzzy cardinality $ir(G) + m$. However, D contains the maximal irredundant set X, hence D is not a fuzzy irredundant set. By Theorem 2.1.13, D properly contains a minimal dominating set and hence $\gamma(G) < ir(G) + m$. ∎

Corollary 2.4.10 *For any fuzzy graph G, $ir(G) > \gamma(G)/2$.*

Proof By Theorem 2.4.9,

$$\gamma(G) < ir(G) + m, \text{ for every values of } m \text{ such that } 0 < m \le ir(G)$$
$$< ir(G) + ir(G)$$
$$< 2ir(G).$$

Therefore, $ir(G) > \gamma(G)/2$. ∎

Theorem 2.4.11 *Let* $G = (\sigma, \mu)$ *be a fuzzy graph. If* $ir(G) = ir(G) + s$, *where* s *is the sum of weights of the nodes* $x_1, x_2, x_3, \ldots, x_k$ *such that* $0 < s \le k$, *where* k *is an integer with* $k \ge 0$, *then* G *has* $k + 1$ *induced subgraphs isomorphic to* P_4 *with node sequences* (a_i, b_i, c_i, d_i), $i = 1, 2, \ldots, k + 1$, *where* $\cup_{i=1}^{k+1} \{b_i, c_i, d_i\}$ *is a set of* $3k + 3$ *nodes and for each* $j = 1, 2, \ldots, k + 1$, $a_j \notin \cup_{i=1}^{k+1} \{c_i, d_i\}$.

Proof Let X be a smallest maximal fuzzy irredundant set and U be the set of nodes, which is not fuzzy dominated by X. Let $X_u = \{x \in X \mid P_N(x, X) \subseteq N(u)\}$ and $Z = \cup_{u \in U} X_u$. For each $x \in Z$ choose a node $x' \in P_N(x, X)$ such that (u, x') is a strong edge. Let B be a strong edge fuzzy bipartite subgraph of G with edge set $\cup_{u \in U} \{(u, x') \mid x \in X\}$ and $Q = \{(u, x_i') \mid i = 1, 2, 3, \ldots, \beta\}$ be a strong maximum matching in B. We claim that $Y = \{x_1, x_2, x_3, \ldots, x_\beta\}$ satisfies $Y \cap X_u \ne \emptyset$ for each $u \in U$, where fuzzy cardinality of Y is t, $0 < t \le \beta$. Suppose $Y \cap X_{u'}$ is a strong edge of B not incident with any strong edge of Q, which is a contradiction to Q is a maximal matching. Hence, m is a value defined as in Theorem 2.4.9, we have

$$\beta \ge t \ge m > \gamma(G) - ir(G)$$
$$= s, \text{ (by hypothesis) which is true for every value of } s, \; 0 < s \le k$$
$$= k.$$

That is $\beta \ge k + 1$.

For each $x \in Z$, there exists $u \in U$ such that $P_N(x, X) \subseteq N(u)$.

Therefore, $x \notin P_N(x, X)$ and there exists $y_k \in X$ such that (x, y_k) is a strong edge. Now, we set $(a_i, b_i, c_i, d_i) = (y_{x_i}, x_i, x_i', u_i)$ for $i = 1, 2, \ldots, k+1$. The result now follows from Corollary 2.4.8. ∎

Corollary 2.4.12 *If* $G = (\sigma, \mu)$ *is a fuzzy graph which does not have two induced subgraphs isomorphic to a strong fuzzy path* P_4 *with node sequences* (a_i, b_i, c_i, d_i), $i = 1, 2$, *where* $b_1, b_2, c_1, c_2, d_1, d_2$ *are distinct and for* $i = 1, 2$, $a_i \notin \{c_1, c_2, d_1, d_2\}$, *then* $\gamma(G) = ir(G)$.

Theorem 2.4.13 *If the fuzzy graph* $G = (\sigma, \mu)$ *has no induced fuzzy subgraph isomorphic to a strong fuzzy graph of the six underlying graphs* G_i, $i = 1, 2, \ldots, 6$, *given Fig. 2.3, then* $irr(G) = \gamma(G)$.

Proof Let us assume $ir(G) < \gamma(G)$. Let X be a maximal fuzzy irredundant set with maximum cardinality. Let U be a set of nodes which is not dominated by X. Here U is nonempty, otherwise X would be a fuzzy dominating set and $\gamma(G) \le ir(G)$. Every node in U makes some node in X which is non isolated and fuzzy redundant,

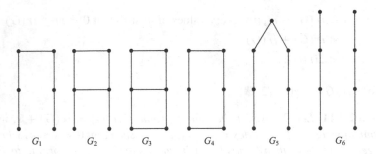

Fig. 2.3 Graphs G_i, $i = 1, 2, \ldots, 6$

then $X \cup \{z'\}$, where z' is private neighbor of z, is a fuzzy dominating set of G. But it is not a minimal fuzzy dominating set, because it strictly contains a maximal fuzzy irredundant set and is therefore not fuzzy irredundant set, which is not possible. For every node x of X, which is not isolated in X, let us define $U_x = \{u \mid x$ is fuzzy irredundant in $X \cup \{u\}\}$. The set does not form a chain under inclusion, otherwise the greatest of them would be U, which is contrary to the fact that all the nodes of U do not make the same node of X fuzzy redundant. Therefore, there are two sets U_x and U_y, each of them having a node not included in the other. Let $u_x \in U_x \backslash U_y$. Then x has a private neighbor of x' which is not a strong neighbor to u_y. Similarly, y has a private neighbor y' which is not a strong neighbor to u_x. Let a_1 a_2 be the strong neighbor to x and y, respectively, in X. If x and y are strong neighbors to a in X, but are not strong neighbor, then we shall take $a_1 = a_2 = a$. Then $\{a_1, x, x', u_x\}$ and $\{a_2, y, y', u_y\}$ are two strong paths P_4 called (a_i, b_i, c_i, d_i) with the required condition: the nodes $b_1, b_2, c_1, c_2, d_1, d_2$ are distinct and for $i = 1, 2$, $a_i \notin \{c_1, c_2, d_1, d_2\}$.

Moreover, $i \neq j$, a_i and b_i are neither strong neighbors to b_j (except when a_i coincides with a_j or b_j) nor to c_j (or to d_j) : and c_i is also not strong neighbor to d_j. Therefore, the fuzzy subgraph induced by the two paths P_4 is isomorphic to one of the strong fuzzy of the fifteen underlying graphs of Fig. 2.4.

But the graph G_1 is an induced subgraph of all the other edges, except the graph G_i, $2 \leq i \leq 6$. Hence, if $ir(G) < \gamma(G)$, G contains a fuzzy subgraph isomorphic to strong fuzzy graph of the underlying graph of G_i, $1 \leq i \leq 6$, then $ir(G) = \gamma(G)$.

∎

Theorem 2.1.13 and Corollary 2.4.12 give the following theorem.

Theorem 2.4.14 *Let $G = (\sigma, \mu)$ be a fuzzy graph. If*

(i) G does not have two induced subgraphs isomorphic to a strong path P_4 with node sequences (a_i, b_i, c_i, d_i), $i = 1, 2$, where $b_1, b_2, c_1, c_2, d_1, d_2$ are distinct and for $i = 1, 2$, $a_i \notin \{c_1, c_2, d_1, d_2\}$, and

(ii) does not have a fuzzy subgraph isomorphic to $K_{1,3}$, then

$$ir(G) = \gamma(G) = i(G).$$

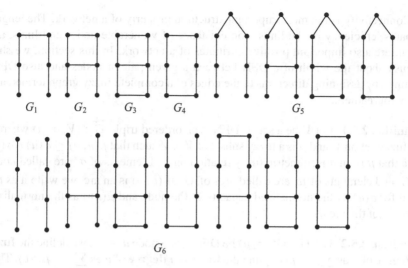

G_1 G_2 G_3 G_4 G_5

G_6

Fig. 2.4 Fuzzy graphs in Theorem 2.4.13

Theorem 2.4.15 *If a fuzzy graph* $G = (\sigma, \mu)$ *has no induced fuzzy subgraph isomorphic to a strong fuzzy graph with underlying graph* $K_{1,3}$, *and also does not have induced fuzzy subgraph isomorphic to a strong fuzzy graph of the six underlying graphs* G_i, $i = 1, 2, \ldots, 6$ *given in Fig. 2.3, then*

$$ir(G) = \gamma(G) = i(G).$$

2.5 Non Deterministic Flow in Fuzzy Networks

We next consider the work in [9]. The figures are from [9].

"The crime of human trafficking affects nearly every country in every region in the world. Human trafficking flows are mainly intraregional. That is, the origin and the destination of those trafficked is within the same region. Victims tend to be trafficked from poor countries to more affluent ones within the region. In this section, we apply techniques of digraph theory to examine the regional flow of trafficking in persons."

Recall that a digraph is said to be strongly connected (or strong) if for every pair of distinct points, x and y, there exists a directed path from x to y and one from y to x. A digraph is said to be unilaterally connected (or unilateral) if for every pair of pints, x and y, there is a directed path from x to y or one from y to x. A digraph is called disconnected if the points can be divided into sets with no line joining any point in one set with a point in the other set. A digraph is called weakly connected (or weak) if it is not disconnected.

Connectivity is the most important structural property of a network. The lengths
of paths, efficiency of the links, non existence of vulnerable nodes and links, and
so on, are also important positive attributes of a network. In this section, we shall
discuss about the width of a DFN between a given pair of nodes and also DFNs
obtained by assigning directions to the edges of a complete fuzzy graph termed as a
fuzzy tournament.

Definition 2.5.1 Let V be a set. A DFN is an ordered triple $\overrightarrow{G}$: (V, σ, μ) where σ
is a fuzzy set on V and μ is a fuzzy subset of $V \times V$ such that $\mu(u, v) \leq \sigma(u) \wedge \sigma(v)$.
Note that μ is an unsymmetric fuzzy relation on V. Elements of σ^* are called nodes
of $\overrightarrow{G}$ and elements of μ^* are called arcs of $\overrightarrow{G}$. If (u, v) is an arc, we write it as uv.
We refer $\sigma(u)$ as the normalized capacity of the node and $\mu(uv)$ as the normalized
capacity of the arc uv.

Definition 2.5.2 Let $\overrightarrow{G}$: (V, σ, μ) be DFN. For a node $u \in \sigma^*$, we define the **fuzzy
in degree** of u as $\sum_{v \in \sigma*} \mu(vu)$ and the **fuzzy out degree** of u as $\sum_{v \in \sigma*} \mu(uv)$. They
are denoted as $d_f^-(u)$ and $d_f^+(u)$, respectively. A node u in a DFN is a **source** if
$d_f^-(u) = 0$, **sink** if $d_f^+(u) = 0$ and an **intermediate node** if both $d_f^-(u) > 0$ and
$d_f^+(u) > 0$.

A DFN $\overrightarrow{G'}$: (V, σ', μ') is said to be a **partial sub DFN** of $\overrightarrow{G}$: (V, σ, μ) if
$\sigma' \subseteq \sigma$ and $\mu' \subseteq \mu$. $\overrightarrow{G'}$ is called a **sub DFN** of G' if $\sigma'(u) = \sigma(u)$ for all $u \in \sigma'^*$
and $\mu'(xy) = \mu(xy)$ for all $xy \in \mu'^*$. A **directed walk** or **di-walk** in a DFN
$\overrightarrow{G}$: (V, σ, μ), is a finite nonempty sequence $(v_0, e_1, v_1, e_2, \ldots, e_k, v_k)$ where terms
are elements of σ^* and μ^* alternatively such that e_i has tail v_{i-1} and head v_i for
$i = 1, 2, \ldots, k$. We denote it by $(v_0, v_1, \ldots, v_k)$. A di-path P in $\overrightarrow{G}$: (V, σ, μ) is a
di-walk $(v_0, v_1, \ldots, v_k)$ such that $v_i \neq v_j$ except for v_0 and v_k referred as the origin
and terminus of P. The number of arcs in the di-path is called its length. A di-path
is said to be a di-cycle if its origin and terminus coincide.

A DFN is said to be **di-connected** if there exists a di-path between any two
elements u and v in σ^* (Either u is connected to v or vice-versa. It is also
termed as unilaterally connected in Digraph theory). Clearly 'di-connected' is an
equivalence relation on σ^*. It decomposes $\overrightarrow{G}$: (V, σ, μ) into di-components
$\overrightarrow{G}(V_1), \overrightarrow{G}(V_2), \ldots, \overrightarrow{G}(V_t)$. The definitions of strongly connected, weakly connected and disconnected fuzzy graphs can be given analogous to the classical case
(See Fig. 2.5).

Let $\overrightarrow{G}$: (V, σ, μ) be a DFN and let P : $(v_0, v_1, v_2, \ldots, v_n)$ be a directed path in
$\overrightarrow{G}$. The **width of the path** P, denoted as $w(P)$ is defined as $\mu(v_0v_1) \wedge \mu(v_1v_2) \wedge
\cdots \wedge \mu(v_{n-1}v_n)$. Also, $\vee\{w(P) : P$ is a $u - v$ di-path$\}$ is called the $u - v$ **width** in
$\overrightarrow{G}$. It is denoted by $w_{\overrightarrow{G}}(u - v)$ or simply $w(u - v)$.

In Fig. 2.5, the width of the $a - c$ di-path abc in the first digraph is $0.5 \wedge 0.6 =
0.5$, where $\wedge$ represents the minimum. Also $b - a$ width in the third digraph is
$0.8 \vee 0.5 = 0.8$.

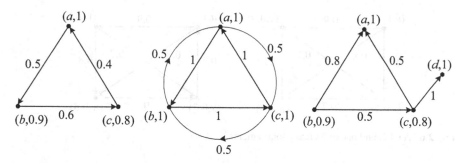

Fig. 2.5 Di-connected, strongly connected and weakly connected DFNs

If $\vec{G}$: (V, σ, μ) is a DFN with $\sigma^* = \{v_1, v_2, \ldots, v_n\}$, then an $n \times n$ matrix $D_f(\vec{G})$ where $D_f(i, j) = w(u_i, u_j)$ if $i \neq j$ and $\sigma(u_j)$ if $i = j$ is called the **width matrix** of the DFN $\vec{G}$: (V, σ, μ).

The width matrix of the first DFN in Fig. 2.5 is,

$$D_f(\vec{G}) = \begin{bmatrix} 1 & 0.5 & 0.5 \\ 0.4 & 0.9 & 0.6 \\ 0.4 & 0.4 & 0.8 \end{bmatrix}$$

Note that the width matrix need not be symmetric like adjacency matrix of a fuzzy graph. As in fuzzy graph theory, by applying max-min composition and taking powers of the directed adjacency matrix, we can reach the width matrix of the DFN as the matrix D^k so that $D^k = D^{k+1}$. In this example, the directed adjacency matrix is

$$A = \begin{bmatrix} 1 & 0.5 & 0 \\ 0 & 0.9 & 0.6 \\ 0.4 & 0 & 0.8 \end{bmatrix}.$$

Here $A^2 = D_f$, where $A = \vec{G}$. Also the indegree and outdegree at any node u_i is given by $d_f^-(u_i) = \{A(1, i) + A(2, i) + \cdots + A(n, i)\} - \sigma(u_i)$ and $d_f^+(u_i) = \{A(i, 1) + A(i, 2) + \cdots + A(i, n)\} - \sigma(u_i)$.

Similar to fuzzy graphs the arcs of a DFN can be divided into three categories. Namely α, β and δ. The only difference is that there will be directed paths in place of paths in fuzzy graphs. But for the completion the definition is formally given below.

Definition 2.5.3 Let $\vec{G}$: (V, σ, μ) be a DFN. An arc xy is said to be α-**strong** if $w_{\vec{G}\setminus xy}(x - y) < \mu(xy)$. It is said to be β- **strong** if $w_{\vec{G}\setminus xy}(x - y) = \mu(xy)$ and δ-**arc** if

$$w_{\vec{G}\setminus xy}(x - y) > \mu(xy),$$

where $\vec{G}\setminus xy$ represents the DFN where the arc xy is removed.

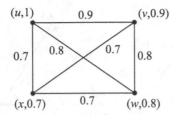 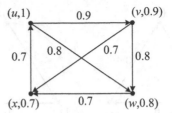

Fig. 2.6 A CFG and one of its fuzzy tournaments

As a consequence, an arc in a DFN is said to be strong if $w_{\overrightarrow{G}\backslash xy}(x-y) \le \mu(xy)$. As in the case of edges in fuzzy graphs, there are no relations with being strong and the μ value of the arc.

As in the classical case, directed fuzzy networks can be constructed by assigning orientation to the edges of a complete graph. We have the definition of a fuzzy tournament as follows.

Definition 2.5.4 Let $G : (\sigma, \mu)$ be a complete fuzzy graph (CFG). The DFN obtained by assigning directions to the edges is called a **fuzzy tournament**.

Example 2.5.5 Consider the following fuzzy tournament (Fig. 2.6) obtained by giving orientation to a CFG on four nodes.

Theorem 2.5.6 *In a fuzzy tournament* $\overrightarrow{G}$, *there always exists a strong spanning di-path.*

Proof First we prove that a fuzzy tournament has no δ-arcs. If possible suppose xy is a δ-arc. Then $q = w_{\overrightarrow{G}}(x-y) = w_{\overrightarrow{G}\backslash xy}(x-y) > \mu(xy) = p$. That is, there exists a directed $x-y$ path say P in $\overrightarrow{G}$ such that $\mu(uv) > \mu(xy)$ for every arc uv in P. Because $\mu(xy) = p$, either $\sigma(x) = p$ or $\sigma(y) = p$. Let w_1 be the first node in P after x and w_2, the last in P before y. Then $\mu(x, w_1) > p$ and $\mu(w_2, y) > p$. But it is not possible, because $\overrightarrow{G}$ is a fuzzy orientation and one of $\sigma(x)$ or $\sigma(y)$ is p.

Next we show that there is a spanning path in $\overrightarrow{G}$. Let $|\sigma^*| = n$. When $n = 1, 2, 3$, we can easily find fuzzy orientations with required property. Now suppose that the result is true for all fuzzy tournaments with $|\sigma^*| < n$. Let $v \in \sigma^*$. Then $\left|\sigma^*(\overrightarrow{G}\backslash v)\right| = n - 1$. Let $S = u_1, u_2, \ldots, u_{n-2}, u_{n-1}$ be the strong spanning di-path of $\overrightarrow{G}\backslash v$. If there exists arc vu_1 or $u_{n-1}v$, we can form the required di-paths by concatenation. Suppose there do not exists such arcs. Because it is a fuzzy orientation, the arcs u_1v and vu_{n-1} should exists. Let w be the last node different from u_{n-1} such that wv is an arc. Let $w = u_i$ so that u_{i+1} do not have this property. Because it is a fuzzy orientation, vv_{i+1} also exists. Thus, $Q = u_1, u_2, \ldots, u_i, v, v_{i+1}, \ldots, u_{n-1}$ is a strong spanning di-path in $\overrightarrow{G}$. ∎

Fig. 2.7 Case 1 of
Theorem 2.5.7

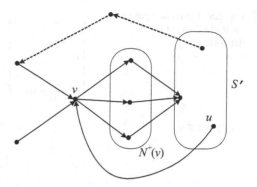

A strong path from u to v is called a **geodesic** if there is no shorter strong path
from u to v.

Let $\vec{G} : (V, \sigma, \mu)$ be a DFN and $S \subseteq \sigma^*$. Then S is called **independent** if there
are no directed arcs between nodes of S.

Theorem 2.5.7 *Let* $\vec{G} : (V, \sigma, \mu)$ *be a DFN. Then there is an independent set*
$S \subseteq \sigma^*$ *of nodes such that for every node* $v \in \sigma^* \backslash S$, *there exists a node* $u \in S$ *such*
that $d(u, v) \leq 2$, *where* $d(u, v)$ *is the minimum of sum of weights of geodesics from*
u *to* v.

Proof Let $\vec{G} : (V, \sigma, \mu)$ be a DFN and suppose $|\sigma^*| = n$. We shall prove the result
by induction on n. If $n = 1$, the result is obvious. Suppose the result is true for
all DFN with $|\sigma^*| < n$. Let $v \in \sigma^*$. Let $\vec{G'} : (V', \tau, \psi)$ be the DFN obtained by
removing v and all its out neighbors. Then because $|\tau^*| < n$, the result is true for $\vec{G'}$
and hence there is an independent set S' satisfying the required property. We have
two different cases.

Case 1: (See Fig. 2.7) In $\vec{G}$, v is an out neighbor of some $u \in S'$. Consider the
following figure.

As seen from the figure, any node in $v \cup N^+(v)$ is reachable from u and the
theorem holds for $S = S'$.

Case 2: For any $u \in S'$, arc uv does not exists.

In this case $S = S' \cup v$ is the required independent set. In both cases, because
$\mu(xy) \leq 1$ for any $xy \in \mu^*$, and hence the result is proved. ∎

The concepts of strong cycle, locamin cycle, multimin cycle, strongest strong
cycle, and so on, can be easily extended to DFN in an analogous way. They will be
called directed strong cycle (DSC), directed locamin cycle (DLC), directed multimin
cycle (DMC) and directed strongest strong cycle (DSSC), respectively. Consider the
DFN given in Fig. 2.8.

In Fig. 2.8, the first DFN is a DSC, DSSC, DLC and DMC. The second is both
DMC and DSC but it is not a DLC or DSSC. The third DFN do not qualify any of
these.

Fig. 2.8 Different types of cycles in directed fuzzy networks

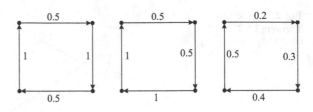

Fig. 2.9 DFN

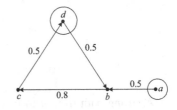

Theorem 2.5.8 *Let $\vec{G} : (V, \sigma, \mu)$ be a strongly connected fuzzy tournament with $|\sigma^*| = n$. Each node of $\vec{G}$ is contained in a DSC of length k where $3 \leq k \leq n$.*

Proof The proof follows from Theorem 10.3 of [2] and the first part of the proof of Theorem 2.5.6. ■

The following definition is a fuzzification of these ideas due to [30] and is presented in Chap. 1. Consider a DFN $\vec{G} = (V, \sigma, \mu)$. Let μ^{-1} be the fuzzy subset of $V \times V$ defined by $\mu^{-1}(y, x) = \mu(x, y)$ for all $(x, y) \in V \times V$. Let Δ be the fuzzy subset of $V \times V$ defined by $\Delta = \mu \cup \mu^{-1}$. Let $\widehat{\mu}, \widehat{\mu^{-1}}$, and $\widehat{\Delta}$ denote the transitive closures of μ, μ^{-1}, and Δ, respectively.

The following definition is a fuzzification of these ideas.

Definition 2.5.9 The grades of membership of a DFN $\vec{G} = (V, \sigma, \mu)$ in U_3, U_2, U_1, and U_0 are defined as follows:

$$\mu_{U_3}(\vec{G}) = \wedge\{\widehat{\mu}(x_i, x_j)\} \mid x_i, x_j \in V\},$$

$$\mu_{U_2}(\vec{G}) = \wedge\{\widehat{\mu}(x_i, x_j) \vee \widehat{\mu}(x_j, x_i)\} \mid x_i, x_j \in V\},$$

$$\mu_{U_1}(\vec{G}) = \wedge\{\widehat{\Delta}(x_i, x_j)\} \mid x_i, x_j \in V\},$$

$$\mu_{U_0}(\vec{G}) = 1 - \wedge\{\widehat{\Delta}(x_i, x_j)\} \mid x_i, x_j \in V\}.$$

Definition 2.5.10 For DFN $\vec{G} = (V, \sigma, \mu)$, let $\vec{G_k}$ be the DFN obtained by the removal of x_k from V. Then x_k is called a **weakening point** for U_i (W_i **point**) if $\mu_{U_i}(\vec{G}) < \mu_{U_i}(\vec{G_k})$; a **neutral point** for u_i (an N_i **point**) if $\mu_{U_i}(\vec{G}) = \mu_{U_i}(\vec{G_k})$; and a **strengthening point** (an S_i point) if $\mu_{U_i}(\vec{G}) > \mu_{U_i}(\vec{G_k})$, $i = 1, 2, 3$.

Consider the following DFN in Fig. 2.9.

In the DFN in Fig. 2.9, a is a strengthening point, b is a weakening point and d is a neutral point as shown in Fig. 2.10.

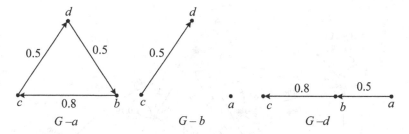

Fig. 2.10 Strengthening, weakening, and neutral points

2.6   Application of DFN in Human Trafficking

In this application, we take a regional study. The regions are the ones defined in [32], namely, Western and Central Europe, Eastern and Central Europe, North America, Central America and the Caribbean, South America, Middle East, Sub-Saharan Africa, Central Europe and the Balkans, East Asia and the Pacific, South Asia.

The following matrix has different regions. The regions and percentage flows are from, United Nations Office on Drugs and Crime, Global Report on Trafficking in Persons 2014, pp.59–86. The ✓ denotes extremely small flow. The regions heading the rows are source regions while those heading the columns are the destination regions.

	WCEur	WSEur	CEur&Bal	EEur&CAsia	NAm&CAm&Car	SAm	EAsia&Pac	SAsia	Sub−Saharan Africa	Mid East
WCEur	0.62	0.13								
WSEur		0.16								
CEur&Bal		0.27	0.79 .09		0.05					
EEur&CAsia	0.04	0.04	0.05	0.99						0.06
NAm&CAm&Car	0.08	✓	✓		0.59	0.04				
SAmerica		0.07			0.03	0.94				
EAsia&Pacific	0.07	0.07	✓		0.25	0.01	0.97			0.33
SAsia		✓			0.07			0.96		0.18
Sub−Saharan Africa		0.16							1.0	0.10
MidEast										0.31

Using the above table, can draw a directed graph as shown in Fig. 2.11.

Consider the directed graph G which is the support of the fuzzy digraph $\overrightarrow{G}$ defined by the above matrix, (see Fig. 2.11). We see that there is no directed path from Western

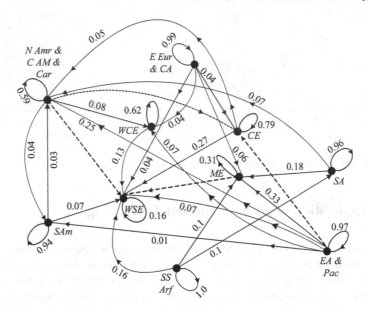

Fig. 2.11 Flow between different regions of the globe

and Southern Europe to the Middle East and no directed path from the Middle East to Western and Southern Europe. Thus, G^* is weakly connected. As long as these two countries remain in a directed graph obtained by deleting points, the resulting digraph can be no more than weakly connected. The digraph obtained by deleting Western and Southern Europe is also weakly connected because there is no directed path from West and Central Europe to the Middle East and no directed path from Middle East to West and Central Europe. The digraph obtained by deleting the Middle east is also weakly connected because there is no directed path from Sub-Saharan Africa to South Asia and no directed path from South Asia to Sub-Saharan Africa. In fact, it follows that every point in G is of type P_{11}. The directed graph obtained by deleting both the Middle East and Western and Southern Europe is disconnected because Sub-Saharan Africa becomes an isolated point.

Nearly every country is affected by the crime of trafficking in persons. Transregional trafficking flows are mainly detected in the rich countries of the Middle East, Western Europe, and North America. These flows often involve victims from the global south mainly East and South Asia and Sub-Saharan Africa. There is a correlation between the affluence of the destination country and the share of victims trafficked there from other regions. Richer countries attract victims from a variety of regions, including from other continents, whereas less affluent countries are mainly affected by domestic or subregional flows.

We now consider the DFN $\overrightarrow{G}$ defined by the matrix immediately above. We see from the discussion above concerning G, the support of $\overrightarrow{G}$, that $\mu_{U_3}(\overrightarrow{G}) = 0 = \mu_{U_2}(\overrightarrow{G})$. It follows that $\mu_{U_1}(\overrightarrow{G}) = 0.05$, else the point Eastern Europe and Central Asia would be isolated.

2.7 Strong Independence and Domination in DFN

We continue with results from [9].

Let DFN $\overrightarrow{G} = (V, \sigma, \mu)$ be a directed fuzzy network and let S be a subset of V. Then S is said to be **strongly independent** if S is independent and for all $s_1, s_2 \in S$, $\nexists$ a directed path in $\overrightarrow{G}$ from s_1 to s_2.

Proposition 2.7.1 *There exists a strongly independent set S with two or more elements if and only if $\mu_{U_2}(\overrightarrow{G}) = 0$.*

Proof Suppose such a set S exists. Let $s_1, s_2 \in S$. Then there does not exist a directed path in $\overrightarrow{G}$ from s_1 to s_2 or one from s_2 to s_1. Thus, $\mu_{U_2}(\overrightarrow{G}) = 0$.

Conversely, suppose $\mu_{U_2}(\overrightarrow{G}) = 0$. Then there exist $s_1, s_2 \in V$ be such that there does not exist a directed path from s_1 to s_2 or one from s_2 to s_1. Hence, $S = \{s_1, s_2\}$ is strongly independent. ∎

In [29], Somasundaram and Somasundaram introduced results on domination to fuzzy graphs. We show how these latter results can be carried over to fuzzy directed graphs and applied to human trafficking.

Let V be a set and P be an anti-reflexive relation on V. Let $x \in V$ and $N(x) = \{y \mid (y, x) \in P\}$. Let $d, x \in V$. Then d is said to **dominate** x with respect to P if $(d, x) \in P$.

Definition 2.7.2 A subset D of V is called a **dominating set** with respect to P if for all $x \in V \backslash D$, there exists $d \in D$ such that $(d, x) \in P$.

Theorem 2.7.3 *Let $D \subseteq V$. Then D is a minimal dominating set with respect to P if and only if D is a dominating set with respect to P and for all $d \in D$ one of the following two conditions hold:*
  (i) $N(d) \cap D = \emptyset$;
  (ii) $\exists c \in V \backslash D$ such that $N(c) \cap D = \{d\}$.

Proof Suppose D is a minimal dominating set with respect to P. Let $d \in D$. Then $D_d = D \backslash \{d\}$ is not a dominating set. Hence, $\exists x \in V \backslash D_d$ such that x is not dominated by any element of D_d, i.e., for all $y \in D_d$, $(y, x) \notin P$. If $x = d$, then (i) holds because P is anti-reflexive. Suppose $x \neq d$. Then $N(x) \cap D = \{d\}$ and so (ii) holds with $c = x$.

Conversely, suppose that D is a dominating set with respect to P and that for all $d \in D$, either (i) and (ii) hold. Let $d \in D$. If (i) holds for d, then $D \backslash \{d\}$ is not a dominating set because there does not exist $d' \in D$ such that $(d', d) \in P$. If (ii) holds for d, then there does not exist $d' \in D\backslash\{d\}$ such that $d' \in N(c)$ and so d' does not dominate c. Hence, $D\backslash\{d\}$ is not a dominating set with respect to P. Thus, D is minimal. ∎

Let $\overrightarrow{G} = (\sigma, \mu)$ be a fuzzy directed graph with vertex set V. Let $P = \text{Supp}(\mu)$. We write $N^{(d)}(x)$ for $N(x)$ and call $N^{(d)}(x)$ the **destination dominating neighborhood** of x. Theorem 2.7.3 holds for this special case.

Let $N^{(s)}(x) = \{y \in V \mid (x, y) \in P\}$ and call $N^{(s)}(x)$ the **source dominating neighborhood** of x. Theorem 2.7.3 holds for this special case.

Suppose P is symmetric. Let G be a fuzzy non-directed graph. The above theorem holds.

Theorem 2.7.4 *Let $\overrightarrow{G}$ be a fuzzy directed graph with no isolated vertices with respect to P. Let D be a destination dominating set with respect to P. Then $V\backslash D$ is a source dominating set with respect to P.*

Proof Let $d \in D$. Because $\overrightarrow{G}$ has no isolated vertices with respect to P, there exists $c \in V$ such that $c \in N^{(d)}(d)$, i.e., $(c, d) \in P$. If $c \notin D$, then d is a source dominated by an element of $V\backslash D$, namely c. If $c \in D$, then (i) of Theorem 2.7.3 does not hold so there exists $c' \in V\backslash D$ such that $c' \in N^{(d)}$, i.e., $(c', d) \in P$. Thus, every element of D is source dominated by some element of $V\backslash D$. ∎

Example 2.7.5 Let $V = (x, y, z\}$ and $P = \{(x, y), (x, z), (z, y)\}$. Then $D = \{x\}$ is a minimal destination dominating set and $V\backslash D = \{y, z\}$ is a source dominating set. However, $\{z\}$ is not a source dominating set because $(y, z) \neq P$. Because $(x, y), (z, y) \in P$, $\{y\}$ is a minimal source dominating set.

Theorem 2.7.6 *If D is an independent destination dominating set with respect to P, then D is both a minimal destination dominating set and a maximal independent set.*

Proof Suppose D is independent and destination dominating. Because D is independent, $D\backslash\{d\}$ is not destination dominating for all $d \in D$ because there does not exist $d' \in D$ such that $(d', d) \in P$. Thus, D is a minimal destination dominating set with respect to P. Also for all $x \in V\backslash D$, $D \cup \{x\}$ is not independent because there exists $d \in D$ such that $(d, x) \in P$. Hence, D is maximal independent. ∎

Example 2.7.7 Let $V = \{x, y, z, w\}$ and $P = \{(z, x), (y, w), (z, w)\}$. Then $D = \{x, y\}$ is a maximal independent set. However, D is not a destination dominating set because $(x, z) \notin P$. Thus, the result which holds for fuzzy undirected graphs does not hold for fuzzy directed graphs.

Example 2.7.8 Let $\overrightarrow{G}$ be the directed graph of Fig. 2.11. Let $D^{(d)} = \{$Sub-Saharan Africa, East Asia and Pacific, E. Europe and C. Asia, S. Asia$\}$. Then $D^{(d)}$ must be a subset of any destination dominating set because every element of $D^{(d)}$ is a source only region. In fact, $D^{(d)}$ can be seen to be destination dominating set and is clearly independent and in fact strongly independent. Thus, by Theorem 2.7.6, $D^{(d)}$ is both a minimal destination dominating set and a maximal independent set. Let $D^{(s)} = \{$Middle East, Western and Southern Europe$\}$. Then it follows that $D^{(s)}$ is both a minimal source dominating set and maximal independent.

Example 2.7.9 Let $S = \{$Sub-Sarahan Africa, East Asia, Eastern Europe and Pacific, South Asia$\}$. Let $u \in S$. Then $N[u] \cap S = \{u\}$ because $N[u] = \{u\}$. Hence, S is irredundant.

Theorem 2.7.10 *Let $D \subseteq V$. Then D is a minimal destination dominating set with respect to P if and only if D is an irredundant and a destination dominating. set.*

Proof Suppose D is a minimal destination dominating set. Suppose D is not irredundant. Then there exists $d \in D$ such that $PN[d, D] = \{v \in V \mid n[v] \cap D = \{d\}\} = \emptyset$. Thus, $\nexists v \in V$ such that $N[v] \cap D = \{d\}$. Hence, for all $v \in V$, either $N[v] \cap D \supset \{d\}$ or $d \notin N[v] \cap D$ so $d \notin N[v]$. If $v \in V$ is such that $N[v] \cap D \supset \{d\}$, then $\exists d' \in D \backslash \{d\}$ such that $(d', v) \in P$ and if v is such that $d \notin N[v]$, then $\exists d'' \in D \backslash \{d\}$ such that $(d'', v) \in P$. Thus, $D \backslash \{d\}$ is a destination dominating set, contradicting the minimality of D.

Conversely, suppose that D is an irredundant and destination dominating set. Suppose D is not minimal destination dominate. Then $\exists d \in D$ such that $D \backslash \{d\}$ is destination dominate. Hence, $\exists d' \in D$ such that $(d', d) \in P$. Thus, $N[d] \supseteq \{d', d\}$. Hence, $N[d'] \cap D \supseteq \{d, d'\} \neq \{d\}$. Thus, it is not the case that for all $d'' \in D, N[d''] \cap D = \{d\}$. Hence, D is not irredundant. ∎

Theorem 2.7.11 *If D is a minimal destination dominating set, then D is a maximal irredundant set.*

Proof Suppose D is a minimal destination dominating set. Then D is irredundant and destination dominating by Theorem 2.7.10. Suppose D is not maximal irredundant. Then there exists $S \subseteq V$ such that S is irredundant and $D \subset S$. Because S must be destination dominate, we have by Theorem 2.7.10 that S is a minimal destination dominating set. However, this is impossible because D is destination dominating. ∎

A dominating set is called **perfect** with respect to P if for all $x \in V \backslash D$ there exist only one $d \in D$ such that $(d, x) \in P$. We see that in our applied example there are no perfect destination dominating sets because every such set must contain $\{$Sub-Saharan Africa, East Asia and Pacific, E. Europe and C. Asia, S. Asia$\}$. Similarly, there are no perfect source dominating sets because Sub-Saharan Africa is source dominated by both Middle East and Western and Southern Europe.

References

1. O. Bobrowicz, C. Choulet, A. Haurat, F. Sandoz, M. Tebaa (1990) A method to build membership functions — Application to numerical/symbolic interface building, in *Proceedings of 3rd Internat. Conference on Information Processing and Management of Uncertainty in Knowledge Based System*, Paris

2. J.A. Bordy, U.S.R. Murty, *Graph Theory with Applications*, 5th edn. (MacMillan Press, New York, 1982)
3. E.J. Cockayne, S.T. Hedetnieme, Towards a theory of domination in graphs. Networks **7**, 247–261 (1977)
4. E.J. Cockayne, O. Favaron, C. Payan, A.G. Thomason, Contribution to the theory of domination, independence, and irredundance in graphs. Discret. Math. **33**, 249–258 (1981)
5. F. Harary, *Graph Theory*, 3rd edn. (Addison Wesley, Reading, MA, 1972)
6. O.T. Manjusha, M.S. Sunitha, Notes on domination in fuzzy graphs. J. Intell. Fuzzy Syst. **27**, 3205–3212 (2014)
7. O.T. Manjusha, M.S. Sunitha, Total domination in fuzzy graphs using strong arcs. Ann. Pure Appl. Math. **9**(1), 23–33 (2014). ISSN: 2279-087X (P), 2279-0888
8. S. Mathew, J.N. Mordeson, Directed fuzzy networks and max-flow min cut theorem. New Math. Nat. Comput. **13**, 219–229 (2017)
9. J.N. Mordeson, S. Mathew, Non determinisitc flow in networks and its application in identification of human trafficking chains. New Math. Nat. Comput. **13**, 231–243 (2017)
10. A. Nagoor Gani and P. Veldivel, On irredundance in fuzzy graphs, Proceedings of the national Conference on Fuzzy Mathematics and Graph Theory, Jamal Mohamed College, Tiruchirappalli, March 2008, 40–43
11. A. Nagoor Gani, V.T. Chandrasekaran, Domination in fuzzy graphs. Adv. Fuzzy Sets Syst. **1**, 17–26 (2006)
12. A. Nagoor Gani, P. Vadiel, A study on domination, independent domination, and irredundance in fuzzy graphs. Appl. Math. Sci. **5**, 2317–2325 (2011)
13. A. Nagoor Gani, P. Veldivel, Relation between the parameters of independent domination and irredundance in fuzzy graph. Int. J. Algorithms Comput. Math. **2**(1), 787–791 (1998)
14. A. Nagoor Gani, P. Veldivel, Fuzzy independent dominating set. Adv. Fuzzy Sets Syst. **2**(1), 99–108 (2007)
15. A. Nagoor Gani, P. Veldivel, On domination, independence, and irredundance in fuzzy graphs. Int. Rev. Fuzzy Math. **3**(2), 191–198 (2008)
16. A. Nagoor Gani, P. Veldivel, Contribution to the theory of domination, independence, and irredundance in fuzzy graphs. Bull. Pure Appl. Sci. **28E**(2), 179–187 (2009)
17. A. Nagoor Gani, P. Veldivel, On the sum of the cardinality of independent and independent dominating S sets in fuzzy graph. Adv. Fuzzy Sets Syst. **4**(2), 157–165 (2009)
18. A. Nagoor Gani, P. Vijayalakshmi, Domination critical nodes in fuzzy graphs. Int. J. Math. Sci. Eng. Appl. (IJMSEA) **5**(1), 295–301 (2011)
19. A. Nagoor Gani, P. Vijayalakshmi, Intensive arcs in domination of fuzzy graphs, Int. J. Contemp. Math Sci. **6**(26), 1303–1309 (2011)
20. A. Nagoor Gani, P. Vijayalakshmi, Fuzzy graphs with equal fuzzy domination and independent domination numbers. Int. J. Eng. Sci. Technol. Dev. **1**(2), 66–68 (2012)
21. A. Nagor Gani, M. Basheer Ahamed, Strong and weak domination in fuzzy graphs. East Asian Math. J. **23**, 1–8 (2007)
22. O. Ore, *Theory of Graphs*, vol. 38, American Mathematical Society Colloquium Publications (Americal Mathematical Society, Providence, RI, 1962)
23. S.S. Rajaram, S. Tidball, Nebraska Sex Trafficking Survivors Speak - A Qualitative Research Study, Submitted to the Women's Fund of Omaha (2016)
24. M. Reha Civanlar, H. Joel Trussel, Constructing membership functions using statistical data. Fuzzy Sets Syst. **18**, 1–13 (1986)
25. I.C. Ross, F. Harary, Identification of strengthening and weakening members of a group. Sociometry **22**, 139–147 (1959)
26. A. Somasundaram, S. Somasundaram, Domination in fuzzy graphs-I, Elsevier Science, 19, 787–791. Discrete Mathematics **33**(1981), 249–258 (1998)
27. A. Somasundaram, Domination in fuzzy graphs-II. J. Fuzzy Math. **13**(2), 281–288 (2005)
28. A. Somasundaram, Domination in product of fuzzy graphs. Int. J. Uncertain. Fuzziness Knowl.-Based Syst. **13**(2), 195–205 (2005)

29. A. Somasundaram, S. Somasundaram, Domination in fuzzy graphs-I. Pattern Recongit. Lett. **19**, 787–791 (1998)
30. E. Takeda, T. Nishida, An application of fuzzy graphs to the problem of group structure. J. Op. Res. Soc. Jpn. **19**, 217–227 (1976)
31. United Nations Office on Drugs and Crime (UNODC) Trafficking in Persons Global Patterns (2006)
32. United Nations Office on Drugs and Crime, Global Report on Trafficking in Persons (2014)

Chapter 3
Fuzzy Incidence Graphs

3.1 Fuzzy Incidence Graphs

We introduce the notion of the degree of incidence of a vertex and an edge in a fuzzy graph in fuzzy graph theory. We concentrate on incidence, where the edge is adjacent to the vertex. We determine results concerning bridges, cutvertices, cutpairs, fuzzy incidence paths, fuzzy incidence tree for fuzzy incidence graphs. In [4, 5], Dinesh introduced the notion of the degree of incidence of a vertex and an edge in fuzzy graph theory. This notion seem to have potential use in a variety of areas involving networks, [12].

There are many ways nations can influence other nations to cooperate in the combating of human trafficking. The purpose of this chapter is to create a mathematical model that can be used to study the effect of states influencing other states. We use a new technique that has recently been created for fuzzy network theory. This technique introduced uses the notion of fuzzy incidence in fuzzy graphs [4, 5, 14].

We discuss some methods how states can influence other states. The discussion is taken from [8, 18]. Many industrialized states fund antitrafficking initiatives in other countries, especially to those countries from which people have been trafficked into their country or region. One major problem for immigration and law enforcement officials is how to distinguish trafficking victims from others. Some victims may be involved in similar income-generating activities, but who have not been victimized. Many methods have been developed at frontiers and for police forces to help identifications. Intelligence about the profile of adults or children who have already been trafficked, aid the police in identifying the characteristics of such persons and using this intelligence to pick out travelers who warrant advice and protection.

We also lay the ground work in this chapter for the further study of human trafficking and illegal immigration in Chap. 5.

We recall the definition of an edge in a graph. Let V be a finite set. Let $\mathcal{E}$ denote the set of all subsets of V with exactly two elements. Let $E \subseteq \mathcal{E}$. If we consider the

© Springer International Publishing AG 2018
J. N. Mordeson et al., *Fuzzy Graph Theory with Applications to Human Trafficking*, Studies in Fuzziness and Soft Computing 365,
https://doi.org/10.1007/978-3-319-76454-2_3

set E to be the set of edges in the graph (V, E), then no element of V has a loop. We write xy for $\{x, y\}$. Then clearly $xy = yx$.

Definition 3.1.1 Let (V, E) be a graph. Then $G = (V, E, I)$ is called an **incidence graph**, where $I \subseteq V \times E$.

We note that if $V = \{u, v\}, E = \{uv\}$ and $I = \{(v, uv)\}$, then (V, E, I) is an incidence graph even though $(u, uv) \notin I$.

Definition 3.1.2 Let $G = (V, E, I)$ be an incidence graph. If $(u, vw) \in I$, then (u, vw) is called an **incidence pair** or simply a **pair**. If

$$(u, uv), (v, uv), (v, vw), (w, vw) \in I,$$

then uv and vw are called **adjacent edges.**

Definition 3.1.3 An **incidence subgraph** H of an incidence graph G is an incidence graph having its vertices, edges, and pairs in G. If H is an incidence subgraph of G, then G is called an incidence **supergraph** of H.

Let $G = (V, E, I)$ be an incidence graph. Let $V' \subseteq V, E' \subseteq E$, and $I' \subseteq I$. Then $G' = (V', E', I')$ is called a **near incidence subgraph** of G if
(i) $u'v' \in E' \Rightarrow u' \in V'$ or $v' \in V'$ and
(ii) $(v', u'v') \in I' \Rightarrow u'v' \in E'$.

Definition 3.1.4 Let $G = (V, E, I)$ be an incidence graph. A sequence

$$v_0, (v_0, v_0v_1), v_0v_1, (v_1, v_0v_1), v_1, \ldots, v_{n-1}, (v_{n-1}, v_{n-1}v_n), v_{n-1}v_n,$$
$$(v_n, v_{n-1}v_n), v_n$$

is called a **walk**. It is **closed** if $v_0 = v_n$. If the pairs are distinct, then it is called an **incidence trail**. If the edges are distinct, then it is called a **trail**. If the vertices are distinct, then it is called a **path**.

$$P_1 : v_0, (v_0, v_0v_1), v_0v_1, (v_1, v_0v_1), v_1, \ldots, v_{n-1}, (v_{n-1}, v_{n-1}v_n), v_{n-1}v_n,$$
$$(v_n, v_{n-1}v_n), v_n, (v_n, v_nv_{n+1}), v_nv_{n+1}$$

$$P_2 : uv_0, (v_0, uv_0), v_0, (v_0, v_0v_1), v_0v_1, (v_1, v_0v_1), v_1, \ldots, v_{n-1},$$
$$(v_{n-1}, v_{n-1}v_n), v_{n-1}v_n, (v_n, v_{n-1}v_n), v_n$$

$$P_3 : uv_0, (v_0, uv_0), v_0, (v_0, v_0v_1), v_0v_1, (v_1, v_0v_1), v_1, \ldots, v_{n-1},$$
$$(v_{n-1}, v_{n-1}v_n), v_{n-1}v_n, (v_n, v_{n-1}v_n), v_n, (v_n, v_nv_{n+1}), v_nv_{n+1}$$

are also incidence walks. The latter is **closed** if $uv_0 = v_n v_{n+1}$. If the vertices are distinct, then they are called **incidence paths**.

A path is a trail and an incidence trail.

Definition 3.1.5 If a walk in an incidence graph is closed, then it is called a **cycle** if the vertices are distinct.

By the definition of a cycle, all pairs of vertices are distinct. Thus, from the definition of a path, if uv is on the path so are (u, uv), (v, uv), but not an incidence pair of the form (u, vw) with $v \neq u \neq w$.

Let (V, E) be a graph and (V, E, I) an incidence graph. Then $I \subseteq V \times E$. We will assume in the following that $I \subseteq \{(u, uv) \mid uv \in E\}$. Let $E^{(i)} = \{(u, uv) \mid uv \in E\}$. (Note that because $uv = vu$, $(v, uv) \in E^{(i)}$.) Although not allowed here, incidence pairs of the form (u, vw), where $v \neq u \neq w$ also have potential applications in network theory. For example, (u, vw) might represent u's influence on vw with respect to flow from v to w. The flow might be human trafficking between countries or illicit flow of drugs, arms, or money between countries, [8, 17].

Definition 3.1.6 An incidence graph in which all pairs of vertices are joined by a path is said to be **connected**.

Definition 3.1.7 An incidence graph having no cycles is called a **forest**. If it is connected, then it is called a **tree**.

Because a tree is connected, all pairs of vertices are connected by a path. By the definition of a path, if uv is on the path so are (u, uv), (v, uv) but no incidence pair of the form (u, vw) with $v \neq u \neq w$ is on the path.

A **component** in an incidence graph is a maximally connected incidence subgraph. Recall that the definition of connectedness uses a path which for incidence graphs involves (u, uv) and (v, uv) for every uv in the path. Thus, the removal of a pair (u, uv) can increase the number of components in an incidence graph. For example, consider the incidence graph $G = (\{u, v\}, \{uv\}, \{(u, uv), (v, uv)\})$. Then G is connected, but $H = (\{u, v\}, \{uv\}, (v, uv)\})$ is not. H has two components, namely $\{u\}$ and $\{v\}$.

Definition 3.1.8 If the removal of an edge in an incidence graph increases the number of connected components, then the edge is called a **bridge**.

Definition 3.1.9 If the removal of a vertex in an incidence graph increases the number of connected components, then the vertex is called a **cutvertex**.

Definition 3.1.10 If the removal of an incidence pair in an incidence graph increases the number of connected components, then the incidence pair is called a **cutpair**.

Consider the incidence graph

$$G = (\{u, v, w\}, \{uv, uw, vw\},$$
$$\{(u, uv), (v, uv), (u, uw), (w, uw), (v, vw), (w, vw)\}).$$

Then (u, uv) is not a cutpair because G remains connected because there is a path from u to v going through w.

Definition 3.1.11 Let $G = (V, E)$ be a graph and σ be a fuzzy subset of V and μ a fuzzy subset of $V \times E$. Let Ψ be a fuzzy subset of $V \times E$. If $\Psi(v, e) \leq \sigma(v) \wedge \mu(e)$ for all $v \in V$ and $e \in E$, then Ψ called a **fuzzy incidence** of G.

Definition 3.1.12 Let $G = (V, E)$ be a graph and (σ, μ) be a fuzzy subgraph of G. If Ψ is a fuzzy incidence of G, then $\widetilde{G} = (\sigma, \mu, \Psi)$ is called a **fuzzy incidence graph (FIG)** of G.

Let (V, E) be a graph and (σ, μ) a fuzzy subgraph of (V, E). Let $G = (V, E, I)$ be an incidence graph. Let (σ, μ, Ψ) be a fuzzy incidence graph. Define $\sigma \cup \mu : V \cup E \rightarrow [0, 1]$ as follows: if $u \in V$, $(\sigma \cup \mu)(u) = \sigma(u)$ and if $uv \in E$, $(\sigma \cup \mu)(uv) = \mu(uv)$. Because $\Psi(u, uv) \leq \sigma(u) \wedge \mu(uv) = (\sigma \cup \mu)(u) \wedge (\sigma \cup \mu)(uv)$, we can consider $(\sigma \cup \mu, \Psi)$ as a fuzzy subgraph of $(V \cup E, E^{(i)})$. That is, the elements of $V \cup E$ are the vertices and the elements of $E^{(i)}$ as the edges. This interpretation will aid in the understanding of the proofs to follow.

Any $x \in V$ is said to be in the support of σ if $\sigma(x) > 0$, $xy \in E$ is said to be in the support of μ if $\mu(xy) > 0$ and $(x, yz) \in V \times E$ is said to be in the support of ψ if $\psi(x, yz) > 0$. The supports of σ, μ and Ψ are denoted as σ^*, μ^* and Ψ^*, respectively.

Definition 3.1.13 Let $xy \in \text{Supp}(\mu)$. Then xy is an edge of the fuzzy incidence graph $\widetilde{G} = (\sigma, \mu, \Psi)$ and if $(x, xy), (y, xy) \in \text{Supp}(\Psi)$, then (x, xy) and (y, xy) are called **pairs**.

Definition 3.1.14 Two vertices v_i and v_j joined by a path in a fuzzy incidence graph are said to be **connected**.

Definition 3.1.15 Let $\widetilde{G} = (\sigma, \mu, \Psi)$ be a fuzzy incidence graph. Define $\Psi^\infty(u, vw)$ to be the incidence strength of the path from u to vw of greatest incidence strength. We shall also use the notation $ICONN_G(u, vw)$ to denote $\Psi^\infty(u, vw)$.

Example 3.1.16 Let $G = (V, E, I)$ be an incidence graph and $\widetilde{G} = (\sigma, \mu, \Psi)$ be a fuzzy incidence graph associated with G, where $V = \{v_1, v_2, v_3, v_4\}$ and

σ	v_1	v_2	v_3	v_4
	0.5	0.7	1.0	0.8

μ	$v_1 v_2$	$v_1 v_3$	$v_2 v_3$	$v_2 v_4$	$v_3 v_4$	$v_1 v_4$
	0.5	0.3	0.7	0.5	0.8	0.5

Ψ	$v_1, (v_1 v_2)$	$v_2, (v_1 v_2)$	$v_1, (v_1 v_3)$	$v_2, (v_2 v_3)$	$v_3, (v_2 v_3)$	$v_2, (v_2 v_4)$
	0.5	0.4	0.3	0.6	0.7	0.5
Ψ	$v_4, (v_2 v_4)$	$v_3, (v_3 v_4)$	$v_4, (v_3 v_4)$	$v_1, (v_1 v_4)$	$v_4, (v_1 v_4)$	
	0.4	0.8	0.7	0.4	0.5	

Fig. 3.1 Representation of a fuzzy incidence graph

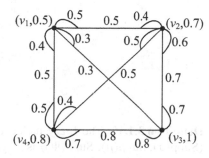

This fuzzy incidence graph can be represented as in Fig. 3.1 given below. In this fuzzy incidence graph,

$$v_1, (v_1, v_1 v_2), v_1 v_2, (v_2, v_1 v_2), v_2, (v_2, v_2 v_3), v_2 v_3, (v_3, v_2 v_3),$$
$$v_3, (v_3, v_3 v_4), v_3 v_4, (v_4, v_3 v_4), v_4, (v_4, (v_2 v_4), v_2 v_4, (v_2, (v_2 v_4), v_2$$

is a walk, but not a path. It is a trail and an incidence trail. The sequence

$$v_1, (v_1, v_1 v_2), v_1 v_2, (v_2, v_1 v_2), v_2, (v_2, v_2 v_3), v_2 v_3, (v_3, v_2 v_3),$$
$$v_3, (v_3, v_3 v_4), v_3 v_4, (v_4, v_3 v_4), v_4$$

is a path, a trail, and an incidence trail. The vertices v_1 and v_4 are connected. The incidence strength of this sequence is 0.4.

We can have another representation for fuzzy incidence graphs, as given in the next example.

Example 3.1.17 Let $G = (V, E, I)$ be an incidence graph and $\widetilde{G} = (\sigma, \mu, \Psi)$ be a fuzzy incidence graph associated with G, which is given in Fig. 3.2. We shall find some of the incidence paths in it $\widetilde{G}$.

Clearly, $v_1, (v_1, v_1 v_2), v_1 v_2, (v_2, v_1 v_2), v_2$ is a path (See the sequence (1)(2)(3)(4) (5)). Thus, v_1 and v_2 are connected.

Definition 3.1.18 The fuzzy incidence graph $\widetilde{G} = (\sigma, \mu, \Psi)$ is a **cycle** if (Supp(σ), Supp(μ), Supp(Ψ)) is a cycle.

Fig. 3.2 Paths in a fuzzy
incidence graph

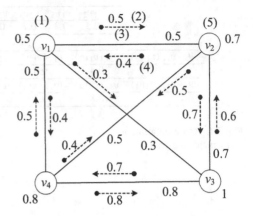

Definition 3.1.19 The fuzzy incidence graph $\widetilde{G} = (\sigma, \mu, \Psi)$ is a **fuzzy cycle** if
$(\mathrm{Supp}(\sigma), \mathrm{Supp}(\mu), \mathrm{Supp}(\Psi))$ is a cycle and there exists no unique $xy \in \mathrm{Supp}(\mu)$
such that $\mu(xy) = \wedge\{\mu(uv) \mid uv \in \mathrm{Supp}(\mu)\}$.

Definition 3.1.20 The fuzzy incidence graph $\widetilde{G} = (\sigma, \mu, \Psi)$ is a **fuzzy incidence
cycle** if it is a fuzzy cycle and there exists no unique $(x, xy) \in \mathrm{Supp}(\Psi)$ such that
$\Psi(x, xy) = \wedge\{\Psi(u, uv) \mid (u, uv) \in \mathrm{Supp}(\Psi)\}$.

Example 3.1.21 Let $\widetilde{G} = (\sigma, \mu, \Psi)$ be the fuzzy incidence cycle with $V = \{v_1, v_2, v_3, v_4\}$ and σ, μ, Ψ defined as follows:

σ	v_1	v_2	v_3	v_4
	0.2	0.3	0.3	0.4

μ	$v_1 v_2$	$v_2 v_3$	$v_3 v_4$	$v_4 v_1$
	0.2	0.2	0.3	0.2

Ψ	$(v_1, v_1 v_2)$	$(v_2, v_1 v_2)$	$(v_2, v_2 v_3)$	$(v_3, v_2 v_3)$
	0.2	0.2	0.2	0.2
Ψ	$(v_3, v_3 v_4)$	$(v_4, v_3 v_4)$	$(v_4, v_4 v_1)$	$(v_1, v_4 v_1)$
	0.3	0.3	0.2	0.2

Consider the walk

$$v_1, (v_1, v_1 v_2), v_1 v_2, (v_2, v_1 v_2), v_2, (v_2, v_2 v_3), v_2 v_3, (v_3, v_2 v_3),$$
$$v_3, (v_3, v_3 v_4), v_3 v_4, (v_4, v_3 v_4), v_4, (v_{34}, v_4 v_1), v_4 v_1, (v_1, v_4 v_1), v_1.$$

Fig. 3.3 Fuzzy incidence
cycle

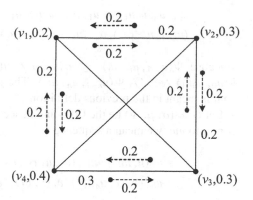

Then it is a fuzzy cycle because there does not exist a unique uv such that $\mu(uv) = 0.2$ the strength of the fuzzy graph (σ, μ). It is also a fuzzy incidence cycle because there does not exist a unique (u, vw) such that $\Psi(u, vw) = 0.2$ the incidence strength of $\widetilde{G}$ (Fig. 3.3).

Definition 3.1.22 The fuzzy incidence graph $\widetilde{G} = (\sigma, \mu, \Psi)$ is a **tree** if (Supp(σ), Supp(μ), Supp(Ψ)) is a tree and is a **forest** if (Supp(σ), Supp(μ), Supp(Ψ)) is a forest.

Definition 3.1.23 Let $\widetilde{G} = (\sigma, \mu, \Psi)$ be a fuzzy incidence graph. Then $\widetilde{H} = (\tau, \nu, \Omega)$ is a **fuzzy incidence subgraph** of $\widetilde{G}$ if $\tau \subseteq \sigma, \nu \subseteq \mu$, and $\Omega \subseteq \Psi$. A fuzzy incidence subgraph $\widetilde{H}$ of $\widetilde{G}$ is a **fuzzy incidence spanning subgraph** of $\widetilde{G}$ if $\tau = \sigma$. A fuzzy incidence subgraph $\widetilde{H} = (\tau, \nu, \Omega)$ is said to be a **subgraph** of $\widetilde{G}$ $\tau(u) = \sigma(u)$ for every $u \in \tau^*$, $\nu(uv) = \mu(uv)$ for every $uv \in \nu^*$, and $\Omega(u, zw) = \Psi(u, vw)$ for every pair $(u, vw) \in \Omega^*$.

The concepts of a spanning subgraph and induced subgraph of a FIG can be defined in the usual way.

Definition 3.1.24 Let $\widetilde{G} = (\sigma, \mu, \Psi)$ be a fuzzy incidence graph.
(i) $\widetilde{G}$ is called a **fuzzy tree** if it has a fuzzy incidence spanning subgraph $\widetilde{F} = (\sigma, \nu, \Omega)$ which is also a tree such that for all $uv \in$ Supp$(\mu)\backslash$Supp(ν), $\mu(uv) < \nu^\infty(uv)$.
(ii) $\widetilde{G}$ is called a **fuzzy forest** if it has a fuzzy incidence spanning subgraph $\widetilde{F} = (\tau, \nu, \Omega)$ which is also a forest such that for all $uv \in$ Supp$(\mu)\backslash$Supp(ν), $\mu(uv) < \nu^\infty(uv)$.

Definition 3.1.25 Let $\widetilde{G} = (\sigma, \mu, \Psi)$ be a fuzzy incidence graph. Define $\Psi^\infty(u, vw)$ to be the incidence strength of the path from u to vw of greatest incidence strength.

Let $u_0 = u$, $u_{n-1} = v$, and $u_n = w$. Then a path from u_0 to u_n would be

$$u_0, (u_0, u_0u_1), u_0u_1, (u_1, u_0u_1), u_1, (u_1, u_1u_2), u_1u_2, u_2, \ldots, u_{n-1},$$

$$(u_{n-1}, u_{n-1}u_n), u_{n-1}u_n, (u_n, u_{n-1}u_n), u_n.$$

[Because $\Psi(u_{i-1}, u_{i-1}u_i) \leq \sigma(u_{i-1}) \wedge \mu(u_{i-1}u_i)$, the strength of the path is $\Psi(u_0, u_0u_1) \wedge \ldots \wedge \Psi(u_{n-1}, u_{n-1}u_n)$. The strength of the strongest such path is what is meant in the previous definition.]

Let $\widetilde{G} = (\sigma, \mu, \Psi)$ be the fuzzy incidence graph and $(u, vw) \in V \times E$. By a path from u to vw, we mean a sequence

$$u, (u, uu_1), uu_1, (u_1, u_1u_2), u_2, \ldots, u_{n-1}, (u_{n-1}, u_{n-1}v),$$

$$u_{n-1}v, (v, u_{n-1}v), v, (v, vw), vw$$

or a sequence

$$u, (u, uu_1), uu_1, (u_1, u_1u_2), u_2, \ldots, u_{n-1}, (u_{n-1}, u_{n-1}w),$$

$$u_{n-1}w, (w, u_{n-1}w), w, (w, wv), vw,$$

where the vertices are distinct. If $u = v$, then a path from u to $vw = uw$ is

$$u, (u, uu_1), uu_1, (u_1, u_1u_2), u_2, \ldots, u_{n-1}, (u_{n-1}, u_{n-1}w),$$

$$u_{n-1}w, (w, u_{n-1}w), w, (w, uw), vw.$$

Another path from u to uw is $u, (u, uw), uw$.

Definition 3.1.26 Let $\widetilde{G} = (\sigma, \mu, \Psi)$ be a fuzzy incidence graph.

(*i*) $\widetilde{G}$ is called a **fuzzy incidence tree** (**FIT**) if it has a fuzzy incidence spanning subgraph $\widetilde{F} = (\sigma, \nu, \Omega)$ which is also a tree such that for all $(u, vw) \in \mathrm{Supp}(\Psi) \backslash \mathrm{Supp}(\Omega)$, $\Psi(u, vw) < \Omega^\infty(u, vw)$.

(*ii*) $\widetilde{G}$ is called a **fuzzy incidence forest** (**FIF**) if $\widetilde{G}$ has a fuzzy incidence spanning subgraph $\widetilde{F} = (\sigma, \nu, \Omega)$ which is also a forest such that for all $(u, vw) \in \mathrm{Supp}(\Psi) \backslash \mathrm{Supp}(\Omega)$, $\Psi(u, vw) < \Omega^\infty(u, vw)$.

Thus, $\widetilde{G}$ is said to be a fuzzy incidence tree if there exists a spanning subgraph $\widetilde{F} = (\tau, \nu, \Omega)$ of $\widetilde{G}$ which is a tree such that for every pair $(u, vw) \in \Psi^* \backslash \Omega^*$, $\Psi(u, vw) < ICONN_{\widetilde{F}}(u, vw)$. In other words, there exists an incidence path in $\widetilde{F}$ from u to vw such that each of its pairs has greater Ψ value than $\Psi(u, vw)$. Note that the existence of an incidence path between any two elements of σ^* is automatically guaranteed in this definition.

Example 3.1.27 Let $\widetilde{G} = (\sigma, \mu, \Psi)$ be a fuzzy incidence graph with $V = \{v_1, v_2, v_3, v_4\}$ and defined as follows:

σ	v_1	v_2	v_3	v_4
	0.2	0.3	0.3	0.4

μ	v_1v_2	v_2v_3	v_3v_4	v_4v_1
	0.2	0.2	0.3	0.2

Ψ	v_1, v_1v_2	v_2, v_1v_2	v_2, v_2v_3	v_3, v_2v_3
	0.2	0.2	0.2	0.2
Ψ	v_3, v_3v_4	v_4, v_3v_4	v_4, v_4v_1	v_1, v_4v_1
	0.3	0.3	0.2	0.2

Consider the walk

$$v_1, (v_1, v_1v_2), v_1v_2, (v_2, v_1v_2), v_2, (v_2, v_2v_3), v_2v_3, (v_3, v_2v_3),$$

$$v_3, (v_3, v_3v_4), v_3v_4, (v_4, v_3v_4), v_4, (v_4, v_4v_1), v_4v_1, (v_1, v_4v_1), v_1.$$

Then $\widetilde{G}$ is a cycle. Now $\widetilde{G}$ has a fuzzy incidence spanning subgraph $\widetilde{F} = (\sigma, \nu, \Omega)$ consisting of

$$v_1, (v_1, v_1v_2), v_1v_2, (v_2, v_1v_2), v_2, (v_2, v_2v_3), v_2v_3, (v_3, v_2v_3),$$

$$v_3, (v_3, v_3v_4), v_3v_4, (v_4, v_3v_4), v_4,$$

having the same membership values as in $\widetilde{G}$ for all vertices, edges, and pairs except for the following:

$$\nu(v_4) = 0.1, \quad \Omega(v_4, v_4v_1) = 0.1, \quad \Phi(v_1, v_4v_1) = 0.1.$$

Then $\widetilde{F}$ is a tree and $v_4v_1 \in \text{Supp}(\mu) \setminus \text{Supp}(\nu)$ is such that $\mu(v_4v_1) < \nu^\infty(v_4v_1)$. Thus, $\widetilde{G}$ is a fuzzy tree.

Furthermore, $\widetilde{H} = (\sigma, \nu, \Omega)$ consisting of

$$v_1, (v_1, v_1v_2), v_1v_2, (v_2, v_1v_2), v_2, (v_2, v_2v_3), v_2v_3, (v_3, v_2v_3),$$

$$v_3, (v_3, v_3v_4), v_3v_4, (v_4, v_3v_4), v_4,$$

having the same membership values as in $\widetilde{G}$ for all vertices, edges, and pairs except for the following:

$$\Omega(v_4, v_4v_1) = 0.1, \Omega(v_1, v_4v_1) = 0.1$$

is a tree and $\Psi(v_4, v_4v_1) < \Omega^\infty(v_4, v_4v_1)$ and $\Psi(v_1, v_4v_1) < \Omega^\infty(v_1, v_4v_1)$. Thus, $\widetilde{G}$ is a fuzzy incidence tree.

Let $(u, uv) \in V \times E$ and $\Psi' = \Psi|_{V \times E \setminus \{(u,uv)\}}$. By Ψ'^∞, we mean the strength of the strongest path from u to uv not including (u, uv). Let $(u, vw) \in V \times E$ with $v \neq u \neq w$ and let $\Psi' = \Psi|_{V \times E \setminus \{(u,vw)\}}$. By Ψ'^∞, we mean the strength of the strongest path from u to vw. (Note that (u, vw) cannot be included by the definition of path.)

Theorem 3.1.28 *Let $\widetilde{G} = (\sigma, \mu, \Psi)$ be a fuzzy incidence graph. Then $\widetilde{G}$ is a fuzzy incidence forest if and only if in any fuzzy incidence cycle of $\widetilde{G}$ there is (x, xy) such that $\Psi(x, xy) < \Psi'^{\infty}(x, xy)$, where $\widetilde{G'} = (\sigma, \mu, \Psi')$ is the fuzzy incidence subgraph obtained by deletion of (x, yz) from $\widetilde{G}$ and $\Psi' = \Psi$ restricted to $(V \times E)\backslash\{(x, xy)\}$ (or $\Psi'(x, xy) = 0$).*

Proof If there is no cycles the result is trivially true. Suppose (x, xy) is an incidence pair in $\widetilde{G}$ that belongs to a fuzzy incidence cycle such that $\Psi(x, xy) < \Psi'^{\infty}(x, xy)$. Let it be the pair with the least value among all such pairs $(u, uv) \in \text{Supp}(\Psi)$. Delete (x, xy). If there are other cycles, remove incidence pairs in a similar way. At each step, the incidence pair deleted will not have lesser incidence than those deleted earlier. After deletion, the remaining fuzzy incidence subgraph is a fuzzy incidence forest $\widetilde{F}$. Therefore, there exists a path $\widetilde{P}$ from x to xy with more incidence strength than $\Psi(x, xy)$ and not containing (x, xy). If incidence pairs deleted earlier are in $\widetilde{P}$, then we can bypass them using a path with more incidence strength.

Conversely, if $\widetilde{G}$ is a fuzzy incidence forest and C any cycle, then by definition, there exists (x, xy) of C not in $\widetilde{F}$ such that $\Psi(x, xy) < \Omega^{\infty}(x, xy) \leq \Psi'^{\infty}(x, xy)$, where $\widetilde{F}$ is as in the definition of a fuzzy incidence forest. ∎

Theorem 3.1.29 *If there is at most one path with the most incidence strength between any vertex and edge of the fuzzy incidence graph $\widetilde{G} = (\sigma, \mu, \Psi)$, then $\widetilde{G}$ is a fuzzy incidence forest.*

Proof Suppose $\widetilde{G} = (\sigma, \mu, \Psi)$ is not a fuzzy incidence forest. Then by Theorem 3.1.28, there exists a cycle C in $\widetilde{G}$ such that $\Psi(x, xy) \geq \Psi'(x, xy)$ for every incidence pair (x, xy) in C. Therefore, (x, xy) is the path with the most incidence strength from x to xy. Let (x, xy) be the incidence pair with the least value for Ψ in C. Then the remaining part of C is a path with the most incidence strength from x to xy. This is a contradiction. Hence, $\widetilde{G}$ is a fuzzy incidence forest. ∎

Theorem 3.1.30 *Let $\widetilde{G} = (\sigma, \mu, \Psi)$ be a cycle. Then $\widetilde{G}$ is a fuzzy incidence cycle if and only if $\widetilde{G}$ is not a fuzzy incidence tree.*

Proof Suppose $\widetilde{G} = (\sigma, \mu, \Psi)$ is a fuzzy incidence cycle. Then there exist at least two incidence pairs $(x_1, x_1y_2), (x_2, x_2y_2)$ with $\Psi(x_1, x_1y_1) = \Psi(x_2, x_2y_2) = \wedge\{\Psi(z, zw) \mid z \in V, zw \in \text{Supp}(\mu), (z, zw) \in \text{Supp}(\Psi)\}$. Let (σ, μ, Ω) be a spanning fuzzy incidence tree in $\widetilde{G}$. Then there exists $u \in V, uv \in E$ such that $\text{Supp}(\Psi)\backslash\text{Supp}(\Omega) = \{(u, uv)\}$. Hence, there does exist a path in (σ, μ, Ω) between u and uv of greater incidence strength than $\Psi(u, uv)$. Thus, $\widetilde{G}$ is not a fuzzy incidence tree.

Conversely, suppose that $\widetilde{G}$ is not a fuzzy incidence tree. Because $\widetilde{G}$ is a fuzzy cycle, it follows that for all (u, uv) in $\text{Supp}(\Psi)$, there is a fuzzy incidence spanning subgraph (σ, μ, Ω) which is a tree and $\Omega(u, uv) = 0, \Omega^{\infty}(u, uv) \leq \Psi(u, uv)$, and $\Omega(x, xy) = \Psi(x, xy)$ for all $(x, xy) \in \text{Supp}(\Psi)\backslash\{(u, vw)\}$. Therefore, Ψ does not attain $\wedge\{\Psi(x, xy) \mid (x, xy) \in \text{Supp}(\Psi)\}$ uniquely. Hence, $\widetilde{G}$ is a fuzzy incidence cycle. ∎

3.2 Cutvertices, Bridges, and Cutpairs

Definition 3.2.1 Let $\widetilde{G} = (\sigma, \mu, \Psi)$ be a fuzzy incidence graph and let $xy \in E$. Then xy is called a **bridge** if there exists $uv \in E\setminus\{xy\}$ such that $\mu'^{\infty}(uv) < \mu^{\infty}(uv)$, where $\mu' = \mu$ restricted to $E\setminus\{xy\}$.

Definition 3.2.2 Let $\widetilde{G} = (\sigma, \mu, \Psi)$ be a fuzzy incidence graph. Let $w \in V$ and E' be the set difference of E and the set of edges with w as an end vertex. Then w is called a **cutvertex** if $\mu'^{\infty}(uv) < \mu^{\infty}(uv)$ for some $uv \in E'$ such that $u \neq w \neq v$, where $\mu' = \mu$ restricted to E'.

Definition 3.2.3 Let $\widetilde{G} = (\sigma, \mu, \Psi)$ be a fuzzy incidence graph. Let $w \in V$ and E' be the set difference of E and the set of edges with w as an end vertex. Then w is called an **incidence cutvertex** if $\Psi'^{\infty}(u, uv) < \Psi^{\infty}(u, uv)$ for some $(u, uv) \in V \times E'$ such that $u \neq w \neq v$, where $\Psi' = \Psi$ restricted to $V \times E'$.

Definition 3.2.4 Let $\widetilde{G} = (\sigma, \mu, \Psi)$ be a fuzzy incidence graph. Then (x, xy) is called an **incidence cutpair** if $\Psi'^{\infty}(u, uv) < \Psi^{\infty}(u, uv)$ for some pair (u, uv) in $\widetilde{G}$, where $\Psi' = \Psi$ restricted to $(V \times E)\setminus\{(x, xy)\}$.

Example 3.2.5 Let $\widetilde{G} = (\sigma, \mu, \Psi)$ be the fuzzy incidence graph consisting of the walk

$$v_1, (v_1, v_1v_2), v_1v_2, (v_2, v_1v_2), v_2, (v_2, v_2v_3), v_2v_3, (v_3, v_2v_3),$$
$$v_3, (v_3, v_3v_4), v_3v_4, (v_4, v_3v_4), v_4, (v_4, v_4v_1), v_4v_1, (v_1, v_4v_1), v_1$$

and σ, μ, Ψ defined as follows:

σ	v_1	v_2	v_3	v_4
	0.2	0.3	0.3	0.4

μ	v_1v_2	v_2v_3	v_3v_4	v_4v_1
	0.2	0.2	0.3	0.2

Ψ	(v_1, v_1v_2)	(v_2, v_1v_2)	(v_2, v_2v_3)	(v_3, v_2v_3)
	0.2	0.2	0.2	0.2

Ψ	(v_3, v_3v_4)	(v_4, v_3v_4)	(v_4, v_4v_1)	(v_1, v_4v_1)
	0.3	0.2	0.2	0.2

Here v_3 and v_4 are cut vertices, v_3v_4 is a bridge, and (v_3, v_3v_4) is a cutpair.

Theorem 3.2.6 *If $\widetilde{G}$ is a fuzzy incidence forest, then the vertex edge pairs of $\widetilde{F}$ (as in the definition of fuzzy incidence forest) are exactly the incidence cutpairs of $\widetilde{G}$.*

Proof Suppose (x, xy) is not in $\widetilde{F}$. Then by definition $\Psi(x, xy) < \Omega^{\infty}(x, xy) \leq \Psi'^{\infty}(x, xy)$, where (σ, μ, Ψ') is the fuzzy incidence spanning subgraph of $\widetilde{G}$ obtained by deleting (x, xy). It has pairs not in $\widetilde{F}$ because $\widetilde{F}$ is a fuzzy incidence forest and has no fuzzy incidence cycles. Therefore, (x, xy) cannot be an incidence cutpair of $\widetilde{G}$. Let (x, xy) be a pair in $\widetilde{F}$. If possible, let (x, xy) not be an incidence cutpair. Then there exists a path $\widetilde{P}$ from x to xy not involving (x, xy) and having incidence strength greater than or equal to $\Psi(x, xy)$. Then some pairs not in $\widetilde{F}$ will be in $\widetilde{P}$ as $\widetilde{F}$ is a fuzzy forest. Any pair of this type say (u, uv) can be replaced by a path $\widetilde{Q}$ in $\widetilde{F}$. This $\widetilde{Q}$ will have incidence strength greater than $\Psi(u, vw)$. Also, $\Psi(u, uv) \geq \Psi(x, xy)$. Therefore, (x, xy) is not in $\widetilde{Q}$. Replacing such (u, uv) by such a path in $\widetilde{F}$ we will get a path in $\widetilde{F}$ from x to xy not containing (x, xy). Thus, we have a cycle in $\widetilde{F}$ which is a contradiction to our assumption. Hence, the vertex edge pairs of $\widetilde{F}$ are just incidence cutpairs of $\widetilde{G}$. ∎

Theorem 3.2.7 *Let $\widetilde{G} = (\sigma, \mu, \Psi)$ be a fuzzy incidence graph and $\widetilde{G}^{*} = (Supp(\sigma), Supp(\mu), Supp(\Psi))$ a cycle. Then an edge is a bridge if and only if it is an edge common to two incidence cutpairs.*

Proof Let e be a bridge. Then there exist edges f and g with e lying on every path with the greatest incidence strength between f and g. Therefore, there exists only one path with the greatest incidence strength joining f and g involving e. Any pair on it will be an incidence cutpair as the removal of any one of them will disconnect the path and reduce the incidence strength.

Conversely, let $e = xy$ be common to two incidence cutpairs (x, xy) and (y, xy). Then they are not the incidence cutpairs with the smallest value for Ψ. A path from f to g not containing (x, xy) and (y, xy) has less incidence strength than $\Psi(x, xy) \wedge \Psi(y, xy)$. Therefore, the path with the most incidence strength from f to g is x, (x, xy), xy, (y, xy), y. Also, the incidence strength of a path with the most incidence strength from f to g is equal to $\Psi(x, xy) \wedge \Psi(y, xy)$. Therefore, e is a bridge. ∎

Theorem 3.2.8 *Let $\widetilde{G}$ be a fuzzy incidence graph. Then the following conditions are equivalent.*

   (i) $\Psi'^{\infty}(u, uv) < \Psi(u, uv)$;
   (ii) (u, uv) is an incidence cutpair;
   (iii) (u, uv) is not the weakest pair of any cycle.

Proof We prove the following three implications by contrapositive.

   $(i) \Rightarrow (ii)$: If (u, uv) is not an incidence cutpair, then $\Psi'^{\infty}(u, uv) = \Psi^{\infty}(u, uv) \geq \Psi(u, uv)$.

   $(ii) \Rightarrow (iii)$: If (u, uv) is the weakest pair in a cycle, then any path involving (u, uv) can be converted into a path not involving (u, uv) but at least as strong by using the rest of the cycle as a path u to v.

$(iii) \Rightarrow (i)$: Suppose that $\Psi'^{\infty}(u, uv) \geq \Psi(u, uv)$. Then there is a path from u to v not involving (u, vu) (not involving vu) that has strength $\geq \Psi(u, vu)$ and the path together with uv (and (v, vu) and u) forms a cycle of which (u, uv) is the weakest pair. ∎

Theorem 3.2.9 *Let* $\widetilde{G} = (\sigma, \mu, \Psi)$ *be the fuzzy incidence graph. If* $\exists t \in (0, 1]$ *such that* $(Supp(\sigma), Supp(\mu), \Psi^t)$ *is a tree, then* (σ, μ, Ψ) *is a fuzzy incidence tree. Conversely, if* (σ, μ, Ψ) *is a cycle and* (σ, μ, Ψ) *is a fuzzy incidence tree, then* $\exists t \in (0, 1]$ *such that* $(Supp(\sigma), Supp(\mu), \Psi^t)$ *is a tree.*

Proof Suppose that t exists such that $(Supp(\sigma), Supp(\mu), \Psi^t)$ is a tree. Let $\Omega = \Psi$ on Ψ^t and $\Omega(x, yz) = 0$ if $(x, yz) \notin \Psi^t$. Then (σ, μ, Ω) is a spanning fuzzy subgraph of (σ, μ, Ψ) such that (σ, μ, Ω) is a fuzzy tree because $(Supp(\sigma), Supp(\mu), Supp(\Omega))$ is a tree. Suppose that $(x, xy) \in Supp(\Psi) \backslash Supp(\Omega)$. Then $(x, xy) \notin \Psi^t$ because $\Psi^t = Supp(\Omega)$. Thus, $\Psi(x, xy) < t \leq \Omega^{\infty}(x, xy)$. Thus, (σ, μ, Ψ) is a fuzzy incidence tree.

Conversely, suppose (σ, μ, Ψ) is a cycle and (σ, μ, Ψ) is a fuzzy incidence tree. Then there exists unique $(x, xy) \in Supp(\Psi)$ such that

$$\Psi(x, xy) = \wedge\{\Psi(u, uv) \mid (u, uv) \in Supp(\Psi)\}.$$

Let t be such that $\Psi(x, xy) < t \leq \wedge\{\Psi(u, uv) \mid (u, uv) \in Supp(\Psi) \backslash \{(x, xy)\}\}$. Then $(Supp(\sigma), Supp(\mu), \Psi^t)$ is a tree. ∎

Theorem 3.2.10 *Let* $\widetilde{G} = (\sigma, \mu, \Psi)$ *be a fuzzy incidence graph and* $(u, uv) \in V \times E$. *If* (u, uv) *is an incidence cutpair, then* $\Psi(u, uv) = \Psi^{\infty}(u, uv)$.

Proof Suppose $\Psi(u, uv) < \Psi^{\infty}(u, uv)$. Then there exists a path from u to uv such that all pairs (x, xy) in the path are such that $\Psi(x, xy) > \Psi(u, uv)$. Hence, $\Psi'^{\infty}(u, uv) > \Psi(u, uv)$. This path together with (u, uv) (and u) forms a cycle in which (u, uv) is the weakest pair and so (u, uv) is not an incidence cutpair. ∎

Theorem 3.2.11 *Let* $\widetilde{G} = (\sigma, \mu, \Psi)$ *be the fuzzy incidence graph. If* w *is a common vertex of at least two cutpairs, then* w *is an incidence cutvertex.*

Proof Let $(w, u_1 w)$ and $(w, u_2 w)$ be incidence two incidence cutpairs. Then there exist u, v such that $(w, u_1 w)$ is on every strongest u-v path. If w is distinct from u and v, then w is an incidence cutvertex. Suppose $w = u$ or $w = v$. Then $(w, u_1 w)$ is on every strongest u-w path or $(w, u_2 w)$ is on every strongest w-v path. Suppose that w is not an incidence cutvertex. Then between every two vertices there exists at least one strongest path not containing w. In particular, there exists at least on strongest path P, joining u_1 and u_2, not containing w. This path together with $(w, u_1 w)$ and $(w, u_2 w)$ (and consequently $u_1 w$, $u_2 w$ and w) form a cycle.

We now consider two cases. First suppose that u_1, w, u_2 is not a strongest path. Then clearly one of $(w, u_1 w)$, $(w, u_2 w)$ or both become the weakest incidence cutpairs of the cycle which contradicts that $(w, u_1 w)$ and $(w, u_2 w)$ are two incidence cutpairs.

Second suppose that u_1, w, u_2 is also a strongest path joining u_1 to u_2. Then the strength of the strongest incidence path from u_1 to u_2 equals $\Psi(w, u_1 w) \wedge \Psi(w, u_2 w)$, the strength of P. Thus, the pairs of P are at least as strong as $\Psi(w, u_1 w)$ and $\Psi(w, u_2 w)$ which implies that $(w, u_1 w), (w, u_2 w)$ or both are the weakest pairs of the cycle, which again is a contradiction. ∎

Theorem 3.2.12 *If $\widetilde{G} = (\sigma, \mu, \Psi)$ is a fuzzy incidence tree and $G^* = (Supp(\sigma), Supp(\mu), Supp(\Psi))$ is not a tree, then there exists at least one incidence pair (u, uv) for which $\Psi(u, uv) < \Psi^\infty(u, uv)$.*

Proof Because $\widetilde{G}$ is a fuzzy incidence tree, there exists a fuzzy incidence spanning subgraph $\widetilde{F} = (\sigma, \nu, \Omega)$ which is a tree such that $\Psi(u, vw) < \Omega^\infty(u, vw)$ for all pairs (u, vw) not in $\widetilde{F}$. Also, $\Omega^\infty(u, uv) \leq \Psi^\infty(u, uv)$. Thus, $\Psi(u, uv) < \Psi^\infty(u, uv)$ for all (u, uv) not in $\widetilde{F}$ and by hypothesis there exists one pair (u, uv) not in $\widetilde{F}$. ∎

Definition 3.2.13 Let $\widetilde{G} = (\sigma, \mu, \Psi)$ be the fuzzy incidence graph. Then $\widetilde{G}$ is said to **fuzzy incidence complete** if for all $(u, vw) \in V \times E$, $\Psi(u, vw) = \sigma(u) \wedge \mu(vw)$.

Note that if $\widetilde{G}$ is fuzzy incidence complete, then $\Psi(u, uv) = \sigma(v) \wedge \mu(uv) = \mu(uv) = \sigma(v) \wedge \mu(uv) = \Psi(v, uv)$.

Theorem 3.2.14 *If $\widetilde{G} = (\sigma, \mu, \Psi)$ is a fuzzy incidence tree, then $\widetilde{G}$ is not fuzzy incidence complete.*

Proof Suppose $\widetilde{G}$ is fuzzy incidence complete. Then $\Psi(u, uv) = \Psi^\infty(u, uv)$ for all (u, uv). Because $\widetilde{G} = (\sigma, \mu, \Psi)$ is a fuzzy incidence tree, $\Psi(u, uv) < \Omega^\infty(u, uv)$ for all (u, uv) not in $\widetilde{F} = (\sigma, \nu, \Omega)$, a fuzzy incidence spanning subgraph of $\widetilde{G}$ which is a tree. Thus, $\Psi^\infty(u, uv) < \Omega(u, uv)$ which is impossible. ∎

Theorem 3.2.15 *Let $\widetilde{G} = (\sigma, \mu, \Psi)$ be a fuzzy incidence tree. Then the internal vertices of a fuzzy incidence spanning subgraph $\widetilde{F}$ which is a tree are the cut vertices of $\widetilde{G}$.*

Proof Let w be a vertex in $\widetilde{G}$ which is not an end vertex of $\widetilde{F}$. Then w is the common vertex of at least two cutpairs in $\widetilde{F}$ which are cutpairs of $\widetilde{G}$ and by Theorem 3.2.11, w is a cutvertex. Also, if w is an end vertex of $\widetilde{F}$, then w is not a cutvertex; else there would exist u, v distinct from w such that w is on every strongest u-v path and one such path certainly lies in $\widetilde{F}$. But because w is an end vertex of $\widetilde{F}$, this is not possible. ∎

Corollary 3.2.16 *Let $\widetilde{G} = (\sigma, \mu, \Psi)$ be a fuzzy incidence tree. Then a cutvertex is the common vertex of at least two cutpairs.*

Theorem 3.2.17 *Let $\widetilde{G} = (\sigma, \mu, \Psi)$ be a fuzzy incidence graph. Then $\widetilde{G}$ is a fuzzy incidence tree if and only if the following conditions are equivalent for all $u, v \in V$:*
(i) (u, uv) is an incidence cutpair.
(ii) $\Psi^\infty(u, uv) = \Psi(u, uv)$.

Proof Let $\widetilde{G} = (\sigma, \mu, \Psi)$ be a fuzzy incidence tree. Suppose that (u, uv) is an incidence cutpair. Then $\Psi^\infty(u, uv) = \Psi(u, uv)$ by Theorem 3.2.10. Let (u, uv) be a pair in $\widetilde{G}$ such that $\Psi^\infty(u, uv) = \Psi(u, uv)$. If $\widetilde{G}^*$ is a tree, then clearly (u, uv) is an incidence cutpair. If $\widetilde{G}^*$ is not a tree, then it follows from Theorem 3.2.12 that (u, uv) is in $\widetilde{F}$ and (u, uv) is an incidence cutpair.

Conversely, assume that (i) and (ii) are equivalent. Construct a maximal fuzzy incidence spanning tree $\widetilde{F} = (\sigma, \mu, \Omega)$ for $\widetilde{G}$. If (u, uv) is in $\widetilde{F}$, $\Psi^\infty(u, uv) = \Psi(u, uv)$ and hence (u, uv) is an incidence cutpair. Now those are the only fuzzy incidence bridges in of $\widetilde{G}$ for if possible let $(u', u'v')$ be an incidence cutpair of $\widetilde{G}$ which is not in $\widetilde{F}$. Consider a cycle C consisting of $(u', u'v')$ and the unique u'-v' path in $\widetilde{F}$. Now incidence pairs of this u'-v' path are incidence cutpairs and so they are not weakest pairs of C and thus $(u', u'v')$ must be the weakest incidence cutpair of C and thus cannot be an incidence cutpair.

Moreover, for all incidence pairs $(u', u'v')$ not in $\widetilde{F}$, we have $\Psi(u', u'v') < \Omega^\infty(u', u'v')$; for if possible let $\Psi(u', u'v') \geq \Omega^\infty(u', u'v')$. But $\Psi(u', u'v') < \Psi^\infty(u', u'v')$, where strict inequality holds because $(u', u'v')$ is not an incidence cutpair. Hence, $\Omega^\infty(u', u'v') < \Psi^\infty(u', u'v')$ which gives a contradiction because $\Omega^\infty(u', u'v')$ is the strength of the unique u'-v' path in $\widetilde{F}$ and $\Psi^\infty(u', u'v') = \Omega^\infty(u', u'v')$. Thus, $\widetilde{F}$ is the required spanning fuzzy incidence subgraph which is a tree and hence $\widetilde{G}$ is a fuzzy incidence tree. ∎

3.3 Connectivity in Fuzzy Incidence Graphs

Connectivity concepts in fuzzy incidence graphs are discussed in this section. The results are based on those in [10]. Fuzzy incidence graphs are important in interconnection networks with influenced flows. Thus, it is of primary importance to analyze their connectivity properties. The existence of a strong path between any node arc pair of a fuzzy incidence graph is established in this section. Also, cut pairs and fuzzy incidence trees are characterized using the concept of strong pairs and several other structural properties of fuzzy incidence graphs are also presented.

A graph represents a relation between the elements of a set. A weighted graph represents the relational strength between the nodes. A fuzzy graph is a weighted graph which gives a normalized relational strength over a fuzzy subset of a set [13, 16]. The nodes in a set V can have certain influences over the elements of $V \times V$. For example, if nodes represents cities and arcs represents the roads connecting cities, we can have a fuzzy graph representing the extent of traffic from one city to another. The city having more population or more heavy vehicles will have more ramps into the city. So if a and b are two cities, and ab is a road connecting them, then (a, ab) could represent the ramp system from the road ab to the city a. In case of an edge uv in an unweighted graph both u and v will have an influence of 1 on uv. Also, for an arc uv from u to v in a directed graph, influence of u on uv, denoted by (u, uv) is one where as (v, uv) is 0. This concept is generalized through fuzzy incidence graphs

(FIG). A fuzzy incidence graph not only gives the extent of the relation between nodes belonging to a set, but also gives the impact or influence of a node on a relation pair.

Note that a fuzzy incidence graph $\widetilde{G}$: (σ, μ, ψ) on $(V, E, V \times E)$ may be considered as a fuzzy graph G^* : $(\sigma \cup \mu, \psi)$, where $\sigma \cup \mu$: $V \cup E \rightarrow [0, 1]$ and ψ : $V \times E \rightarrow [0, 1]$ such that $\psi(u, vw) \leq \sigma(u) \wedge \mu(vw)$. Elements of both σ^* and μ^* may be considered as nodes and incidence pairs as arcs. Even though elements of the form (u, vw) where $u \neq v \neq w$ are permitted in the definition, only pairs of the form (u, uv) will be considered in this section. The fuzzy graph G^* will help to study the connectivity concepts of a fuzzy incidence graph smoothly. Thus, incidence paths in a fuzzy incidence graph can take different forms. $u = u_0$, (u_0, u_0u_1), u_0u_1 is an incidence path of length one, $u = u_0$, (u_0, u_0u_1), u_0u_1, $(u_1, u_0u_1)u_1$ is an incidence path of length two, and so on. Note that an incidence path need not be a path. But it is clear that every path in an incidence graph is an incidence path. $u - v$, $u - vw$, $uv - w$, $uv - xy$ type incidence paths exist in a fuzzy incidence graph. Any path in the corresponding fuzzy graph G^* of $\widetilde{G}$ will be an incidence path in $\widetilde{G}$. The number of pairs in an incidence path is the incidence length of the path and an incidence path is an incidence cycle if it is closed. The number of pairs in a path will be an even number whereas that in an incidence path need not be. uv, (v, uv), v, (v, vx), vx, (x, vx), x, (x, xu), xu, (u, xu), u, (u, uv), uv is also an incidence cycle. The incidence strength of an incidence path is the minimum ψ value of all pairs in it. The incidence strength of u, (u, uv), uv is $\psi(u, uv)$.

In a fuzzy graph $G = (\sigma, \mu)$, an arc uv is said to be **strong** if $\mu(uv) \geq CONN_{G \setminus uv}(u, v)$, where $CONN_{G \setminus uv}(u, v)$ is the strength of connectedness between u and v in the arc deleted fuzzy subgraph $G \setminus uv$ of G [1]. Arc uv is said to be α-**strong** if $\mu(uv) > CONN_{G \setminus uv}(u, v)$ and is β-**strong** if $\mu(uv) = CONN_{G \setminus uv}(u, v)$. An arc which is not strong is termed as a δ-**arc**. Clearly for a δ-arc, $\mu(uv) < CONN_{G \setminus uv}(u, v)$ [7]. Analogous to these definitions in fuzzy graphs, we have the following definitions.

Definition 3.3.1 Let $\widetilde{G} = (\sigma, \mu, \Psi)$ be a fuzzy incidence graph. An incidence pair (x, xy) is said to be a **strong incidence pair** if

$$\psi(x, xy) \geq ICONN_{\widetilde{G} \setminus (x,xy)}(x, xy).$$

In particular, it is called α-**strong incidence pair** if

$$\psi(x, xy) > ICONN_{\widetilde{G} \setminus (x,xy)}(x, xy)$$

and β-**strong incidence pair** if $\psi(x, xy) = ICONN_{\widetilde{G} \setminus (x,xy)}(x, xy)$.

It is clear that a pair is strong if it is either α-strong or β-strong. There are pairs which are not strong in FIGs, which are given in the following definition.

Definition 3.3.2 Let $\widetilde{G} = (\sigma, \mu, \Psi)$ be a fuzzy incidence graph. An incidence pair (x, xy) is said to be a δ -**incidence pair** if

$$\psi(x, xy) < ICONN_{\widetilde{G} \setminus (x,xy)}(x, xy).$$

(x, xy) is said to be δ^*, if it is a δ-incidence pair, but $\psi(x, xy) > \wedge\{\psi(u, uv) \mid (u, uv) \in \psi^*\}$.

Definition 3.3.3 Let $\widetilde{G} = (\sigma, \mu, \Psi)$ be a fuzzy incidence graph. An incidence path ρ in $\widetilde{G}$ is said to be a **strong incidence path** if all pairs of ρ are strong. A **strong incidence cycle** is a closed strong incidence path.

Note that even if an arc uv is strong, the corresponding pairs (u, uv) and (v, uv) need not be strong. Similarly even if (u, uv) and (v, uv) are strong, the arc uv need not be strong. It is explained in the following example.

Example 3.3.4 Consider the fuzzy incidence graphs given in Fig. 3.4a, b.

In Fig. 3.4a, all arcs except arc uw are strong. Indeed uv and vw are α-strong arcs. But all the pairs in Fig. 3.4a are of equal membership value and hence they are all β-strong by definition. In Fig. 3.4b, arc vw is α-strong and the others are β-strong. But the pair (u, uw) is a δ-pair because $\Psi(u, uw) = 0.05$, but ICONN$_{\widetilde{G}\setminus(u,uw)}(u, uw) = 0.1 > 0.05$. In Fig. 3.4a, the incidence path

$$u, (u, uv), uv, (v, uv), v, (v, vw), vw, (w, vw), w, (w, wu), w, (w, wu), wu$$

is a strong $u - wu$ incidence path. There are no strong cycles in Fig. 3.4a or in Fig. 3.4b.

Next we have an easy proposition.

Proposition 3.3.5 *A fuzzy incidence cycle is a strong cycle.*

Proof Let $\widetilde{G} = (\sigma, \mu, \Psi)$ be a fuzzy incidence cycle. Then $G^* = (\sigma^*, \mu^*, \psi^*)$ is a cycle such that there is no unique $xy \in \mu^*$ such that $\mu(x, y) = \wedge\{\mu(u, v) \mid uv \in \mu^*\}$ and there exists no unique pair $(x, xy) \in \psi^*$ such that $\psi(x, xy) = \wedge\{\psi(u, uv) \mid (u, uv) \in \psi^*\}$. We next show that each arc and pair in $\widetilde{G}$ are strong. Clearly every arc in a fuzzy cycle is strong. Now let (x, xy) be a pair in $\widetilde{G}$. Suppose pair (x, xy) is not strong. Then it is a δ-incidence pair. By definition, $\psi(x, xy) < ICONN_{\widetilde{G}\setminus(x,xy)}(x, xy)$. Because G^* is a cycle, every pair in the incidence subgraph $\widetilde{G}\setminus(x, xy)$ will have membership value greater than $ICONN_{\widetilde{G}\setminus(x,xy)}(x, xy)$. It is a contradiction to our assumption that $\widetilde{G}$ is a fuzzy incidence cycle. Thus, (x, xy) is strong. ∎

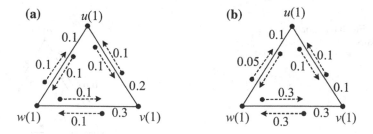

Fig. 3.4 a arc (u,w) not strong, **b** Pair (u,uw) not strong

Proposition 3.3.6 *Let* $\widetilde{G} = (\sigma, \mu, \Psi)$ *be fuzzy incidence graph and* $\widetilde{H} = (\tau, \nu, \Omega)$, *a fuzzy incidence subgraph of* $\widetilde{G}$. *For any pair* $(u, uv) \in \Omega^*$,

$$ICONN_{\widetilde{H}}(u, uv) \leq ICONN_{\widetilde{G}}(u, uv).$$

Proof The proof follows from the fact that, several $u - uv$ paths in $\widetilde{G}$ may be absent or have less strength in $\widetilde{H}$. ∎

Proposition 3.3.7 *Let* $\widetilde{G} = (\sigma, \mu, \Psi)$ *be a fuzzy incidence graph. A pair* (x, xy) *such that* $\psi(x, xy) = \vee\{\psi(u, uv) \mid (u, uv) \in \psi^*\}$ *is a strong pair.*

Proof Incidence strength of any $x - xy$ path in $\widetilde{G}$ will be less than or equal to $\psi(x, xy)$. If (x, xy) is the only pair with $\psi(x, xy) = \vee\{\psi(u, uv) \mid (u, uv) \in \psi^*\}$, then every other $u - uv$ path in $\widetilde{G}$ will have incidence strength less than $\psi(x, xy)$ and hence $ICONN_{\widetilde{G}\backslash(x,xy)}(x, xy) < \psi(x, xy)$. Thus, (x, xy) is an α-strong pair. If (x, xy) is not unique, the maximum possible value for incidence strength of a path in $\widetilde{G}\backslash(x, xy)$ will be $\psi(x, xy)$. So if there exists an $x - xy$ path in $\widetilde{G}\backslash(x, xy)$ with incidence strength $\psi(x, xy)$, then (x, xy) is β-strong. Otherwise (x, xy) is α-strong. ∎

The converse of Proposition 3.3.7 is not always true.

Example 3.3.8 Let $\widetilde{G} = (\sigma, \mu, \Psi)$ be the fuzzy incidence graph given in Fig. 3.5.

In this FIG, all pairs are strong even the weakest ones. Clearly it is a strong fuzzy incidence cycle.

Proposition 3.3.9 *Let* $\widetilde{G} = (\sigma, \mu, \psi)$ *be a FIG. If* $\psi(u, uv) = \wedge\{\sigma(u), \mu(uv)\}$, *then the pair* (u, uv) *is strong.*

Proof Consider the fuzzy incidence subgraph $\widetilde{G}\backslash(u, uv)$. If $\widetilde{G}\backslash(u, uv)$ is disconnected, then (u, uv) is a fuzzy incidence cut pair. Then, $\psi(u, uv) = ICONN_{\widetilde{G}}(u, uv) \geq ICONN_{\widetilde{G}\backslash(u,uv)}(u, uv)$. Hence, by definition (u, uv) is a strong pair. If $\widetilde{G}\backslash(u, uv)$ is connected, there exist pairs (u, ux) for some $x \neq u$, such that (u, ux) and (v, vu) belongs to a path ρ from u to uv in $\widetilde{G}\backslash(u, uv)$. Hence, the incidence strength of the path ρ is at most $\wedge\{\psi(u, ux), \psi(v, uv)\} \leq \wedge\{\sigma(u), \mu(ux), \sigma(v), \mu(uv)\} \leq \wedge\{\sigma(u), \mu(uv)\} = \psi(u, uv)$. That is, $\psi(u, uv) \geq ICONN_{\widetilde{G}\backslash(u,uv)}(u, uv)$. ∎

Fig. 3.5 FIG having only
strong pairs

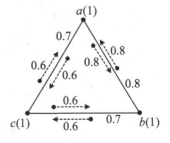

The converse of Proposition 3.3.9 is not true. In Example 3.3.8 (Fig. 3.5), the pair (b, bc) is strong. But $\psi(b, bc) \neq \wedge\{\sigma(b), \mu(bc)\}$.

Proposition 3.3.10 *In a fuzzy incidence graph, every cutpair is a strong pair.*

Proof Let $\widetilde{G} = (\sigma, \mu, \Psi)$ be a fuzzy incidence graph. Let $(u, uv) \in \psi^*$ be an incidence cutpair. Then by definition

$$ICONN_{\widetilde{G}\backslash(u,uv)}(u, uv) < ICONN_{\widetilde{G}}(u, uv).$$

If possible suppose that (u, uv) is not a strong pair. Then, $\psi(u, uv) < ICONN_{\widetilde{G}\backslash(u,uv)}$ (u, uv). Let ρ be a strongest $u - uv$ incidence path in $\widetilde{G}\backslash(u, uv)$. Then ρ together with the pair (u, uv) in $\widetilde{G}$ form an incidence cycle whose weakest pair is (u, uv). By Theorem 3.2.8, it is not possible because (u, uv) is a cutpair. ∎

The **incidence strength** of an incidence path P, IStrength(P), is defined to be $\vee\{\psi(x, xy) \mid (x, xy) \text{ is in } P\}$.

Proposition 3.3.11 *Let $\widetilde{G} = (\sigma, \mu, \psi)$ be a fuzzy incidence graph. An incidence pair $(u, uv) \in \psi^*$ is a strong pair if and only if $\psi(u, uv) = ICONN_{\widetilde{G}}(u, uv)$.*

Proof Suppose that the pair $(u, uv) \in \psi^*$ is strong. Because $\rho : u, (u, uv), uv$ is an incidence path from u to uv, $ICONN_{\widetilde{G}}(u, uv) \geq \psi(u, uv)$. If there exists no incidence path other than ρ from u to uv, the result is trivial. Now consider another path ς from u to uv in $\widetilde{G}$. If (u, uv) is a pair in ς, then $IStrength(\varsigma) \leq \psi(u, uv)$. If (u, uv) is not a pair of ς, ς is an incidence path in $\widetilde{G}\backslash(u, uv)$, whose incidence strength is less than or equal to $ICONN_{\widetilde{G}\backslash(u,uv)}(u, uv)$. Because the pair (u, uv) is strong, $ICONN_{\widetilde{G}\backslash(u,uv)}(u, uv) \leq \psi(u, uv)$. Thus, in any case the incidence strength of an incidence path in $\widetilde{G}$ is at most $\psi(u, uv)$. Thus, $ICONN_{\widetilde{G}}(u, uv) \leq \psi(u, uv)$.

Conversely, if $\psi(u, uv) = ICONN_{\widetilde{G}}(u, uv)$, then by Proposition 3.3.6, $\psi(u, uv) \geq ICONN_{\widetilde{G}\backslash(u,uv)}(u, uv)$. Hence, (u, uv) is a strong node arc pair. ∎

Note that in a fuzzy incidence graph, a path is a sequence of nodes, arcs and incidence arcs. Hence, paths connecting nodes to nodes are also important. A $u - v$ path will be of the form

$$u = u_0, (u_0, u_0u_1), u_0u_1, (u_1, u_0u_1), \ldots (u_{n-1}, u_{n-1}u_n), u_{n-1}u_n,$$
$$(u_n, u_{n-1}u_n), u_n = v$$

Note that in a path any arc should have two pairs to support. It is trivial that there exists a strong path between any two elements of σ^*. A new fuzzy graph $\widetilde{G}^*$ can be constructed from any fuzzy incidence graph $\widetilde{G}$ by subdividing each element xy in μ^* and assigning the incidence membership values of the concerned pairs to the newly formed arcs. The node set of the new fuzzy graph is exactly same as that of $\widetilde{G}$. Being a fuzzy graph, any two nodes of $\widetilde{G}^*$ are connected by a strong path [1]. This path can be restructured to get a strong path in $\widetilde{G}$.

The incidence path from a node u to an arc vw have to be considered differently as mentioned before. Such incidence paths exists between any node arc pair as seen from the following theorem.

Theorem 3.3.12 $\widetilde{G} = (\sigma, \mu, \psi)$ *be a connected fuzzy incidence graph. There exists a strong incidence path from any element* $u \in \sigma^*$ *to any element* $vw \in \mu^*$ *in* $\widetilde{G}$.

Proof Because $\widetilde{G}$ is connected, there exists an incidence path $\rho : u = u_0, (u_0, u_0 u_1)$, $u_0 u_1, (u_1, u_0 u_1), \ldots, u_n = v, (v, vw), vw, (w, vw)$ (u_n can be w also) between u and vw. Note that $\psi(u_i, u_{i+1}) > 0$ for every $i = 0, 1, 2, \ldots, n - 1$ and $\psi(v, vw), \psi(w, vw) > 0$. If pair $(u_{j-1}, u_{j-1} u_j)$ is not strong, then it is a δ-pair. So by definition

$$\psi(u_{j-1}, u_{j-1} u_j) < ICONN_{\widetilde{G} \backslash (u_{j-1}, u_{j-1} u_j)}(u_{j-1}, u_{j-1} u_j).$$

Let ρ be the incidence path from u to vw in $\widetilde{G} \backslash (u_{j-1}, u_{j-1} u_j)$ with incidence strength $ICONN_{\widetilde{G} \backslash (u_{j-1}, u_{j-1} u_j)}(u_{j-1}, u_{j-1} u_j)$. Note that every arc in ρ will have ψ value greater than $\psi(u_{j-1}, u_{j-1} u_j)$. If ρ is strong, we are done. Otherwise find δ-pairs in ρ and replace them by incidence paths whose strength are more than their ψ value. This process cannot be continued indefinitely. Eventually, we can find an incidence path ξ from u to vw so that every pair in ξ is strong. ∎

Theorem 3.3.13 (Characterization of incidence cutpairs.) *Let* $\widetilde{G} = (\sigma, \mu, \psi)$ *be a fuzzy incidence graph.* $(u, uv) \in \psi^*$ *is an incidence cutpair if and only if it is* α-strong.

Proof Let $\widetilde{G} = (\sigma, \mu, \psi)$ be a fuzzy incidence graph and let $(u, uv) \in \psi^*$ be an incidence cutpair of $\widetilde{G}$. By Theorem 3.2.8,

$$\psi(u, uv) > ICONN_{\widetilde{G} \backslash (u, uv)}(u, uv),$$

showing that (u, uv) is α-strong.

Conversely, suppose that (u, uv) is α-strong. Then by definition, it follows that $u, (u, uv), uv$ is the unique strongest incidence path from u to uv. The removal of (u, uv) reduces the incidence strength between u and uv and hence is an incidence cutpair. ∎

Example 3.3.14 Let $\widetilde{G} = (\sigma, \mu, \Psi)$ be the fuzzy incidence graph given in Fig. 3.6.

The spanning subgraph $\widetilde{F} = (\tau, \nu, \Omega)$ of $\widetilde{G}$ is the FIG with the same membership value for all nodes, arcs,, and pairs, except for $\Omega(v_4, v_4 v_1) = 0.15$ and $\Omega(v_1, v_1 v_4) = 0$. Then $(\tau^*, \nu^*, \Omega^*)$ is a tree. $(v_1, v_1 v_4) \in \Psi^* \backslash \Omega^*$. There exists an incidence path namely,

$$\xi \; : \quad v_4 v_1, (v_4, v_4 v_1), v_4, (v_4, v_3 v_4), v_4 v_3, (v_3, v_3 v_4), v_3, (v_3, v_3 v_2),$$
$$v_3 v_2, (v_2, v_3 v_2), v_2, (v_2, v_2 v_1), v_2 v_1, (v_1, v_2 v_1), v_1$$

Fig. 3.6 Fuzzy incidence tree

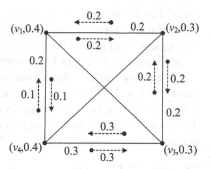

with $\mathrm{ICONN}_{\widetilde{F}}(v_1, v_4 v_1) = 0.15 > 0.1 = \Psi(v_1, v_4 v_1)$. Thus, $\widetilde{G}$ is a fuzzy incidence tree. Note that the underlying fuzzy graph $G = (\sigma, \mu)$ is not a fuzzy tree. Thus, the concepts of fuzzy incidence graph and fuzzy graph are somewhat different.

Now we discuss strong pairs of a fuzzy incidence tree (FIT). According to Theorem 3.1.28, in each incidence cycle of a fuzzy incidence tree (FIT) there exists a δ-pair. Note that two δ-pairs do not exist in any such incidence cycle. Removal of all such δ-pairs from cycles make the fuzzy incidence tree $\widetilde{G}$, a tree $\widetilde{F}$. The following theorem guarantees that the strong pairs of $\widetilde{G}$ are nothing but the pairs in $\widetilde{F}$ as in the case of fuzzy trees.

Theorem 3.3.15 $\widetilde{G} = (\sigma, \mu, \psi)$ be a fuzzy incidence tree. A pair in ψ^* is a strong pair if and only if it is a pair of the tree $\widetilde{F} = (\tau, \nu, \Omega)$ in the definition of a fuzzy incidence tree.

Proof Let $\widetilde{G} = (\sigma, \mu, \psi)$ be a fuzzy incidence tree and let $\widetilde{F} = (\tau, \nu, \Omega)$ be the tree in the definition. Let (u, uv) be a strong pair. If (u, uv) is not in $\widetilde{F}$, then $\psi(u, uv) < \mathrm{ICONN}_{\widetilde{F}}(u, uv)$. Because (u, uv) is strong, $\psi(u, uv) \geq \mathrm{ICONN}_{\widetilde{G} \setminus (u,uv)}(u, uv)$. Because (u, uv) is not in $\widetilde{F}$, $\widetilde{F}$ is a subgraph of $\widetilde{G} \setminus (u, uv)$ and hence $\psi(u, uv) \geq \mathrm{ICONN}_{\widetilde{G} \setminus (u,uv)}(u, uv) \geq \mathrm{ICONN}_{\widetilde{F}}(u, uv)$, a contradiction.

Conversely, suppose that $(u, uv) \in \Omega^*$. By Theorem 3.2.6, (u, uv) is an incidence cutpair. By Theorem 3.3.13, (u, uv) is strong. ∎

Thus, the spanning tree given in the definition of a fuzzy incidence tree is uniquely determined. The node arc pairs of $\widetilde{F}$ are nothing but the strong pairs of $\widetilde{G}$. Also, note that the sum of ψ membership values of the strong pairs of $\widetilde{G}$ will be greatest for $\widetilde{F}$, among all other spanning trees of $\widetilde{G}$. Thus, here onwards, we can refer $\widetilde{F}$, as the unique maximum spanning tree associated with $\widetilde{G}$. Also, a pair (u, uv) in a fuzzy incidence tree is strong if and only if it is an incidence cutpair. Thus, $\widetilde{F}$ may be obtained by removing all δ-pairs belonging to cycles in $\widetilde{G}$. Using these ideas, we have the following proposition.

Proposition 3.3.16 *A connected fuzzy incidence graph is a fuzzy incidence tree if and only it has no β-strong pairs.*

Proof Let $\widetilde{G} = (\sigma, \mu, \psi)$ be a connected fuzzy incidence graph. If $\widetilde{G}$ is fuzzy incidence tree, then there exists a unique $\widetilde{F} = (\tau, \nu, \Omega)$ such that every pair in $\widetilde{F}$ is an incidence cutpair and hence is α-strong. By definition of a fuzzy incidence tree, all other pairs of $\widetilde{G}$, not in $\widetilde{F}$ are δ-pairs. Thus, $\widetilde{G}$ has no β-strong pairs.

Conversely suppose that $\widetilde{G}$ has no β-strong pairs. We show that $\widetilde{G}$ is a fuzzy incidence tree. If $\widetilde{G}$ has no cycles, clearly $\widetilde{G}$ is a fuzzy incidence tree. Now assume that $\widetilde{G}$ has cycles. Let ρ be a cycle in $\widetilde{G}$. Then ρ contains only α-strong and δ pairs. All pairs of ρ cannot be α strong by definition. Hence, there exists a unique δ-pair in every cycle of $\widetilde{G}$ and by Theorem 3.1.28, $\widetilde{G}$ is a fuzzy incidence tree. ∎

Theorem 3.3.17 *Let $\widetilde{G} = (\sigma, \mu, \psi)$ be a connected fuzzy incidence graph. $\widetilde{G}$ is a fuzzy incidence tree if and only there exists a unique strong incidence path between any node arc pair. In particular, this path will be an α-strong incidence path.*

Proof Suppose $\widetilde{G}$ is a fuzzy incidence tree. By Theorem 3.3.12, there exists a strong incidence path ρ between any node $u \in \sigma^*$ and arc $vw \in \mu^*$. By Theorem 3.3.13, this incidence path ρ lies entirely in the associated maximum spanning tree $\widetilde{F}$. Because $\widetilde{F}$ is a tree, such a path is unique. Also, because there are no β-strong pairs in $\widetilde{F}$, this path will be an α-strong incidence path.

Conversely, assume that $\widetilde{G}$ is a fuzzy incidence graph such that there exists a unique α-strong incidence path between any node arc pair. If possible suppose $\widetilde{G}$ is not a fuzzy incidence tree. Then there exists a cycle ρ in $\widetilde{G}$ such that $\psi(u, uv) \geq ICONN_{\widetilde{G}\setminus(u,uv)}(u, uv)$ for every node arc pair (u, uv) of ρ. That is, every node arc pair (u, uv) of ρ is a strong pair, which is a contradiction to the assumption that there exists a unique strong incidence path between a node arc pair. ∎

It is clear that a strongest incidence path need not be a strong incidence path and vice versa. A strongest incidence path having no δ-pairs and arcs is a strong path. Next we have the following proposition.

Proposition 3.3.18 *Let $\widetilde{G} = (\sigma, \mu, \psi)$ be a fuzzy incidence graph and let ρ be a strong incidence path between u and uv. Then ρ is a strongest $u - uv$ incidence path in the following cases.*

(i) ρ contains only α strong pairs.
(ii) ρ is a unique strong $u - uv$ incidence path.
(iii) All $u - uv$ incidence paths in $\widetilde{G}$ are of equal strength.

A complete fuzzy incidence graph $\widetilde{G} = (\sigma, \mu, \psi)$ is such that $\psi(u, uv) = \sigma(u) \wedge \mu(uv)$ for every $(u, uv) \in \psi^*$. It is clear that $\psi(u, uv) = \psi(v, uv)$ for every $u, v \in \sigma^*$. Consider the following example.

Example 3.3.19 Let $\widetilde{G} = (\sigma, \mu, \Psi)$ be the fuzzy incidence subgraph shown in Fig. 3.7. It is a complete fuzzy incidence graph. It can be seen that the underlying fuzzy subgraph of a complete graph fuzzy incidence graph is not complete.

Fig. 3.7 Complete fuzzy
incidence graph

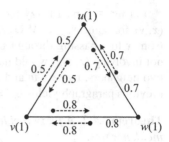

Proposition 3.3.20 *A complete fuzzy incidence graph has no δ pairs.*

Proof If possible assume that a complete incidence fuzzy graph $\widetilde{G} = (\sigma, \mu, \psi)$ has
a δ-pair say (u, uv). By definition, $\psi(u, uv) < ICONN_{\widetilde{G} \backslash (u,uv)}(u, uv)$. That is, there
exists a stronger incidence path ρ from u to uv in $\widetilde{G}$. Let q be its incidence strength.
If $\psi(u, uv) = \psi(v, uv) = p$, then $p < q$. Every pair in ρ will have incidence strength
more than p. Let w be the first node in ρ after u. Then $\psi(u, uw) > p$. This is not
possible as $\wedge\{\sigma(u), \mu(uw)\} = p$. Thus, $\widetilde{G}$ has no δ-pairs. Hence, every node arc pair
of a complete fuzzy incidence graph is strong. ∎

3.4   Fuzzy End Nodes in Fuzzy Incidence Graphs

We consider results from [2] in this section.
   A fuzzy incidence cycle $\widetilde{G}$ is called **multimin** if $\widetilde{G}$ has more than one weakest
incidence pair. $\widetilde{G}$ is called locamin if every element of $V \cup E$ lies on a weakest
incidence pair. We sometimes use the term node for an element from $V \cup E$. Because
a fuzzy incidence cycle has at least three nodes, locamin implies multimin. An
incidence pair (x, xy) is called a **weakest incidence pair** if $\Psi(x, xy) = \wedge\{\Psi(u, uv) \mid (u, uv) \in \text{Supp}(\Psi)\}$.

Theorem 3.4.1 *A fuzzy incidence cut node has at least two strong neighbors.*

Proof Let $z \in V$ be a fuzzy incidence cut vertex. Then $ICONN_G(x, y) > ICONN_{G \backslash z}$
(x, y) for some $x, y \in (V \backslash \{z\}) \cup E$. Then there is a strongest incidence path P from
x to y passing through z, say $x, \ldots, uz, z, vz, \ldots, y$. If (z, uz) is not strong, we
have $\Psi(z, uz) < ICONN_{G \backslash (z,uz)}$. Thus, there is an incidence path Q from uz to z, not
involving (z, uz), that is stronger than $\Psi(z, uz)$. Let $u'z$ be the node just preceding z on
Q. Because the strength of Q is at most $\Psi(z, u'z)$, we must have $\Psi(z, u'z) > \Psi(z, uz)$.
If $(z, u'z)$ is not strong, then this argument can be repeated. However, it cannot be
repeated indefinitely. Therefore, we eventually find a u^*z such that (z, u^*z) is strong.
Similarly, we can find a v^*z such that (z, v^*z) is strong. If $u^* = v^*$, we would have
a path from x to $u^*z = v^*z$ to y that is stronger than P, so that deleting z would not
reduce $ICONN(x, y)$, a contradiction. Hence, z has at least two strong neighbors.

Let $wz \in E$ be a fuzzy incidence cutedge. Then $\text{ICONN}_G(x, y) > \text{ICONN}_{G \backslash wz}$ (x, y) for some $x, y \in V \cup (E \backslash \{wz\})$. Then there is a strongest incidence path P from x to y passing through wz, say $x, \ldots, z, wz, w, \ldots, y$. A path from wz to z not involving (z, wz) would have a $u' \in V$ just preceding wz. Because wz has only two neighbors, namely w and z, it follows that $u' = w$. Thus, a procedure as in the previous paragraph yields $u^* = w$ and $v^* = z$. ∎

Theorem 3.4.2 *A fuzzy incidence cycle is multimin if and only if it is not a fuzzy incidence tree.*

Proof Suppose that (σ, μ, Ψ) is a fuzzy incidence cycle. Then there exist distinct pairs $(x_1, x_1y_1), (x_2, x_2y_2) \in \text{Supp}(\Psi)$ such that $\Psi(x_1, x_1y_1) = \Psi(x_2, x_2y_2) = \wedge\{\Psi(u, uv) \mid (u, uv) \in\text{Supp}(\Psi)\}$. If (u, uv) is in any incidence spanning tree of (σ, μ, Ψ), then $\text{Supp}(\Psi) \backslash \text{Supp}(\Omega) = \{(u, uv)\}$ for some $u, v \in V$ because (σ, μ, Ψ) is an incidence cycle. Hence, there does not exist an incidence path in (σ, μ, Ω) between u and uv of greater strength than $\Psi(u, uv)$. Thus, (σ, μ, Ψ) is not a fuzzy incidence tree. Conversely, suppose that (σ, μ, Ψ) is not a fuzzy incidence tree. Because (σ, μ, Ψ) is an incidence cycle, we have for all $(u, uv) \in\text{Supp}(\Psi)$ that (u, uv) is in a fuzzy spanning incidence subgraph of (σ, μ, Ψ) which is an incidence tree and $\Omega^\infty(u, uv) \leq \Psi(u, uv)$, where $\Omega(u, uv) = 0$ and $\Omega(x, xy) = \Psi(x, xy)$ for all $(x, xy) \in\text{Supp}(\Psi) \backslash \{(u, uv)\}$. Hence, Ψ does not attain $\wedge\{\Psi(x, xy) \mid (x, xy) \in\text{Supp}(\Psi)\}$ uniquely. Thus, (σ, μ, Ψ) is a fuzzy incidence cycle. ∎

It follows that in a multimin fuzzy incidence cycle, every incidence pair, even a weakest incidence pair is strong. This is used to prove the following result.

Theorem 3.4.3 *A fuzzy incidence cycle is multimin if and only if it has no fuzzy incidence nodes.*

Proof In a multimin fuzzy incidence cycle, every pair is strong. Hence, for any vertex z or edge wz both of the pairs on which z or wz lie are strong, so z or wz cannot be a fuzzy end node or a fuzzy end edge, respectively. Conversely, if a fuzzy incidence cycle has only one weakest pair (x, xy), it follows that xy is not a strong neighbor of x or vice versa. Hence, only the other neighbors of x and xy are strong, so that x is a fuzzy end vertex and xy is a fuzzy end edge. ∎

Theorem 3.4.4 *A multimin fuzzy incidence cycle is locamin if and only if it has no fuzzy incidence cut nodes.*

Proof If (u, uv) is a weakest pair of a fuzzy incidence cycle G, then clearly u and uv cannot be incidence cut nodes of G. Hence, if G is locamin, no incidence node of G can be an incidence cut node because every incidence node of G lies on a weakest pair. Conversely, suppose G is not locamin. Let x, xy, y be three consecutive incidence nodes of G such that neither (x, xy) nor (y, xy) is a weakest pair. Then deleting xy reduces $\text{CONN}_G(x, y)$ so that xy is an incidence cut node. Consider the consecutive incidence nodes xy, x, xz such that neither (x, xy) nor (x, xz) is a weakest pair. Then deleting x reduces $\text{ICONN}_G(xy, xz)$. ∎

Theorem 3.4.5 *A nontrivial fuzzy incidence tree G has at least two fuzzy incidence end nodes.*

Proof Let G' be a spanning fuzzy incidence tree of G. Because the support of G' is a nontrivial tree, it has at least two end nodes. We will prove that these nodes are fuzzy incidence end nodes of G. Suppose $z \in V$ is an end vertex of G' and not a fuzzy incidence end vertex of G. Then z has at least two strong incidence neighbors xz, yz so that $\Psi(x, xz) \geq \text{ICONN}_{G\backslash(z,xz)}(z, xz)$ and $\Psi(y, yz) \geq \text{ICONN}_{G\backslash(z,yz)}(z, yz)$. Because z is an end vertex of G', at most one of (z, xz) and (z, yz) can be pairs of G'. Suppose (z, xz) is not a pair of G'. By definition of a fuzzy incidence tree, this implies that $\Psi(z, xz) < \text{ICONN}_{G'}(z, xz)$. Hence,

$$ICONN_{G\backslash(z,xz)}(z, xz) \leq \Psi(z, xz) < ICONN_{G'}(z, xz).$$

But G' is a fuzzy subgraph of G and because (z, xz) is not a pair of G', G' is also a fuzzy incidence subgraph of $G\backslash(z, xz)$. This implies that

$$ICONN_{G'}(z, xz) \leq_{G\backslash(z,xz)} (z, xz),$$

a contradiction.

Suppose $wz \in E$ is a fuzzy incidence end edge in the support of G'. Now w and z are the only neighbors of wz. This implies that wz is a fuzzy incidence end node of G. Hence, either w or z is also a fuzzy incidence end node of G. ∎

Theorem 3.4.6 *A fuzzy incidence cycle G is multimin if and only if it has at least one incidence node which is neither a fuzzy incidence cut node nor a fuzzy incidence endnode.*

Proof By Theorem 3.4.3, if G is multimin it has no fuzzy incidence end nodes. Thus, a fuzzy incidence node that lies on a weakest pair of G cannot be a fuzzy incidence cut node. Conversely, if G is not multimin it has a unique weakest pair. Hence, the nodes that lie on this pair must be fuzzy incidence end nodes. Thus, all the other nodes of G must be incidence cut nodes. Hence, every incidence node of G is either a fuzzy incidence end node or a fuzzy incidence cut node. ∎

Corollary 3.4.7 *A fuzzy incidence cycle G is a fuzzy incidence tree if and only if every incidence node of G is either a fuzzy incidence cut node or a fuzzy incidence endnode.*

Proof The proof follows from Theorems 3.4.2 and 3.4.6. ∎

It follows by Theorem 3.2.14 that in a fuzzy incidence tree, every incidence node is either an incidence cut node or an incidence end node. Indeed the incidence end nodes of the fuzzy incidence spanning tree of G must be fuzzy incidence end nodes and the other nodes must be fuzzy incidence cut nodes. The converse is not true even if every node of G is either a fuzzy incidence end node or a fuzzy incidence cut node. For example, let G be the crisp cycle x, (x, xy), xy, (y, xy), y, (y, yz), yz, (z, yz), z, (z, xz),

$xz, (x, xz), x$ together with the ux, yv, wz and pairs $(x, ux), (y, yv), (z, zw)$. Then x, xy, y, yz, z, xz are incidence cutnodes and ux, yv, zw are incidence end nodes, but G is not an incidence tree.

3.5 Human Trafficking and Fuzzy Influence Graphs

Trafficking in persons is the most heinous and condemnable organized crime of our time. The eradication of this crime must involve the full cooperation of all. In this section, we provide a mathematical method to model such cooperation. We use a new concept in fuzzy graph theory, namely that of influence. We characterize fuzzy influence cutpairs because their removal increases the number of connected components of a fuzzy network and thus weakens the potential flow in the network.

The following discussion is based on [20].

The United Nations Global Initiative to Fight Human Trafficking (UN.GIFT) aims to mobilize state and non-state actors to eradicate human trafficking by focusing on three main points:

(1) The reduction of both the vulnerability of potential victims and the demand for exploitation;

(2) Ensuring adequate protection and support of victims;

(3) Supporting the prosecution of the criminals involved while respecting fundamental human rights of all persons.

Joining forces, pooling knowledge, expanding the scope and number of stakeholders and cooperating across borders must occur if we hope to eradicate human trafficking. There are many ways nations can influence other nations to cooperate in the combating of human trafficking. The purpose of this section is to create a mathematical model that can be used to study the effect of states influencing other states. We use a new technique that has recently been created fuzzy network theory. This technique introduced the notion of fuzzy incidence in fuzzy graphs [4, 5].

United States rates countries annually with regard to their involvement in human trafficking. The United States places countries into four tiers, [17]. Tier 1 countries fully comply with the Trafficking Victims Protection Act (TVPA) passed by Congress in 2000 and strengthened in 2003. Tier 2 countries do not fully comply with the minimum standards for eliminating trafficking, but are making efforts to bring trafficking into compliance. Tier 2 countries that require special scrutiny because of higher increasing numbers of victims are placed in the Tier 2 Watch List. Countries that neither satisfy the minimum standards nor show a significant effort to come into compliance are placed in Tier 3. Some experts believe that countries are influenced by the rating they are given. These experts believe that this type of rating exerts social pressure on governments.

Let (σ, μ) be a fuzzy subgraph of G and Ψ a fuzzy incidence of G. Then (σ, μ, Ψ) is a fuzzy incidence graph. If the elements of V are states, the elements of E edges, then $\Psi(u, vw)$ can be interpreted in many ways. One way is to consider $\Psi(u, vw)$

as the degree to which state u influences a reduction of the flow of traffic between states v and w.

We discuss some methods how states can influence other states. The discussion is taken from [8, 18]. Many industrialized states fund antitrafficking initiatives in other countries, especially to those countries from which people have been trafficked into their country or region. This funding should be pooled in order to provide a more effective way to use the funds. One major problem for immigration and law enforcement officials is how to distinguish trafficking victims from others. Some victims may be involved in similar income-generating activities, but who have not been victimized. Many methods have been developed at frontiers and for police forces to help identifications. Intelligence about the profile of adults or children who have already been trafficked, aid the police in identifying the characteristics of such persons and using this intelligence to select travelers who warrant advice and protection. For example, the Philippine Centre on Transnational Crime developed an 11-point checklist to help identify such people and a further 11-point checklist to identify possible traffickers.

Border posts offer an opportunity to identify victims of transnational trafficking and to intercept them. The United Nations Children's Fund (UNICEF) developed a checklist for immigration officers (border officials) to assess whether a child is at risk of being trafficked which lists even indicators to show a child might be in the process of being trafficked.

Traffickers move their victims from place to place within a country. Consequently, effective response to trafficking requires the collaboration of multiple countries. Bilateral agreements on law enforcement cooperation and mutual legal assistance have been made.

Certain agencies have adopted action oriented plans to improve cooperation to combat trafficking. The include the Association of Southeastern Asian Nations (ASEAN), the Economic Community of West African States (ECOWAS) and the Organization for Security and Cooperation in Europe (OSCE).

If $G : (\sigma, \mu)$ is a fuzzy graph on $G : (V, E)$, then an incidence fuzzy graph $\widetilde{G} : (\sigma, \mu, \psi)$ may be considered as a fuzzy graph $G^* : (\sigma \cup \mu, \psi)$, where $\sigma \cup \mu :$ $V \cup E \to [0, 1]$ and $\psi : V \times E \to [0, 1]$ such that $\psi(u, vw) \leq \sigma(u) \wedge \mu(vw)$ [12]. Even though the elements of the form (u, vw) where $u \neq v \neq w$ are permitted in the definition, only pairs of the form (u, uv) have been studied in the literature. Clearly for an arc $uv \in \mu^*$, u as well as v will have some influence on uv. In case of an edge in an unweighted graph both u and v will have an influence of 1 on uv. Also, for an arc u from v in a directed graph, influence of u on uv, denoted by (u, uv) is one where as $(v, uv) = 0$. This concept is generalized through fuzzy influence graphs.

Consider a fuzzy incidence graph $\widetilde{G}(\sigma, \mu, \psi)$. $\psi(u, vw) > 0$ implies that the node u, which is not an end node of the arc vw has an influence on vw. It is quiet natural that there are shortcuts in any procedure or process dominating usual incidence paths. In a triangle abc with all vertex, edge weights and incidence values are ones, $a, (a, ab), ab, (b, ab), b, (b, bc), bc$ is a $a - bc$ incidence path. If there is a shortcut pair (a, bc), then the path $a, (a, bc), bc$ is an influence path.

3.6 Fuzzy Influence Graphs

Now, we introduce fuzzy influence graphs. As mentioned before, the pairs of the form (u, vw) where $u \neq v \neq w$ are considered in fuzzy influence graphs. Clearly, $\psi(u, vw)$ represents the nontrivial influence of u on vw and any such incidence make a FIG a fuzzy influence graph. The figures in this section are from [15].

Definition 3.6.1 Let $\widetilde{G} = (\sigma, \mu, \Psi)$ be a fuzzy incidence graph (FIG). A path P in $\widetilde{G}$ is said to be **influenced** or an **influence path** if it contains at least one influence pair. That is, a pair of the form (u, vw), $u \neq v \neq w$.

If (u, vw) is an influence pair, we say that vw is influenced by u. If vw is influenced by u, one can reach v, w or vw from u using the influence.

Definition 3.6.2 Let $\widetilde{G} = (\sigma, \mu, \Psi)$ be a fuzzy incidence graph (FIG). $\widetilde{G}$ is said to be a **fuzzy influence graph** if it contains at least one influence pair. Or equivalently if it contains a fuzzy influence path. The **influence of a path** $i(P)$ is the minimum of non zero ψ values of all influence pairs in it. As in the case of incidence graphs, we can define **strength of influence** or **influence connectivity** between two nodes as $ICONN_{\widetilde{G}}^*(u, v) = \vee\{i(P) \mid$ where P is an influence path between u and v and $\vee$ represents the maximum}. If there is no influence or incidence path between a pair $u, v \in \sigma^*$, then $ICONN_{\widetilde{G}}^*(u, v) = 0$.

Definition 3.6.3 Let $\widetilde{G} = (\sigma, \mu, \Psi)$ be a fuzzy influence graph. A pair (u, vw) is said to be a **strong influence pair** if

$$\psi(u, vw) \geq ICONN_{\widetilde{G}\setminus(u,vw)}^*(u, vw).$$

(u, vw) is called a **strongest influence pair** or an α **-strong influence pair** if $\psi(u, vw) > ICONN_{\widetilde{G}\setminus(u,vw)}^*(u, vw)$. If an influence pair is not strong, we call it a weak influence pair and a strong pair which is not α-strong is termed as a β-**strong pair**. An influence pair (u, vw) is said to be an **effective influence pair** if $\psi(u, vw) = \sigma(u) \wedge \mu(vw)$.

Example 3.6.4 Consider the fuzzy influence graph in Fig. 3.8.

For the fuzzy influence graph given in Fig. 3.8, (u, vw) is an effective pair. Note that vw is influenced by u. The path $P : u, (u, vw), vw, (v, vw), w$ is an influenced $u - w$ path. $ICONN_{\widetilde{G}}^*(u, v) = ICONN_{\widetilde{G}}^*(v, w) = 0.5$. Note that if the influence (u, vw) is absent from this fuzzy influence graph, it becomes a fuzzy incidence graph.

Definition 3.6.5 Let $\widetilde{G} = (\sigma, \mu, \Psi)$ be a fuzzy influence graph.

(i) A node $x \in \sigma^*$ is said to be an **influence cutnode** if $ICONN_{\widetilde{G}\setminus x}^*(s, t) < ICONN_{\widetilde{G}}^*(s, t)$ for some $s, t \in \sigma^* \cup \mu^*$.

(*ii*) An edge $xy \in \mu^*$ is said to be an **influence bridge** if $ICONN^*_{\widetilde{G} \setminus xy}(s, t) < ICONN^*_{\widetilde{G}}(s, t)$ for some $s, t \in \sigma^* \cup \mu^*$.

(*iii*) A pair (x, yz) is said to be an **influence cutpair** if $ICONN^*_{\widetilde{G} \setminus (x,yz)}(s, t) < ICONN^*_{\widetilde{G}}(s, t)$ for some $s, t \in \sigma^* \cup \mu^*$.

In Fig. 3.8, u an influence cutnode, vw is an influence bridge. (v, vw) and (w, vw) are influence pairs.

Consider the following example.

Example 3.6.6 Let $\widetilde{G_1}$ and $\widetilde{G_2}$ be the fuzzy influence graphs given in Fig. 3.9a, b.

In Fig. 3.9a, $\psi(u, vw) = 0.3$. $ICONN^*_{\widetilde{G} \setminus (u,vw)}(u, vw) < \psi(u, vw)$. Thus, (u, vw) is a strongest influence pair. Equivalently it is an α-pair. In fact, it is an effective influence pair. In Fig. 3.9b, (u, vw) is a β-strong influence pair and (w, uv) is a weak influence pair. Next we have the following proposition.

Proposition 3.6.7 *An effective influence pair is a strong influence pair.*

Proof Let $\widetilde{G} = (\sigma, \mu, \Psi)$ be fuzzy influence graph with an effective influence pair (u, vw). Then $\psi(u, vw) = \sigma(u) \wedge \mu(vw)$. Suppose (u, vw) not strong. Then, $\psi(u, vw) < ICONN_{\widetilde{G} \setminus (u,vw)}(u, vw)$. Therefore, there exists a $u - vw$ incidence path ρ in $\widetilde{G}$ so that incidence strength of $\rho > \psi(u, vw)$. Let $\rho = u =$

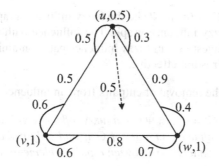

Fig. 3.8 Fuzzy influence graph with a single influence pair

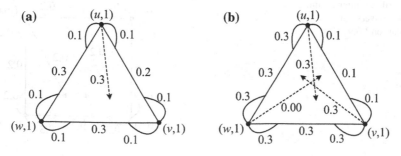

Fig. 3.9 **a** α-influence pair, **b** β-strong and weak influence pairs

u_0, (u_0, u_0u_1), u_0u_1, (u_1, u_0u_1), ..., vw. Then $\psi(u_0, u_0u_1) > \wedge\{\sigma(u = u_0), \mu(vw)\}$. This is a contradiction because $\psi(u, uv) \leq \sigma(u) \wedge \mu(uv)$ for any incidence pair. ∎

Proposition 3.6.8 *Every strongest influence pair is a cutpair.*

Proof The follows from the definitions of strongest influence pair and an influence cutpair. ∎

A strongest influence pair and a strongest influence path are different. Also, an effective influence pair need not be a strongest pair. But we have the following result.

Proposition 3.6.9 *Let $\tilde{G} = (\sigma, \mu, \Psi)$ be a fuzzy influence graph. If (u, vw) is an effective influence pair, then there exists a strongest influence path ρ in $\tilde{G}$ containing (u, vw).*

Proof Consider the $u - vw$ incidence path $\xi = u$, (u, vw), vw. No incidence pair incident at u can have incidence value more than $\sigma(u)$ and no incident pair incident at vw can have incidence value more than $\mu(vw)$. So any incidence path in $\tilde{G}$ from u to vw other than ξ will have incidence strength at most $\sigma(u) \wedge \mu(vw)$. Thus, ξ becomes a strongest influence path containing the pair (u, vw). ∎

The converse of Proposition 3.6.9 is not always true. Consider the following example.

Example 3.6.10 Let $\tilde{G} = (\sigma, \mu, \Psi)$ be the fuzzy influence graph given in Fig. 3.10.
Figure 3.10 is a fuzzy influence graph. The influence path v_2, (v_2, v_3v_4), v_3v_4, (v_4, v_3v_4), v_4 is a strongest $v_2 - v_4$ fuzzy influence path containing the influence pair (v_2, v_3v_4). But the pair is not effective.

Next we consider the removal of cutpairs from an influence path.

Theorem 3.6.11 *Let $\tilde{G} = (\sigma, \mu, \Psi)$ be a fuzzy influence graph with $|\sigma^* \cup \mu^*| \geq 6$. A pair (u, vw) is an influence cutpair of $\tilde{G}$ if and only if the removal of (u, vw) reduces the influence connectivity between a pair of elements $s, t \in \sigma^* \cup \mu^*$.*

Fig. 3.10 Counter example for the converse of Proposition 3.6.9

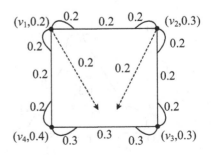

Proof Let $\widetilde{G} = (\sigma, \mu, \Psi)$ be a fuzzy influence graph. Suppose (u, vw) is an influence cut pair of $\widetilde{G}$. Then by definition, $ICONN^*_{\widetilde{G}\backslash(u,vw)}(s, t) < ICONN^*_{\widetilde{G}}(s, t)$ for some $s, t \in \sigma^* \cup \mu^*$. Also, among all $u - vw$ influence paths, the path $P : u, (u, vw), vw$ will have the maximum influence strength, namely $\psi(u, vw)$. Let $s = uv$ and $t = vw$. Then any $uv - vw$ incidence path in $\widetilde{G}$ contains a $u - vw$ sub path P^* such that $i(P^*) < \psi(u, vw)$. Clearly the removal of (u, vw) reduces the influence connectivity between uv and vw and the conclusion follows. It follows that uv and vw need not be the only pairs having this property.

Conversely, suppose for some $s, t \in \sigma^* \cup \mu^*$, $ICONN^*_{\widetilde{G}\backslash(u,vw)}(s, t) < ICONN^*_{\widetilde{G}}(s, t)$. By definition, it follows that (u, vw) is an influence cut pair. ∎

Definition 3.6.12 A complete fuzzy incidence graph $\widetilde{G} = (\sigma, \mu, \Psi)$ is said to be a **complete influence graph** if each $s \in \sigma^*$ influences every $t \in \mu^*$ such a way that $\psi(s, t) = \wedge\{\sigma(s), \mu(t)\}$.

Consider the following example.

Example 3.6.13 Let $\widetilde{G} = (\sigma, \mu, \Psi)$ be a fuzzy influence graph given in Fig. 3.11.

Representation of complete influence graphs is difficult due to large number of possible influences. The support of an influence graph on four vertices is shown in Fig. 3.12.

An influence path becomes strong if all the influence pairs present are strong. Hence, we have the following theorem.

Theorem 3.6.14 *Let $\widetilde{G} = (\sigma, \mu, \Psi)$ be a complete influence graph. Then there exists a strong influence path between any two elements of $\sigma^* \cup \mu^*$.*

Proof First we show that there exists an influence path between any two elements s, t of $\sigma^* \cup \mu^*$. If $s, t \in \sigma^*$, then s will have an influence over some element of μ^* say xy. Then the influence path $s, (s, xy), xy, y$ together with any incidence path from y to t, forms a new influence path between s and t. Similarly if $s \in \sigma^*$ and $t \in \mu^*$, then we can concatenate one influence and another incidence path to get the new influence path. Suppose both $s, t \in \mu^*$. We can find an influence pair at the end node of s or at the end node of t and hence an influence path between s and t.

Fig. 3.11 A complete fuzzy influence graph

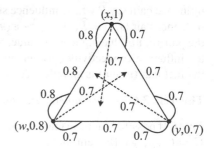

Fig. 3.12 Support of a
complete fuzzy influence
graph on four nodes

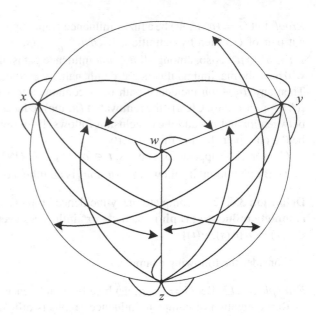

We next show that a complete influence graph has no weak influence pairs. If possible suppose (u, vw) is a weak influence pair. Then by definition, $\psi(u, vw) < ICONN^*_{\widetilde{G}\setminus(u,vw)}(u, vw)$. That is, there exists an influence path P between u and vw of higher influence strength say d than $\psi(u, vw)$. Every influence pair in P will have ψ value greater than or equal to d. Also, in particular, $\sigma(u) \geq d$ and $\mu(vw) \geq d$. $\widetilde{G}$ being a complete fuzzy influence graph, $\psi(u, vw) = \wedge\{\sigma(u), \mu(vw)\} \geq d$, which is a contradiction, and hence we conclude that every influence pair of $\widetilde{G}$ is strong. ∎

Theorem 3.6.15 *Let $\widetilde{G} = (\sigma, \mu, \Psi)$ be a fuzzy influence graph. Then for any influence cutpair (u, vw), $\psi(u, vw) = ICONN^*_{\widetilde{G}}(u, vw)$.*

Proof From Theorem 3.6.11, it follows that

$$\psi(u, vw) \geq ICONN^*_{\widetilde{G}\setminus(u,vw)}(u, vw).$$

If possible suppose that $\psi(u, vw) > ICONN^*_{\widetilde{G}\setminus(u,vw)}(u, vw)$. Then there exists an influence path say $\widetilde{P}$ whose influence strength is greater than $\psi(u, vw)$ and for every influence pair (s, t) of $\widetilde{P}$, $\psi(s, t) > \psi(u, vw)$ and we can find an influence cycle with the pair (u, vw) having least ψ value. So the removal of pair (u, vw) does not affect the influence connectivity between any $s, t \in \sigma^* \cup \mu^*$ and we get a contradiction to the fact that (u, vw) is an influence cut pair. ∎

Theorem 3.6.16 *A complete fuzzy influence graph has no influence cutnodes.*

Proof If possible suppose that u is an influence cutnode in the fuzzy influence graph $\widetilde{G} = (\sigma, \mu, \Psi)$. By definition,

Fig. 3.13 Fuzzy influence tree

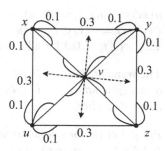

$$ICONN^*_{\widetilde{G}\setminus u}(s, t) < ICONN^*_{\widetilde{G}}(s, t), \ \ s, t \in \sigma^* \cup \mu^*.$$

That is, every strongest $s - t$ influence path has to pass through u. Being a complete fuzzy influence, the paths $s, (s, t), st, (t, st), t$ is a strongest $s - t$ influence path when $s, t \in \sigma^*$, $s, (s, t), t$ is a strongest $s - t$ influence path when $s \in \sigma^*, t \in \mu^*$ and $s, (x, s), x, (x, t), t$ is an $s - t$ influence path when $s, t \in \mu^*$. Here x is the element of $\sigma^* \setminus \{u\}$ influencing the element t. Note that $\widetilde{G} \setminus u$ also will be a complete fuzzy influence. None of the above mentioned influence paths contains $u \in \sigma^*$ and hence u cannot be an influence cutnode of $\widetilde{G}$. ∎

Definition 3.6.17 A fuzzy influence graph is said to be a **fuzzy influence block** if it has no influence cutnodes.

By Theorem 3.6.16, it follows that all complete fuzzy influences are influence blocks. Figure 3.9a is an influence block.

As in fuzzy incidence graphs, we can characterize blocks in fuzzy influence graphs by the following theorem. Because the proof is similar to that of incidence blocks, only statement is given below.

Theorem 3.6.18 Let $\widetilde{G} = (\sigma, \mu, \Psi)$ be a fuzzy influence graph. $\widetilde{G}$ is a fuzzy incidence block if and only if any two elements of $\sigma^* \cup \mu^*$, where (s, t) is not an influence cutpair are joined by two internally disjoint strongest influence paths.

Definition 3.6.19 Let $\widetilde{G} = (\sigma, \mu, \Psi)$ be a fuzzy influence graph. $\widetilde{G}$ is said to be an **influence tree** if it has an influence subgraph $\widetilde{F} = (\sigma, \tau, \eta)$ such that $\widetilde{F}^*$ is a tree and for all pairs $(u, vw) \in \psi^* \setminus \eta^*$, $ICONN^*_{\widetilde{F}}(u, vw) > \psi(u, vw)$.

Because incidence edges are trivial influence edges, every incidence tree is trivially an influence tree. Treating elements of σ^* and μ^* as vertices and pairs as edges, we can obtain a fuzzy graph realization $\widetilde{F}'$ of the incidence tree $\widetilde{F}$ in the definition. It can be observed that when $\widetilde{G}$ is an influence tree, $\widetilde{F}'^*$ is a tree. Also, the influence constraint in the definition shows that this tree is unique. Here is an example for a non trivial influence tree (Fig. 3.13).

Example 3.6.20 Consider the fuzzy influence graph $\widetilde{G} = (\sigma, \mu, \Psi)$ given in Fig. 3.13. Let $\sigma(s) = 1$ for every $s \in \sigma^*$, $\mu(vy) = \mu(vx) = \mu(vz) = \mu(vu) = 0.9$,

$\mu(xy) = \mu(yz) = \mu(zu) = \mu(ux) = 0.3$, $\psi(x, xy) = \psi(x, xu) = \psi(z, zy) = \psi(y, ux) = \psi(y, yz) = \psi(u, ux) = 0.1$. On the eighth cycle note that $\psi(z, zu) = 0$. Take $\psi(s, t) = 0.3$ for all other influence pairs. Then $\widetilde{G}$ is a fuzzy influence tree.

Theorem 3.6.21 *Let $\widetilde{G} = (\sigma, \mu, \Psi)$ be a fuzzy influence tree. A pair $(u, vw) \in \psi^*$ is a strongest influence pair if (u, vw) is a pair of its incidence subgraph $\widetilde{F} = (\sigma, \tau, \eta)$ in the definition.*

Proof Let (u, vw) be a strongest influence pair in $\widetilde{G}$. By definition, $\psi(u, vw) > ICONN^*_{\widetilde{G}\setminus(u,vw)}(u, vw)$. If possible assume that $(u, vw) \in \psi^*\setminus\eta^*$. Then,

$$ICONN^*_{\widetilde{F}\setminus(u,vw)}(u, vw) = ICONN^*_{\widetilde{F}}(u, vw) > \psi(u, vw).$$

But being a subgraph, $ICONN^*_{\widetilde{F}\setminus(u,vw)}(u, vw) \le ICONN^*_{\widetilde{G}\setminus(u,vw)}(u, vw)$ and

$$ICONN^*_{\widetilde{G}\setminus(u,vw)}(u, vw) > \psi(u, vw),$$

which is a contradiction. ∎

3.7 Social Influence

Suppose we have a set of n players. The players may be persons or states or regions for example. We let t denote units of time, $t = 0, 1, 2, \ldots$. The following model describes a method by which players weigh and integrate their own attitudes and the attitudes of others on an issue, [6]:

$$y_j^{(t+1)} = a_{jj} \sum_{k=1}^{n} w_{jk} y_k^{(t)} + (1 - a_{jj}) y_j^{(0)},$$

$j = 1, 2, \ldots, n$ and $t = 0, 1, \ldots$, where the n group members' time position on an issue at time t are $y_1^{(t)}, \ldots, y_n^{(t)}$ and where their initial position is given at $t = 0$. Group member's individual susceptibilities to interpersonal influences are $a_{jj}, j = 1, 2, \ldots, n$. The relative interpersonal influence of each group member k on j is w_{jk}, where $0 \le w_{jk} \le 1$ $(j, k = 1, \ldots, n, j \ne k)$, $w_{jj} = 1 - a_{jj}, j = 1, \ldots, n$ and $\sum_{k=1}^{n} w_{jk} = 1$ for all j. It follows that for each j, $\sum_{k=1}^{n} a_{jj} w_{jk} + (1 - a_{jj}) = 1$ so that j's attitude at time $t + 1$ is formed as a weighted average of the attitudes of others and self at time t and j's initial attitude.

Proposition 3.7.1 *Suppose $\exists a \in [0, 1]$ such that for all j, $a_{jj} = a$. Suppose also that $\sum_{j=1}^{n} w_{jk} = 1$ for $j = 1, \ldots, n$. Then $\sum_{k=1}^{n} y_k^{(t+1)} = \sum_{k=1}^{n} y_k^{(0)}$ for $t = 0, 1, \ldots$.*

Proof Suppose that $\sum_{k=1}^{n} y_k^{(t)} = \sum_{k=1}^{n} y_k^{(0)}$, the induction hypothesis. We have that

$$y_j^{(t+1)} = a \sum_{k=1}^{n} w_{jk} y_k^{(t)} + (1-a) y_j^{(0)}$$

and so

$$\sum_{j=1}^{n} y_j^{(t+1)} = \sum_{j=1}^{n} a \sum_{k=1}^{n} w_{jk} y_k^{(t)} + \sum_{j=1}^{n} (1-a) y_j^{(0)}$$

$$= a \sum_{j=1}^{n} \sum_{k=1}^{n} w_{jk} y_k^{(t)} + (1-a) \sum_{j=1}^{n} y_j^{(0)}$$

$$= a \sum_{k=1}^{n} \sum_{j=1}^{n} w_{jk} y_k^{(t)} + (1-a) \sum_{j=1}^{n} y_j^{(0)}$$

$$= a \sum_{k=1}^{n} (1) y_k^{(t)} + (1-a) \sum_{j=1}^{n} y_j^{(0)}$$

$$= a \sum_{k=1}^{n} y_k^{(0)} + (1-a) \sum_{j=1}^{n} y_j^{(0)} \quad \text{the induction hypothesis}$$

$$= \sum_{k=1}^{n} y_k^{(0)}.$$

∎

Corollary 3.7.2 *Suppose that* $w_{jk} = \frac{1}{n}$ *for* $j, k = 1, \ldots, n$. *Then* $\sum_{k=1}^{n} \frac{1}{n} y_k^{(t+1)} = \sum_{k=1}^{n} \frac{1}{n} y_k^{(0)}$, $t = 0, 1, \ldots$.

Proof The proof follows from the Proposition 3.7.1 with $a = \frac{n-1}{n}$ because $1 - a = \frac{1}{n} = w_{jj}$. ∎

Corollary 3.7.3 *Suppose that* $w_{jk} = \frac{1}{n}$ *for* $j, k = 1, \ldots, n$. *Then* $y_k^{(t)} = y_k^{(1)}$, $k = 1, \ldots, n$, $t = 1, 2, \ldots$.

Proof $y_j^{(1)} = \left(1 - \frac{1}{n}\right) \sum_{k=1}^{n} \frac{1}{n} y_k^{(0)} + \frac{1}{n} y_j^{(0)} = \left(1 - \frac{1}{n}\right) \sum_{k=1}^{n} \frac{1}{n} y_k^{(t-1)} + \frac{1}{n} y_j^{(0)} = y_j^{(t)}$. ∎

Example 3.7.4 Let $n = 3$ and let W denote the 3×3 matrix $[w_{jk}]$, where $W = \begin{bmatrix} 1/3 & 1/2 & 1/6 \\ 1/6 & 1/3 & 1/2 \\ 1/2 & 1/6 & 1/3 \end{bmatrix}$

Let $y_1^{(0)} = \frac{3}{4}$, $y_2^{(0)} = \frac{1}{2}$, and $y_3^{(0)} = \frac{1}{4}$. Then

$$y_1^{(1)} = \frac{2}{3}\left(\frac{1}{3}\frac{3}{4} + \frac{1}{2}\frac{1}{2} + \frac{1}{6}\frac{1}{4}\right) + \frac{1}{3}\frac{3}{4} = \frac{2}{3}\frac{13}{24} + \frac{1}{4} = \frac{11}{18},$$

$$y_2^{(1)} = \frac{2}{3}\left(\frac{1}{6}\frac{3}{4} + \frac{1}{3}\frac{1}{2} + \frac{1}{2}\frac{1}{4}\right) + \frac{1}{3}\frac{1}{2} = \frac{2}{3}\frac{5}{12} + \frac{1}{6} = \frac{8}{18},$$

$$y_3^{(1)} = \frac{2}{3}\left(\frac{1}{2}\frac{3}{4} + \frac{1}{6}\frac{1}{2} + \frac{1}{3}\frac{1}{4}\right) + \frac{1}{3}\frac{1}{4} = \frac{2}{3}\frac{13}{24} + \frac{1}{12} = \frac{8}{18}.$$

We see that $\sum_{j=1}^{3} y_j^{(0)} = \frac{1}{2} = \sum_{j=1}^{3} y_j^{(1)}$, yet $y_j^{(1)} \neq y_j^{(0)}$ for $j = 1, 2, 3$. However, by Corollary 3.7.3, we will have $y_j^{(1)} = y_j^{(t)}$, $t = 1, 2, \ldots$.

Suppose we have n countries, $C_1, \ldots, C_n$. There are four main vulnerabilities for a country to be involved in human trafficking, V_1, V_2, V_3, and V_4. Let $M^{(0)}$ be the $4 \times n$ matrix $[y_{ij}^{(0)}]$, where $y_{ij}^{(0)}$ denotes C_j's vulnerability for V_i. One country may try to influence another with respect to human trafficking. For example, one country may place certain sanctions on another to encourage it to pass certain anti-trafficking laws. Fix i, i.e., focus on V_i. We have by the above that under certain assumptions the row average of the ith row of $M^{(t)}$ remains the same over time. Without these assumptions we might try to determine if the $M^{(t)}$ converge over time.

Let $\mu_j^{(t+1)}(F_i)$ denote the degree of importance that expert j places on factor $i, j = 1, \ldots, n$; $i = 1, \ldots, m$, at time t, where $0 \leq \mu_j^{(t)} \leq 1$. Suppose $\mu_j^{(0)}$ is given and

$$\mu_j^{(t+1)}(F_i) = a_{ii} \sum_{k=1}^{n} w_{ik} \mu_k^{(t)}(F_i) + (1 - a_{ii})\mu_j^{(0)}(F_i),$$

where $j = 1, \ldots, n$; $i = 1, \ldots, m$.

Let $C_j^{(0)} = \sum_{i=1}^{m} \mu_j^{(0)}(F_i), j = 1, \ldots, n$, i.e., $C_j^{(0)}$ is sum of the elements of the jth column at time $t = 0, j = 1, \ldots, n$.

Theorem 3.7.5 Let $a_{jj} = 1 - \frac{1}{n}, j = 1, \ldots, n$, and $w_{ij} = \frac{1}{n}, i = 1, \ldots, m; j = 1, \ldots, n$. Suppose that $C_j^{(0)} = C_k^{(0)}, j, k = 1, \ldots, n$. Then $C_j^{(t)} = C_j^{(0)}$ for $j = 1, \ldots, n$; $t = 0, 1, \ldots$.

Proof Suppose $C_j^{(t)} = C_j^{(0)}$ for $j = 1, \ldots, n$; $t \geq 0$, the induction hypothesis. Then

$$C_j^{(t+1)} = \sum_{i=1}^{m} \mu_j^{(t+1)}(F_i) = \sum_{i=1}^{m} \frac{n-1}{n}\frac{1}{n} \sum_{k=1}^{n} \mu_k^{(t)}(F_i) + \frac{1}{n} \sum_{i=1}^{m} \mu_j^{(0)}(F_i)$$

$$= \frac{n-1}{n^2} \sum_{k=1}^{n} \sum_{i=1}^{m} \mu_k^{(t)}(F_i) + \frac{1}{n} C_j^{(0)}$$

$$= \frac{n-1}{n^2} \sum_{k=1}^{n} C_k^{(t)} + \frac{1}{n} C_j^{(0)} = \frac{n-1}{n^2} n C_j^{(t)} + \frac{1}{n} C_j^{(0)}$$

$$= \frac{n-1}{n} C_j^{(0)} + \frac{1}{n} C_j^{(0)} = C_j^{(0)}.$$

∎

Corollary 3.7.6 *Let* $a_{jj} = 1 - \frac{1}{n}$, $j = 1, \ldots, n$, *and* $w_{ij} = \frac{1}{n}$, $i = 1, \ldots, m$; $j = 1, \ldots, n$. *Suppose that* $C_j^{(0)} = 1$. *Then* $C_j^{(t)} = 1$ *for* $t = 0, 1, \ldots$.

Hence, we see that if $a_{jj} = 1 - \frac{1}{n}$, $j = 1, \ldots, n$, and $w_{ij} = \frac{1}{n}$, $i = 1, \ldots, m$; $j = 1, \ldots, n$, then a Guiasu table remains a Guiasu table for $t = 0, 1, \ldots$. Also, it is known that if the columns of the AHP table add to the same number, then the AHP equation and the Guiasu equation are the same, [16, p. 127].

Lemma 3.7.7 *Let* n *be a positive integer and* $y_j \in \mathbb{R}$, $j = 1, \ldots, n$. *Then*

$$\sum_{j=1}^{n} y_j = \sum_{j=1}^{n} \left[\frac{n-1}{n} \sum_{k=1}^{n} \frac{1}{n} y_k + \frac{1}{n} y_j \right].$$

Proof We have

$$\sum_{j=1}^{n} \left[\frac{n-1}{n} \sum_{k=1}^{n} \frac{1}{n} y_k + \frac{1}{n} y_j \right] = \frac{n-1}{n} \sum_{j=1}^{n} \sum_{k=1}^{n} \frac{1}{n} y_k + \sum_{j=1}^{n} \frac{1}{n} y_j$$

$$= \frac{n-1}{n} n \sum_{k=1}^{n} \frac{1}{n} y_k + \sum_{j=1}^{n} \frac{1}{n} y_j$$

$$= \frac{n-1}{n} \sum_{k=1}^{n} y_k + \frac{1}{n} \sum_{j=1}^{n} y_j$$

$$= \sum_{j=1}^{n} y_j.$$

∎

Theorem 3.7.8 *Let* $a_{jj} = 1 - \frac{1}{n}$, $j = 1, \ldots, n$, *and* $w_{ij} = \frac{1}{n}$, $i = 1, \ldots, m$; $j = 1, \ldots, n$. *Then*

$$\sum_{j=1}^{n} \mu_j^{(t)}(F_i) = \sum_{j=1}^{n} \mu_j^{(0)}(F_i)$$

for $t = 0, 1, \ldots$.

Proof Clearly the result is true for $t = 1$, Assume the result is true for t, i.e., $\sum_{j=1}^{n} \mu_j^{(t)}(F_i) = \sum_{j=1}^{n} \mu_j^{(0)}(F_i)$, the induction hypothesis. Then

$$\sum_{j=1}^{n} \mu_j^{(t+1)}(F_i) = \sum_{j=1}^{n} \left[\frac{n-1}{n} \sum_{k=1}^{n} \frac{1}{n} \mu_k^{(t)}(F_i) + \frac{1}{n} \mu_j^{(0)}(F_i) \right]$$

$$= \sum_{j=1}^{n} \left[\frac{n-1}{n} \sum_{k=1}^{n} \frac{1}{n} \mu_k^{(0)}(F_i) + \frac{1}{n} \mu_j^{(0)}(F_i) \right]$$

$$= \sum_{j=1}^{n} \mu_j^{(0)}(F_i)$$

by the induction hypothesis and the Lemma 3.7.7. ∎

The theorem says that even though the $\mu_j^{(0)}$ may change with t, the row averages remain the same. Hence, the AHP equation remains the same as t varies. (The Guiasu equation also remains the same.)

The next result shows that after $t = 0$, the AHP table remains the same.

Corollary 3.7.9 *Let $a_{jj} = 1 - \frac{1}{n}$, $j = 1, \ldots, n$, and $w_{ij} = \frac{1}{n}$, $i = 1, \ldots, m$; $j = 1, \ldots, n$. Then $\mu_j^{(t)}(F_i) = \mu_j^{(1)}(F_i)$ for $t = 1, 2, \ldots$, $i = 1, \ldots, m$; $j = 1, \ldots, n$.*

Proof

$$\mu_j^{(1)}(F_i) = \left(1 - \frac{1}{n} \right) \sum_{k=1}^{n} \frac{1}{n} \mu_k^{(0)}(F_i) + \frac{1}{n} \mu_j^{(0)}(F_i)$$

$$= \left(1 - \frac{1}{n} \right) \sum_{k=1}^{n} \frac{1}{n} \mu_k^{(t)}(F_i) + \frac{1}{n} \mu_j^{(0)}(F_i) \text{(by Theorem 3.7.8)}$$

$$= \mu_j^{(t+1)}(F_i)$$

for $t = 1, 2,$ ∎

Example 3.7.10 In this example, players 1 and 2 change player's preference of x over y. Even if player 1 is a dictator, his choice can be influenced. Let $X = \{x, y\}$, $N = \{1, 2, 3\}$ and let $\rho_1^{(0)}, \rho_2^{(0)}, \rho_3^{(0)}$ be fuzzy relations on x at time $t = 0$. Let $a_{11} = 0.8$ so $w_{11} = 0.2$. Let $w_{12} = w_{13} = 0.4$. Suppose that $\rho_1^{(0)}(x, y) = 1, \rho_1^{(0)}(y, x) = 0, \rho_2^{(0)}(x, y) = 0, \rho_2^{(0)}(y, x) = 1, \rho_3^{(0)}(x, y) = 0, \rho_3^{(0)}(y, x) = 1$. Then

$$\rho_1^{(1)}(x, y) = 0.8(0.2(1) + 0 + 0) + 0.2(1) = 0.36$$
$$\rho_1^{(1)}(y, x) = 0.8((0 + 0.4(1) + 0.4(1)) + 0.2(0) = 0.64.$$

Example 3.7.11 The $t = 0$ and the $t = 1$ AHP tables below are different, even though the row averages remain the same.

	$y = 0$	E_1	E_2	Row Avg
AHP	F_1	9.5	0.4	9.9/2
	F_2	0.5	0.6	1.1/2
	Col Total			11/2

	$t = 0$	E_1	E_2	Row Avg
Guiasu	F_1	0.95	0.4	1.35/2
	F_2	0.05	0.6	0.65/2
	Col Total			1

where the Guiasu table for $t = 0$ is obtained from the $t = 0$ AHP table.

Let $w_{ij} = \frac{1}{2}$, $i, j = 1, 2$. Then using the formula

$$\mu^{(1)}(F_i) = \frac{1}{2}\left(\sum_{j=1}^{2} \frac{1}{2}\mu^{(0)}(F_i)\right) + \frac{1}{2}\mu^{(0)}(F_i), \ i = 1, 2$$

for both $t = 0$ tables, we have

	$y = 0$	E_1	E_2	Row Avg
AHP	F_1	7.225	2.675	9.9/2
	F_2	0.525	0.575	1.2/2
	Col Total	7.750	3.250	11/2

	$t = 0$	E_1	E_2	Row Avg
Guiasu	F_1	0.8125	0.5375	0.675
	F_2	0.1875	0.4625	0.325
	Col Total	1	1	1

Consider the AHP table at time $t = 1$. If we normalize the columns of this table, we obtain

$t = 1$	E_1	E_2	Row Avg
F_1	0.932	0.823	0.8775
F_2	0.68	0.177	0.1225
Col Total	1	1	1

We see that we obtain two different Guiasu tables for $t = 1$.

3.8   Fuzzy Incidence Blocks and Illegal Migration

Incidence blocks in fuzzy incidence graphs are studied in this section. The results are based on [11]. The figures are from [11]. Incidence relations play very important roles in several natural and human made networks like road, pipe, power and data flow networks. Incidence blocks are characterized using strongest strong incidence paths.

Violence against illegal border crossers occurs regularly around land and sea borders. Criminal acts committed against illegal immigrants include kidnapping, robbery, extortion, sexual violence, and death at the hands of cartels, smugglers, and even corrupt government officials. Individuals also die due to heat exhaustion, dehydration, and drowning. As a government attempts to clamp down on narcotics operations and other criminal activities, criminal organizations turn to other sources of income such as human smuggling and sex trafficking. Much of the discussion to follow is based on [11, 20].

"We focus our discussion on illegal routes from Latin America to the United States. Of the 400,235 individuals that Mexico's National Immigration Institute (INM) estimates enter Mexico every year illegally, approximately 150,000 or 37% intend to cross over in to the United States. At Mexico's southern border there is a dangerous journey of some 2000 miles to the United States. In 2009, along the deadliest areas of the border, such as Arizona's Sonoran Desert, the risk of death for illegal border-crossers was one and a half times greater than it had been in 2004, and 17 times greater than it had been in 1998. Not only is heat a danger, but several varieties of rattle snakes, tarantulas, Gila monsters, and killer bees inhabit the desert. It is estimated that 400 kidnapping involving some 22,000 individuals may be occurring in Mexico on an annual basis. Not only are criminal organization involved in kidnapping of illegal immigrants, but state, local, or federal authorities were also responsible for or complicit in 9% of the kidnappings.

Those who attempt to cross dangerous parts of the Mexican-U.S border face other dangers. One particular dangerous journeys is riding freight trains. Many illegal immigrants ride on the top of train cars to avoid detection. They not only risk falling asleep or losing their grip and rolling off, but also being knocked off by tunnels or passing branches. Similar dangers occur when they stow away by clinging to couplings or shock absorbers between cars and axles. Many are knocked off or injured by rocks that are kicked up under the train, or by falling asleep. Many are killed or lose limbs when they fall onto the tracks. When people hide in sealed train cars they risk dying from heat or suffocation. Illegal immigrants also face risk when they decide to journey to the U.S. by sea. Rafts, small boats, styrofoam vessels and inner tubes make up the vessel of choice by a large number of illegal immigrants. These vessels present many dangers such as capsizing and drowning, dehydration, exhaustion, and lack of food. Train routes taken by illegal immigrants can be found in [19]. Routes through Central America to the U.S. including routes by sea can be found in [21]. Routes into Europe can be found in [9].

Illegal immigrants from around the world migrate to South America to travel through Mexico to the United States. For example, immigrants from Delhi, India travel to Istanbul to the jungles of Guatemala (See [7]). A Spanish-language news report states that over 700 Africans have illegally traveled through Panama on their way to the United States this year."

The concept of a fuzzy incidence block will be discussed in this section. According to Definition 3.2.3, a node $w \in \sigma^*$ is called an incidence cutvertex if $ICONN_{\widetilde{G} \setminus \{w\}}(u, uv) < ICONN_{\widetilde{G}}(u, uv)$ for some $(u, uv) \in \Psi^*$ such that $u \neq w \neq v$. Thus, we have the definition of fuzzy incidence block as follows.

Definition 3.8.1 Let $\widetilde{G} = (\sigma, \mu, \Psi)$ be a fuzzy incidence graph. $\widetilde{G}$ is said to be a **fuzzy incidence block** if $\widetilde{G}$ has no incidence cutvertices.

Definition 3.8.1 is illustrated in the following example.

Example 3.8.2 Consider the following fuzzy incidence graphs with $|\sigma^*| = 3$ given in Fig. 3.14.

In Fig. 3.14a, all arcs are strong. Indeed, ab and bc are β-strong whereas ac is α-strong. Also, all the pairs in Fig. 3.14a are strong. In Fig. 3.14b, arc ac is α-strong and other arcs are β-strong. All the pairs are strong. The node c in Fig. 3.14a is an incidence cutnode. But Fig. 3.14b is an incidence block.

There are different types of pairs in an IFG. We have the following definition.

Definition 3.8.3 Let $\widetilde{G} = (\sigma, \mu, \Psi)$ be a fuzzy incidence graph. Then (x, xy) is called an **incidence bond** if

$$ICONN_{\widetilde{G} \setminus (x,xy)}(u, uv) < ICONN_{\widetilde{G}}(u, uv)$$

for some pair (u, uv) in $\widetilde{G}$ such that either u is different from x or uv is different from xy. (x, xy) is called an **incidence cutbond** if

$$ICONN_{\widetilde{G} \setminus (x,xy)}(u, uv) < ICONN_{\widetilde{G}}(u, uv)$$

for some pair (u, uv) in $\widetilde{G}$ such that u is different from x and uv is different from xy.

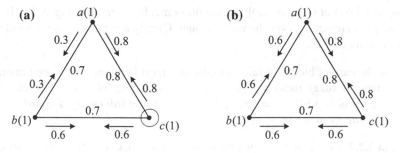

Fig. 3.14 **a** FIG with incidence cut vertex, **b** FIG incidence block

Clearly in an incidence block, no two incidence cutpairs share a common node. However, the incidence cutpairs of incidence blocks are special as seen from the following proposition.

Proposition 3.8.4 *Let $\widetilde{G} = (\sigma, \mu, \Psi)$ be a fuzzy incidence graph with at least one incidence cutpair. Then the removal of any incidence cutpair (u, uv) reduces the incidence strength between u and uv alone. That is, an incidence block has no incidence bonds or incidence cutbonds.*

Proof Let $\widetilde{G} = (\sigma, \mu, \Psi)$ be a fuzzy incidence graph and (u, uv) be an incidence cutpair of $\widetilde{G}$. If possible suppose that the removal of (u, uc) reduces the incidence strength between x and yz, where $z \neq u$ and $yz \neq uv$. Then every strongest $x - xy$ path passes through the pair (u, uv). Hence, it follows that removal of u from $\widetilde{G}$ reduces the incidence strength between x and yz, a contradiction to the assumption that $\widetilde{G}$ is an incidence block.

Now if the removal of (u, uv) reduces the incidence strength between x and uv in $\widetilde{G}$, where $x \neq u$, then also every strongest $x - uv$ incidence path passes through u which becomes a fuzzy incidence cutnode, a contradiction.

If the removal of (u, uv) reduces the incidence strength between u and xy in $\widetilde{G}$, where $xy \neq uv$, then v becomes a fuzzy incidence cutnode, which also is not possible. ∎

Proposition 3.8.5 *Let $\widetilde{G} = (\sigma, \mu, \Psi)$ be a fuzzy incidence graph. A pair (x, xy) such that $\Psi(x, xy) = \vee\{\Psi(u, uv) \mid (u, uv) \in \Psi^*\}$ is a strong pair.*

Proof The incidence strength of any $x - xy$ path in $\widetilde{G}$ will be less than or equal to $\Psi(x, xy)$. If (x, xy) is the unique pair with $\Psi(x, xy) = \vee\{\Psi(u, uv) \mid (u, uv) \in \Psi^*\}$, then every other $u - uv$ path in $\widetilde{G}$ will have incidence strength less than $\Psi(x, xy)$ and hence $CONN_{\widetilde{G}\backslash(x,xy)}(x, xy) < \Psi(x, xy)$. Thus, (x, xy) is an α-strong pair. If (x, xy) is not unique, then the maximum possible value for the incidence strength of a path in $\widetilde{G}\backslash(x, xy)$ will be $\Psi(x, xy)$. Thus, there exists an $x - xy$ path in $\widetilde{G}\backslash(x, xy)$ with incidence strength $\Psi(x, xy)$. Hence, (x, xy) is β-strong, otherwise it is α-strong. ∎

The converse of Proposition 3.8.5 is not true as can be seen by the following example.

Example 3.8.6 Let $\widetilde{G} = (\sigma, \mu, \Psi)$ be the fuzzy incidence graph in Fig. 3.14a. In this FIG, all pairs are strong, even the weakest ones. Clearly it is a strong fuzzy incidence cycle. But not an FIB.

As in the case of blocks in fuzzy graphs, strongest incidence paths are important in the study of fuzzy incidence blocks as seen from the following theorem. Two incidence paths in a FIG $\widetilde{G} = (\sigma, \mu, \Psi)$ are said to be **internally disjoint**, if they have no elements of σ^* or μ^* in common.

Lemma 3.8.7 *Let $\widetilde{G} = (\sigma, \mu, \Psi)$ be an incidence block. Let $\widetilde{M} = (\tau, \sigma, \nu)$ be a maximum spanning tree of $\widetilde{G}$. Then each pair of $\widetilde{M}$ is a strong pair.*

The following result provides equivalent necessary conditions for a fuzzy incidence graph to be an incidence block.

Theorem 3.8.8 *Let* $\widetilde{G} = (\sigma, \mu, \Psi)$ *be an incidence block. Then the following conditions hold and are equivalent.*

(i) *Any two elements* $s, t \in \sigma^* \cup \mu^*$ *lie on a common strong incidence cycle.*

(ii) *Any element* $s \in \sigma^* \cup \mu^*$ *and a string pair of* $\widetilde{G}$ *lie on a common strong incidence cycle.*

(iii) *Any two strong pairs of* $\widetilde{G}$ *lie on a common strong incidence cycle.*

(iv) *For any two given* $s, t \in \sigma^* \cup \mu^*$ *and a strong pair in* $\widetilde{G}$, *there exists a strong incidence path joining s and t containing the pair.*

(v) *For every three distinct elements* $s, t, r \in \sigma^* \cup \mu^*$ *there exist strong incidence paths joining any two of them containing the third.*

(vi) *For every three distinct elements* $s, t, r \in \sigma^* \cup \mu^*$ *there exist strong paths joining any two of them which does not contain the third.*

Proof (i) Suppose that $\widetilde{G}$ is FIB. By Lemma 3.8.7, there always exists a strong path between s and t. Let $s, t \in \sigma^* \cup \mu^*$ be such that there exists only one strong incidence path between s and t, say P. Two cases arise.

1. P has exactly one pair which is strong. That is, P is of the form $s, (s, t), t$ where $t \in \mu^*$ and $s \in \sigma^*$.

2. There exists an $s - t$ incidence path of length more than one in $\widetilde{G}$. That is, P will have more than one strong pair.

Suppose that 1 holds, i.e., P is of the form $s, (s, t), t$. Because (s, t) is not on any strong incidence cycle, pair (s, t) is in every maximum spanning tree of $\widetilde{G}$ and hence it is an incidence cutpair. If s is an end node in all maximum spanning trees of $\widetilde{G}$, then clearly it is a fuzzy incidence end node of $\widetilde{G}$ and hence the other end node of t is an incidence cutnode of $\widetilde{G}$ and vice versa, contradicting our assumption that $\widetilde{G}$ is an incidence block.

Now suppose that s is an end node in some maximal spanning tree T_1 and the other end of the arc t say w is an end node in some other maximal spanning tree T_2. Let u' be a strong incidence neighbor of s in T_2. Because s is an end node and w is an internal node in T_1, there exists a strong incidence path P in T_1 from u to u' passing through w. The incidence path P together with the strong arc su' forms a strong incidence cycle in $\widetilde{G}$, a contradiction.

Suppose that 2 holds, i.e., there exists an $s - t$ incidence path of length more than one in $\widetilde{G}$. Because there is an unique strong incidence $s - t$ path P in $\widetilde{G}$, P belongs to all maximum spanning incidence trees. Thus, all internal nodes in P are internal nodes in all the maximum spanning incidence trees and hence all of them are fuzzy incidence cutnodes in $\widetilde{G}$, a contradiction to the assumption that $\widetilde{G}$ is an FIB. Thus, condition (1) holds in an incidence block. Next we show that each of the given conditions are equivalent.

(i) $\Rightarrow$ (ii): Suppose that any pair of elements $s, t \in \sigma^* \cup \mu^*$ lie on a strong incidence cycle. To prove that any element $s \in \sigma^* \cup \mu^*$ and a strong pair (u, uv) of $\widetilde{G}$ lie on a common strong incidence cycle. Let C be a strong incidence cycle containing a and arc uv. If (u, uv) is in C, then there is nothing to prove. Now suppose that (u, uv) is a pair not in C. Let C_1 be a strong cycle containing s and uv. Let P_1 and P_2 be the strong $s - u$ incidence paths in C and P_1' and P_2' the strong $s - uv$ incidence paths in C_1. Let x_1 be the node at which P_1' leaves P_1. Then clearly $s \ldots (P_1) \ldots x_1 \ldots (P_1') \ldots (u, uv) \ldots (P_2)s$ is a strong cycle containing s and (u, uv). If $x = s$, then $s \ldots (P_1') \ldots (u, uv) \ldots (P_2') \ldots s$ is the required cycle. In case $x_1 = u$, let x_2 be the node at which P_2' leaves P_2. Then $s \ldots (P_1) \ldots (u, uv) \ldots (P_2') \ldots x_2 \ldots (P_2)s$ is the required strong cycle. If $x_2 = s$, then $s \ldots (P_2') \ldots (u, uv) \ldots (P_1) \ldots s$ is the required strong incidence cycle. Because P_1 and P_2 are internally disjoint, both x_1 and x_2 cannot be the same as u.

$(ii) \Rightarrow (iii)$: Suppose that each $s \in \sigma^* \cup \mu^*$ and strong pair of $\widetilde{G}$ lie on a common strong cycle. To prove that any two strong pairs lie on a common strong incidence cycle. Let (u, uv) and (x, xy) be two strong pairs. of $\widetilde{G}$. Let P_1 and P_2 be two internally disjoint strong incidence paths between u and x and Q_1 and Q_2 be two internally disjoint strong incidence paths between uv and xy. If P_1, P_2, Q_1, and Q_2 are internally disjoint, then $u(u, uv)uv \ldots (P_1) \ldots xy(x, xy) \ldots (Q_2) \ldots u$ is a strong incidence cycle containing (u, uv) and (x, xy). If Q_1 and Q_2 intersect P_1 or P_2, then a strong incidence cycle containing (u, uv) and (x, xy) can be extracted from the parts of the four paths P_1, P_2, Q_1, and Q_2.

$(iii) \Rightarrow (iv)$: Consider any two $s, t \in \sigma^* \cup \mu^*$ and a strong pair (u, uv) in $\widetilde{G}$. Let (s, sx') be a strong pair at s and (t, ty') a strong pair at t. Now there exists a strong cycle C_1 containing (s, sx') and (u, uv) and C_2 a strong cycle containing (t, ty') and (u, uv). Now $s(s, sx') \ldots (C_1) \ldots (u, uv) \ldots (C_2) \ldots (t, ty')y'$ is a strong $s - t$ path containing the pair (u, uv).

$(iv) \Rightarrow (v)$: Let $\widetilde{G}$ be a FIB. Let $s, t, r \in \sigma^* \cup \mu^*$. Let (t, tt') be a strong pair at t. Then s and r are distinct elements of s, r of $\sigma^* \cup \mu^*$ and (t, tt') is a strong pair of $\widetilde{G}$. By (iv), there exists a strong incidence path from s to r containing the pair (t, tt').

$(v) \Rightarrow (vi)$: Let $s, t, r \in \sigma^* \cup \mu^*$. Let P be a strong incidence path from s to r containing t. Then clearly the $s - t$ strong sub incidence path say P' does not contain r.

$(vi) \Rightarrow (i)$: Let $s, t \in \sigma^* \cup \mu^*$. Let r be a third element in $\sigma^* \cup \mu^*$. Let P_1 be the strong incidence path joining s and t not containing r. Let P_2 be the strong incidence path joining s and r not containing t and let P_3 be the strong incidence path joining r and t not containing s. Then $P_1 \cup P_2 \cup P_3$ will contain a strong incidence cycle containing s and t. ∎

Theorem 3.8.9 (A Characterization of incidence blocks in incidence fuzzy graphs) *Let $\widetilde{G} = (\sigma, \mu, \Psi)$ be an incidence fuzzy graph with at least 6 elements in $\sigma^* \cup \mu^*$. Then $\widetilde{G}$ is an incidence block if and only if any two elements s, t of $\sigma^* \cup \mu^*$ such that (s, t) is not a cutpair, there exists an incidence cycle containing the elements s and t which is formed by two strongest strong $s - t$ incidence paths.*

Proof Suppose $\widetilde{G} = (\sigma, \mu, \Psi)$ is an incidence block. Consider a maximum spanning tree $\widetilde{T}$ of $\widetilde{G}$. By Lemma 3.8.7, every pair in an MST is a strong pair. Also, every $s - t$ path in $\widetilde{T}$ is a strongest $s - t$ incidence path in $\widetilde{G}$, where $s, t \in \sigma^* \cup \mu^*$. Thus, between any two elements of $\sigma^* \cup \mu^*$ there exists a strongest incidence path. Let P be a strongest strong incidence $s - t$ path in $\widetilde{G}$. Assume that P is the unique $s - t$ strongest incidence path in $\widetilde{G}$. Then P should belong to all maximum spanning trees of $\widetilde{G}$. Also, note that the length of P is at least two. Then all internal nodes of P are internal nodes of every maximum spanning tree and they are all fuzzy incidence cut nodes, contradicting the fact that $\widetilde{G}$ is a FIB. Thus, it follows that the strongest strong $s - t$ incidence path P does not belong all maximum spanning trees. Hence, there exists a maximum spanning tree T_1 not containing P. Let P_1 be a strongest strong $x - y$ incidence path in T_1. The strongest strong incidence path P_1 together with P gives an incidence cycle in $\widetilde{G}$ containing s and t as required. Note that P and P_1 should be internally disjoint because otherwise the common nodes of P and P_1 become incidence fuzzy cut nodes of $\widetilde{G}$.

Conversely, suppose that for any two elements s, t of $\sigma^* \cup \mu^*$ such that (s, t) is not a cut pair there exists an incidence cycle containing the elements s and t which is formed by two strongest incidence pairs. We have to prove that $\widetilde{G}$ is an FIB. On the contrary assume that w is an incidence cutnode of $\widetilde{G}$. Then $ICONN_{\widetilde{G} \setminus w}(x, y) < ICONN_{\widetilde{G}}(x, y)$ for some $x, y \in \sigma^*$. That is, all strongest $x - y$ incidence paths pass through w which is a contradiction. to the assumption. $\blacksquare$

Definition 3.8.10 Let $\widetilde{G} = (\sigma, \mu, \Psi)$ be a fuzzy incidence cycle. Then $\widetilde{G}$ is said to be a **strong incidence cycle** (SIC) if all its pairs are strong. $\widetilde{G}$ is said to be **locmin** if there exists a pair with minimum Ψ value incident at every element of $\sigma^* \cup \mu^*$. $\widetilde{G}$ is said to be a **strongest strong incidence cycle** (SSIC) if $\widetilde{G}$ can be written as the union of two strongest $u - uv$ incidence paths, provided (u, uv) is not a cutpair of $\widetilde{G}$.

Note that when $\widetilde{G}$ is a FIG and C is a strongest strong incidence cycle in $\widetilde{G}$, (u, uv) can be a cutpair of $\widetilde{G}$. But it cannot be a pair of C.

Example 3.8.11 Consider the following incidence cycle in Fig. 3.15.

Note that the above FIG is a strongest strong incidence cycle, which is both a locamin incidence cycle and a strong incidence cycle. Motivated by this example, we have the following result.

Fig. 3.15 A strongest strong incidence cycle

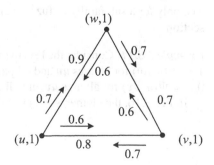

Proposition 3.8.12 *Let $\widetilde{G} = (\sigma, \mu, \Psi)$ be a FIG and let $G = (\sigma, \mu)$ be the corresponding underlying fuzzy graph. If G^* is a cycle, then the following statements are equivalent.*

(i) Let $\widetilde{G} = (\sigma, \mu, \Psi)$ is a FIB;
(ii) Let $\widetilde{G} = (\sigma, \mu, \Psi)$ is SSIC;
(iii) Let $\widetilde{G} = (\sigma, \mu, \Psi)$ is locamin.

Proof $(i) \Rightarrow (ii)$: Suppose that $\widetilde{G}$ is a fuzzy incidence block. Then for any two elements s, t of $\sigma^* \cup \mu^*$ such that (s, t) is not a cutpair, there exists an incidence cycle containing the elements s and t which is formed by two strongest strong $s - t$ incidence paths. Then by definition, $\widetilde{G}$ is an SSIC.

$(ii) \Rightarrow (iii)$: Suppose that $\widetilde{G}$ is an SSIC. If possible suppose that $\widetilde{G}$ is not locamin. Then there exists some node w such that w is not on a weakest pair (w, wv) of $\widetilde{G}$ for any $v \in \sigma^*$. Let (x, xw) and (w, wy) be the two pairs from Ψ^* incident on w which are not weakest pairs. This implies that the path $x(x, xw)w(w, wy)y$ is the unique strongest incidence $x - y$ path in $\widetilde{G}$, a contradiction of the assumption that $\widetilde{G}$ is an SSIC.

$(iii) \Rightarrow (i)$: Suppose that $\widetilde{G}$ is locamin. That is, for any element s of $\sigma^* \cup \mu^*$, there is a weakest pair incident on s. Let $s, t \in \sigma^* \cup \mu^*$. If (s, t) is a weakest pair, then (s, t) and $\widetilde{G} \backslash (s, t)$ are two strongest strong paths joining s and t. If (s, t) is a pair, but not weakest, it becomes a cutpair. If $s, t \in \sigma^* \cup \mu^*$, because $\widetilde{G}$ is a locamin, there exists two strongest $s - t$ incidence paths in $\widetilde{G}$. Hence, it follows that $\widetilde{G}$ is an incidence block. ■

Now we discuss a sufficient condition for an FIG to be an FIB.

Theorem 3.8.13 *Let $\widetilde{G} = (\sigma, \mu, \Psi)$ be a FIG. If any two elements of $\sigma^* \cup \mu^*$ lie on a common SSIC, then $\widetilde{G}$ is a block.*

Proof Let $\widetilde{G} = (\sigma, \mu, \Psi)$ be a FIG satisfying the condition of the theorem. Clearly, $\widetilde{G}$ is connected. Let $w \in \sigma^*$. For any two elements $u, uv \in \sigma^* \cup \mu^*$ such that $w \neq u \neq v$, there exists an SSIC, say C, containing u and uv. If (u, uv) is not strong, there exists two internally disjoint strongest $u - v$ incidence paths in $\widetilde{G}$ belonging to C. Because w is arbitrary, it follows that $\widetilde{G}$ is a block. ■

The converse of Theorem 3.8.13 is not true as seen from the next example. It is true only for a sub family of fuzzy incidence graphs which are discussed in the next section.

Example 3.8.14 Consider the FIG in Fig. 3.16 with $\sigma(u) = 1$ for all $u \in \sigma^*$, $\mu(uv) = 1 = \Psi(u, uv)$ for all arcs uv and all pairs (u, uv) in the outer rectangle and $\mu(uv) = 0.3 = \Psi(u, uv)$ for all arcs uv and all pairs (u, uv) in the inner quadrilateral.

In Fig. 3.16, the elements b and ec do not belong to an SSIC. But $\widetilde{G}$ is a FIB.

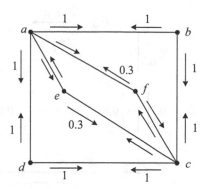

Fig. 3.16 Counter example
for the converse of
Theorem 3.8.13

3.9 Cyclically Strong Incidence Graphs

The incidence strength of an incidence cycle is the minimum of the Ψ values of
its pairs. In the above example, there are strong incidence cycles with two different
incidence strengths passing through the pair $a, c \in \sigma^* \cup \mu^*$. As in [11], we have the
following definition of cycle connectivity in FIGs.

Definition 3.9.1 Let $\widetilde{G} = (\sigma, \mu, \Psi)$ be a fuzzy incidence cycle. The **cycle connectivity** between two elements $s, t \in \sigma^* \cup \mu^*$ is defined as the maximum of the incidence
strengths of strong cycles passing through s and t. It is denoted as $C_{\widetilde{G}}(s, t)$. The cycle
connectivity of $\widetilde{G}$ denoted by $C(\widetilde{G})$ is defined as $\vee \{C_{\widetilde{G}}(s, t) \mid s, t \in \sigma^* \cup \mu^*\}$.

For any FIG, $0 \leq C(\widetilde{G}) \leq 1$. In Example 3.8.14 (Fig. 3.16), $C_{\widetilde{G}}(e, fc) = 0.3$ and
$C_{\widetilde{G}}(a, c) = 1$ and the cycle connectivity of $\widetilde{G}$ equals 1.

Also, the cycle connectivity is a measure of connectedness in a FIG and $C_{\widetilde{G}}(s, t) \leq$
$ICONN_{\widetilde{G}}(s, t)$ for any $s, t \in \sigma^* \cup \mu^*$. Next we have the definition for a new class of
FIGs.

Definition 3.9.2 Let $\widetilde{G} = (\sigma, \mu, \Psi)$ be a fuzzy incidence graph. Then $\widetilde{G}$ is said to
be **cyclically strong fuzzy incidence graph** (CSFIG) if the incidence strength of
each strong incidence cycle passing through every pair $s, t \in \sigma^* \cup \mu^*$ is $C_{\widetilde{G}}(s, t)$.

Example 3.9.3 Consider the FIG in Fig. 3.17.

It follows that for a pair $s, t \in \sigma^* \cup \mu^*$, the strengths of all strong cycles passing
through s and t are the same and are equal to $C_{\widetilde{G}}(s, t)$. Thus, the FIG in Fig. 3.17 is a
CSFIG.

Next we have an obvious characterization for a fuzzy incidence tree (FIG).

Theorem 3.9.4 Let $\widetilde{G} = (\sigma, \mu, \Psi)$ be a fuzzy incidence graph. Then $\widetilde{G}$ is a fuzzy
incidence tree if and only if $C_{\widetilde{G}}(s, t) = 0$ for every pair $s, t \in \sigma^* \cup \mu^*$.

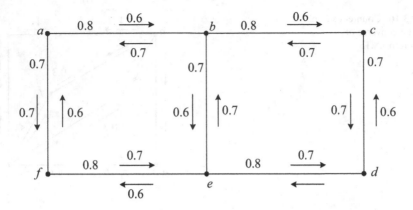

Fig. 3.17 A cyclically strong fuzzy incidence graph

A complete fuzzy incidence graph (CFIG) $\widetilde{G} = (\sigma, \mu, \Psi)$ is such that $\Psi(u, uv) = \sigma(u) \wedge \mu(uv)$ for every $(u, uv) \in \Psi^*$. It is clear that $\Psi(u, uv) = \Psi(v, uv)$ for every $u, v \in \sigma^*$. Note that a complete fuzzy incidence graph has no δ -pairs. Next we have a result related to with the cycle connectivity of a complete fuzzy incidence graph.

Theorem 3.9.5 *Let $\widetilde{G} = (\sigma, \mu, \Psi)$ be a complete fuzzy incidence graph. Then for any $s, t \in \sigma^* \cup \mu^*$, $C_{\widetilde{G}}(s, t) = \vee\{\wedge\{\sigma \cup \mu(s), \sigma \cup \mu(t), \sigma(w)\} \mid w \in \sigma^*\}$.*

Proof Let $\widetilde{G} = (\sigma, \mu, \Psi)$ be a complete fuzzy incidence graph. Because it has no δ-pairs, all pairs of $\widetilde{G}$ are strong and hence all incidence cycles in $\widetilde{G}$ are strong incidence cycles. Now let $s, t \in \sigma^* \cup \mu^*$. Without loss of generality assume that $s \in \sigma^*$ and $t \in \mu^*$. Let C be a strong cycle containing s and t. Let it be

$$C : s = v_1(v_1, v_1v_2)v_1v_2(v_2, v_1v_2)v_2 \ldots v_k(v_k, t)t \ldots s.$$

Clearly, the incidence strength (*is*) of C is $\Psi(v_1, v_1v_2) \wedge \Psi(v_2, v_1v_2) \wedge \ldots$. Because $\widetilde{G}$ is complete, $is(C) = \wedge\{\sigma(v_1), \mu(v_1v_2), \sigma(v_2), \mu(v_2v_3, \sigma(v_3), \ldots\}$. Because $\mu(v_iv_j) \leq \sigma(v_i) \wedge \sigma(v_j)$, it follows that $is(C) = \wedge\{\sigma(v_1), \sigma(v_2), \ldots\}$, where the $v_i \in C$. Thus, it follows that for any $s, t \in \sigma^*\mu^*$, $C_{\widetilde{G}}(s, t) = \vee\{\wedge\{\sigma\cup\mu(s), \sigma\cup\mu(t), \sigma(w)\} \mid w \in \sigma^*\}$ ∎

From the Theorem 3.9.6, it can be seen that the concept of strong incidence paths and strongest incidence paths coincide in a cyclically strong fuzzy graph which is an incidence block.

Theorem 3.9.6 *Let $\widetilde{G} = (\sigma, \mu, \Psi)$ be a cyclically strong fuzzy incidence graph which is a block. Then any $s - t$ incidence path such that (s, t) is not an incidence cutpair is a strongest $s - t$ incidence path and hence any strong incidence cycle in $\widetilde{G}$ is a strongest strong incidence cycle.*

Proof Let $\widetilde{G} = (\sigma, \mu, \Psi)$ be a cyclically strong fuzzy incidence graph which is a block. Let $s, t \in \sigma^* \cup \mu^*$ be such that (s, t) is not an incidence cutpair. Let P be a strong $s - t$ incidence path in $\widetilde{G}$. If P is not a strongest $s - t$ incidence path, then because $\widetilde{G}$ is a cyclically strong fuzzy incidence graph, there exists two internally disjoint strongest strong $s - t$ paths P_1 and P_2. Then $P_1 \cup P_2$ is a strong incidence cycle with incidence strength less than that of the cycle $P_1 \cup P_2$. Both these incidence cycles pass through s and t and hence it contradicts the fact that all incidence cycles passing through s and t are of the same incidence strength. Thus, P must be a strongest strong $u - v$ path.

Let C be a strong incidence cycle in $\widetilde{G}$. Let $s, t \in \sigma^* \cup \mu^*$ be such that (s, t) is not an incidence cutpair. Then by the first part both these $s - t$ incidence paths in C are strongest $u - v$ incidence paths. Thus, $\widetilde{G}$ is a strongest strong cycle. ∎

Thus, from Theorem 3.9.6, any strong incidence cycle can be replaced with a strongest strong incidence cycle in a block in cyclically strong fuzzy incidence graphs. Hence, we have the following characterization for blocks in CSFIGs.

Theorem 3.9.7 *Let* $\widetilde{G} = (\sigma, \mu, \Psi)$ *be a cyclically strong fuzzy incidence graph. Then the following statements are equivalent.*

(i) $\widetilde{G}$ *is a FIB.*

(ii) Every two elements $s, t \in \sigma^* \cup \mu^*$ *lie on a common strongest strong incidence cycle.*

(iii) Each element $s \in \sigma^* \cup \mu^*$ *and a strong pair in* $\widetilde{G}$ *lie on a common strongest strong incidence cycle.*

(iv) Any two strong pairs of $\widetilde{G}$ *lie on a common strongest strong incidence cycle.*

(v) For any two given elements $s, t \in \sigma^* \cup \mu^*$ *such that* (x, xy) *in* $\widetilde{G}$, *there exists a strongest strong* $s - t$ *incidence path containing the pair* (x, xy).

(vi) For every three distinct elements u_i *in* $\sigma^* \cup \mu^*$, $i = 1, 2, 3$, *such that* $u_i u_j$, $i \neq j$, *is not an incidence cutpair, there exist strongest strong incidence paths joining any two of them containing the third.*

(vii) For every three distinct elements u_i *in* $\sigma^* \cup \mu^*$, $i = 1, 2, 3$, *such that* $u_i u_j$, $i \neq j$, *is not an incidence cutpair, there exist strongest strong incidence paths joining any two of them not containing the third.*

Proof Most of the proof follows from Theorems 3.8.8 and 3.9.6.

$(vii) \Rightarrow (i)$ For any element r of $\widetilde{G}$, and for every pair of elements $s, t \in \sigma^* \cup \mu^*$ other than r, there exists a strongest $s - t$ incidence path not containing r. That is, r is not in every strongest $s - t$ path for all pairs of nodes s and t and hence r is not an incidence cut node. Because r is arbitrary, it follows that $\widetilde{G}$ is an FIB. ∎

3.10 Illegal Migration

We continue with results from [11]. Let $\widetilde{G} = (\sigma, \mu, \Psi)$ be a fuzzy incidence graph. For all $x, y \in V$, we let $\Psi(x, xy)$ denote the degree of safety for illegal immigrants to use x as a source, travel on xy, and arrive at destination y. Let $P : u, (u, u_1)$, $uu_1, u_1, (u_1, u_1u_2), \ldots, (u_n, v), (v, vw), w$ denote an incidence path from u to w. Then $\Psi^\infty(u, uw)$ denotes the strongest of all such incidence paths. Hence, such a path is a safest path to travel. Suppose (x, xy) is a fuzzy cutpair. Then there exists (u, uv) such that $\Psi'^\infty(u, uv) < \Psi^\infty(u, uv)$, where Ψ' is determined by the deletion of (x, xy) from $\widetilde{G}$. Thus, the removal of (x, xy) would make the network less safe. Hence, because a fuzzy incidence block has no fuzzy incidence cutpairs, there are no pairs (x, xy) whose removal would decrease the safety of the network.

Let c be an involutive complement of $[0, 1]$, i.e., $c : [0, 1] \to [0, 1]$ be such that $c(0) = 1, c(1) = 0$, for all $a, b \in [0, 1]$, $a \leq b$ implies $c(a) \geq c(b)$, and for all $a \in [0, 1]$, $c(c(a)) = a$. Let Ω be the fuzzy subset of $V \times E$ defined by for all $(x, xy) \in V \times E$, $\Omega(x, xy) = c(\Psi(x, xy))$. Then Ω can be interpreted as providing the degree of risk for an illegal immigrant to travel from x to y along xy. For all $a, b, f, g \in [0, 1]$,

$$c((a \wedge b) \vee (f \wedge g)) = c(a \wedge b) \wedge c(f \wedge g) = (c(a) \vee c(b)) \wedge (c(f) \vee c(g)).$$

It thus follows that for P as above,

$$\begin{aligned}
\Omega^\infty(x, xy) &= c(\Psi^\infty(x, xy)) \\
&= c(\vee\{\Psi(u, uu_1) \wedge \Psi(u_1, u_1u_2) \wedge \cdots \wedge \Psi(v, vw) \mid P \text{ incidence} \\
&\quad \text{path from } x \text{ to } y\}) \\
&= \wedge\{c(\Psi(u, uu_1)) \vee c(\Psi(u_1, u_1u_2)) \vee \cdots \vee c(\Psi(v, vw)) \mid \\
&\quad P \text{ incidence path from } x \text{ to } y\} \\
&= \wedge\{\Omega(u, uu_1) \vee \Omega(u_1, u_1u_2) \vee \cdots \vee \Omega(v, vw) \mid P \text{ incidence} \\
&\quad \text{path from } x \text{ to } y\}.
\end{aligned}$$

Thus, $\Omega^\infty(x, xy)$ is the smallest risk over all incidence paths yielding the most risk.

A result for Ψ has a complementary result for Ω. This complementary result holds for Ω without writing a formal proof because the proof is a complementary proof. This can also be shown for intuitionistic fuzzy incidence graphs, where $0 \leq \Psi(x, xy) + \Omega(x, xy) \leq 1$ rather than Ω being a complement of Ψ, [3]. We can also conclude that if (x, xy) is a fuzzy incidence cutpair, then the removal of (x, xy) would make the network riskier because (x, xy) would be a fuzzy incidence cutpair for Ω in a complementary sense.

Fuzzy incidence graphs can be used as a good model for a nondeterministic network with supporting links. The dynamism of the network can be then effectively studied. Block structure always avoids links in the network.

References

1. K.R. Bhutani, A. Rosenfeld, Strong arcs in fuzzy graphs. Inf. Sci. **152**, 319–322 (2003)
2. K.R. Bhutani, A. Rosenfeld, Fuzzy end notes in fuzzy graphs. Inf. Sci. **152**, 323–326 (2003)
3. C. Caniglia, B. Cousino, S.-C. Cheng, D.S. Malik, J.N. Mordeson, Intuition fuzzy graphs: weakening and strengthening members of a group. J. Fuzzy Math. **24**, 87–102 (2016)
4. T. Dinesh, Ph.D. thesis, Kannur University, Kerala, India
5. T. Dinesh, Fuzzy incidence graph - an introduction. Adv. Fuzzy Sets Syst. **21**, 33–48 (2016)
6. N.E. Friedkin, E.C. Johnson, *Social Influence Network Theory: A Sociological Examination of Small Group Dynamics.* Structural Analysis in the Social Sciences, vol. 33 (Cambridge University Press, Cambridge, 2011)
7. H. Irfan, From India to the U. S. via the jungles of Guatemala: investigation exposes route taken by human traffickers, Daily Mail, India, 20 Jan 2012
8. D. Kar, D. Cartwright-Smith, Illicit financial flows from developing countries, Global Financial Integrity, A Program of the Center for International Policy, 2002–2006, pp. 1–67
9. S. Mathew, M.S. Sunitha, Types of arcs in a fuzzy graph. Inf. Sci. **179**, 1760–1768 (2009)
10. S. Mathew, J.N. Mordeson, Connectivity concepts in fuzzy incidence graphs. Inf. Sci. **382–383**, 326–333 (2017)
11. S. Mathew, J.N. Mordeson, Fuzzy incidence blocks and their applications in illegal migration problems. New Math. Nat. Comput. **13**(3), 245–260 (2017). https://doi.org/10.1142/S1793005717400099
12. J.N. Mordeson, Fuzzy incidence graphs. Adv. Fuzzy Sets Syst. **22**, 121–133 (2016)
13. J.N. Mordeson, P.S. Nair, *Fuzzy Graphs and Fuzzy Hypergraphs*, vol. 46 (Physica-Verlag, Heidelberg, 2000)
14. J.N. Mordeson, S. Mathew, Fuzzy end nodes in fuzzy incidence graphs. New Math. Nat. Comput. **13**, 13–20 (2017)
15. S. Mathew, J.N. Mordeson, Fuzzy influence graphs. New Math. Nat. Comput. **13**, 311–325 (2017)
16. A. Rosenfeld, Fuzzy graphs, in *Fuzzy Sets and Their Applications*, ed. by L.A. Zadeh, K.S. Fu, M. Shimura (Academic Press, London, 1975), pp. 77–95
17. C.H. Smith, Tier rankings and the fight against human trafficking, Congress of the United States, Subcommittee on Africa, Global Health, Global Human Rights, and Int'l Organizations, 2013
18. United Nations on Drug and Crime, UN.GIFT United Nations Initiative to Fight Human Trafficking, An Introduction to Human Trafficking: Vulnerability, Impact, and Action, 2008
19. R.D. Villedas, Central American Migrants and "La Besta": The Route, Dangers, and Government Response, Migration Information Source, 10 Sept 2014
20. R. Walser, J.B. McNeill, J. Zuckerman, The human tragedy of illegal immigration: greater efforts needed to combat smuggling and violence, Backgrounder, Heritage Foundation No. 2568, 22 June 2011, pp. 1–18
21. S. Worrall, An anthropologist unravels the mysteries of Mexican migration, National Geographic, 6 Dec 2015

Chapter 4
Networks

4.1 Network Models

In the first three sections, we devote our attention to the problem of maximizing the flow through a network. We present a fuzzy version of a max flow, min cut theorem.

Definition 4.1.1 A transport network (or simply a network) is a simple weighted, directed graph G satisfying the following conditions.

(i) There is exactly one vertex in G, called the **source**, having no incoming edges.

(ii) There is exactly one vertex in G, called the **sink**, having no outgoing edges.

(iii) The weight C_{ij} of the directed edge (i, j) is a positive real number called the **capacity** of (i, j).

(iv) The graph G is unilaterally connected.

Definition 4.1.2 A **flow** in a network $G = (V, E)$ is a function F of E into the positive real numbers such that

(i) $F(i, j) \leq C_{ij}$ for all $(i, j) \in E$,

(ii) and for all $j \in V \backslash \{a, z\}$, $\sum_{i \in V} F(i, j) = \sum_{i \in V} F(j, i)$, where a denotes the source and z denotes the sink.

If F is a flow in a network $G = (V, E)$, we let F_{ij} denote $F(i, j)$ for all (i, j) in E. We call F_{ij} the **flow in edge** (i, j). For all $j \in V$, we call $\sum_{i \in V} F_{ij}$ the **flow into** j and we call $\sum_{i \in V} F_{ji}$ the **flow out** of j. The property (ii) of Definition 4.1.2 is called **conservation of flow**.

Theorem 4.1.3 Let F be a flow in a network G. Then $\sum_{i \in V} F_{ai} = \sum_{i \in V} F_{iz}$.

Proof Clearly

$$\sum_{j \in V} \left(\sum_{i \in V} F_{ij} \right) = \sum_{j \in V} \left(\sum_{i \in V} F_{ji} \right).$$

© Springer International Publishing AG 2018

J. N. Mordeson et al., *Fuzzy Graph Theory with Applications to Human Trafficking*, Studies in Fuzziness and Soft Computing 365, https://doi.org/10.1007/978-3-319-76454-2_4

Thus,

$$
\begin{aligned}
0 &= \sum_{j \in V} \left(\sum_{i \in V} F_{ij} - \sum_{i \in V} F_{ji} \right) \\
&= \sum_{i \in V} F_{iz} - \sum_{i \in V} F_{zi} + \sum_{i \in V} F_{ia} - \sum_{i \in V} F_{ai} \\
&\quad + \sum_{j \in V \setminus \{a,z\}} \left(\sum_{i \in V} F_{ij} - \sum_{i \in V} F_{ji} \right) \\
&= \sum_{i \in V} F_{iz} - \sum_{i \in V} F_{ai}
\end{aligned}
$$

by Definition 4.1.2(ii).   ∎

Let F be the flow in a network G. Then the value $\sum_{i \in V} F_{ai} = \sum_{i \in V} F_{iz}$ is called **the value of the flow** F.

4.2 A Maximal Flow Algorithm

If G is a network, a maximal flow in G is a flow with maximum value. There may be several flows having the same maximum value. We present an algorithm for finding a maximal flow in this section. The basic idea is to start with some initial flow and iteratively increase the value of the flow until no larger value is possible. We can take the initial flow to be the flow in which the flow in each edge is zero. To increase the value of a given flow, a path must be found from the source to the sink and increase the flow along this path.

Let $P = (v_0, v_1, \ldots, v_n)$, $v_0 = a$, $v_n = z$, be an undirected path from a to z. If an edge e in P is directed from v_{i-1} to v_i, we say that e is **properly oriented** with respect to P; otherwise, we say that e is **improperly oriented** with respect to P. If we find a path P from the source to the sink in which every edge in P is properly ordered and the flow in each edge is less than the capacity of the edge, it is possible to increase the value of the flow.

Theorem 4.2.1 *Let P be a path from a to z in a network G satisfying the following properties:*
(i) For each properly ordered edge (i, j) in P, $F_{ij} < C_{ij}$;
(ii) For each improperly ordered edge (i, j) in P, $0 < F_{ij}$.
Let $\Delta = \min X$, where X consists of the numbers $C_{ij} - F_{ij}$ for properly oriented edges (i, j) in P and F_{ij} for improperly oriented edges (i, j) in P. Define

$$
F_{ij}^* = \begin{cases} F_{ij} \text{ if } (i, j) \text{ is not in } P, \\ F_{ij} + \Delta \text{ if } (i, j) \text{ is properly oriented in } P, \\ F_{ij} - \Delta \text{ if } (i, j) \text{ is improperly oriented in } P. \end{cases}
$$

Then F^* is a flow whose value is Δ greater than the value of F.

An algorithm to find a maximal flow is based on the following ideas:

(1) Start with any flow, for example the flow in which the flow in every edge is 0.

(2) Search for a path satisfying the conditions of Theorem 4.2.1. If no such path exists, stop; the flow is maximal.

(3) Increase the flow through the path by Δ, where Δ is defined in Theorem 4.2.1.

Algorithm 4.2.2 Let the vertices of G be ordered as follows: $a = v_0, \ldots, v_n = z$.

(1) Initialize flow: set $F_{ij} = 0$ for each edge (i, j).

(2) Label source: Label vertex (a, ∞).

(3) Sink labeled? If the sink is labeled, go to Step 6.

(4) Next label vertex: Choose the not yet examined labeled vertex v_i with the smallest index i. If none, stop, the flow is maximal; otherwise, set $v = v_i$.

(5) Label adjacent vertices. Let (α, Δ) be the label of v. Examine each edge of the form (v, w), (w, v) in the order (v, v_0), (v_0, v), (v, v_1), (v_1, v), $\ldots$, where w is unlabeled. For an edge of the form (v, x), if $F_{vw} < C_{vw}$, label w $(v, \min\{\Delta, C_{vw} - F_{vw}\})$; if $F_{vw} = C_{vw}$, do not label w. For an edge of the form (w, v), if $F_{wv} > 0$, label w $(v, \min\{\Delta, F_{wv}\})$, if $F_{wv} = 0$ do not label w. Go to step (3).

(6) Revise flow: Let (γ, Δ) be the label of z. Let $w_0 = z$, $w_1 = \gamma$. If the label of w_i is (γ', Δ'), set $w_{i+1} = \gamma'$. Continue until $w_k = a$. Now

$$P : a = w_k, w_{k-1}, \ldots, w_1, w_0 = z$$

is a path from a to z. Change the flow of the edges in P as follows: If the edge e in P is properly oriented, increase the flow in e by Δ : otherwise decrease the flow in e by Δ. Remove all labels from vertices and go to step (2).

4.3 The Max Flow, Min Cut Theorem

In this section, we show that at the termination of Algorithm 4.2.2, the flow in the network is maximal. Let G be a network and let F denote the flow at the termination of Algorithm 4.2.2. Let P denote the set of labeled vertices and $\overline{P}$ the set of unlabeled vertices. The source a is in P and the sink z is in $\overline{P}$. Let $S = \{(v, w) \in E \mid v \in P$ and $w \in \overline{P}\}$. Then S is called a **cut** and the sum of the capacities of the edges in S is called the **capacity of the cut**.

Definition 4.3.1 A **cut** $(P, \overline{P})$ in G consists of a set of vertices and the complement $\overline{P}$ of P with $a \in P$ and $z \in \overline{P}$.

Theorem 4.3.2 *Let F be a flow in G and let $(P, \overline{P})$ be a cut in G. Then the capacity of $(P, \overline{P})$ is greater than or equal to the value of F, i.e.,*

$$\sum_{i \in P} \sum_{j \in \overline{P}} C_{ij} \geq \sum_{i \in V} F_{ai}. \tag{4.1}$$

Proof Clearly, $\sum_{j \in P} \sum_{i \in P} F_{ji} = \sum_{j \in P} \sum_{i \in P} F_{ij}$. Thus,

$$
\sum_{i \in V} F_{ai} = \sum_{j \in P} \sum_{i \in V} F_{ji} - \sum_{j \in P} \sum_{i \in V} F_{ij}
$$

$$
= \sum_{j \in P} \sum_{i \in P} F_{ji} - \sum_{j \in P} \sum_{i \in \overline{P}} F_{ji}
$$

$$
- \sum_{j \in P} \sum_{i \in P} F_{ij} - \sum_{j \in P} \sum_{i \in \overline{P}} F_{ij}
$$

$$
= \sum_{j \in P} \sum_{i \in \overline{P}} F_{ji} - \sum_{j \in P} \sum_{i \in \overline{P}} F_{ij}
$$

$$
\leq \sum_{j \in P} \sum_{i \in \overline{P}} F_{ji} \leq \sum_{j \in P} \sum_{i \in \overline{P}} C_{ji}.
$$
∎

A **minimal cut** is a cut having minimum capacity.

Theorem 4.3.3 (Max Flow, Min Cut Theorem) *Let F be a flow in G and let $(P, \overline{P})$ be a cut in G. If the equality (4.1) holds, then the flow is maximal and the cut is minimal. Moreover, equality holds in (4.1) if and only if*
  (i) $F_{ij} = C_{ij}$ for $i \in P$, $j \in \overline{P}$, and
  (ii) $F_{ji} = 0$ for $i \in \overline{P}$, $j \in P$.

Proof That the flow is maximal and the cut is minimal follows immediately from Eq. (4.1). From the proof of Theorem 4.3.2, it is immediate that the equality in (4.1) holds if and only if $\sum_{j \in P} \sum_{i \in \overline{P}} F_{ji} = 0$ and $\sum_{j \in P} \sum_{i \in \overline{P}} F_{ji} = \sum_{j \in P} \sum_{i \in \overline{P}} C_{ji}$. Consequently, the equivalence statement holds. ∎

Theorem 4.3.4 *At termination, Algorithm 4.2.2 produces a maximal flow.*

Proof Let P be the set of labeled vertices and $\overline{P}$ the set of unlabeled vertices at the termination of the algorithm. Consider an edge (i, j), where $i \in P$ and $j \in \overline{P}$. Because i is labeled, we have $F_{ij} = C_{ij}$ else j would have labeled at step 5. By Theorem 4.3.3, the flow at the termination of Algorithm 4.2.2 is maximal. ∎

4.4 Reduction Nodes and Arcs

Directed fuzzy networks are introduced in this section. They are normalized node capacitated networks and provide a good platform to model different types of complicated flows in nature. A directed fuzzy network version of Menger's theorem and the celebrated Max flow Min cut theorem are also provided. Because the maximum flow through minimum number of directed internally disjoint paths is important in quality of service(QoS) problems in networking, the results in this section can be applied to a wide variety of problems.

The problem of finding *maximum flow* in a network and flow between pairs of nodes in a network have been studied widely in [1]. Most of the studies were oriented in an algorithmic way. Problems related with *maximum bandwidth* in [5], *widest path* in [4], *bottleneck paths* in [13] are examples. There are no specific mathematical formulations for a fuzzy network in the literature. The maximum flow problems in networks can be effectively studied using Zadeh's logic [16, 17] as most of the complicated networks like internet and power grids obey fuzzy principles [3]. See also [6, 12, 15].

In this section, the concept of networks is extended to a fuzzy situation. A fuzzy graph represents an undirected network with normalized capacities. When we provide directions to all arcs in a fuzzy graph, we get a directed fuzzy network (DFN). When the exact capacity function and exact flows are not available, such a network is useful. For example, the exact data of humans trafficked between different parts of the globe is not a well defined function, but can be estimated as a fuzzy membership function based on the available data. Thus, it can be modeled as a DFN.

Basic concepts from DFNs are introduced in this section and the concepts of strength reducing sets of nodes and arcs are introduced (Definitions 4.5.2 and 4.5.3) in Sect. 4.5. Strength reducing sets of nodes are characterized in Theorem 4.5.5 and strength reducing sets of arcs in Theorem 4.5.6. The fuzzy generalization for the node version of the classical Menger's theorem is proved in Theorem 4.5.7 and it's arc version is stated without proof in Theorem 4.5.8. The flow problems in DFNs are studied in Sect. 4.6. Max-flow min cut theorem for DFN is presented in Theorem 4.6.8.

We recall some definitions. A *path* P of length n is a sequence of distinct nodes $u_0, u_1, \ldots, u_n$ such that $\mu(u_{i-1}, u_i) > 0, i = 1, 2, \ldots, n$ and the degree of membership of a weakest arc is its strength. The *strength of connectedness* between two nodes x and y is defined as the maximum of the strengths of all paths between x and y and is denoted by $CONN_G(x, y)$ [9, 11]. An $x - y$ path P is called a *strongest* $x - y$ path if its strength equals $CONN_G(x, y)$. An f-graph $G : (\sigma, \mu)$ is connected if for every x, y in σ^*, $CONN_G(x, y) > 0$. Through out, we assume that G is connected. An arc of an f-graph is called *strong* if its weight is at least as great as the connectedness of its end nodes when it is deleted. If (x, y) is a strong arc, then x and y are strong neighbors.

The *degree* of a node v is defined as $d(v) = \sum_{u \neq v} \mu(u, v)$. The *minimum degree* of G is $\delta(G) = \wedge\{d(v) \mid v \in \sigma^*\}$ and the *maximum degree* of G is $\Delta(G) = \vee\{d(v) \mid v \in \sigma^*\}$.

In a network, the capacity of a node is assumed as infinity in the sense that any flow coming to that node will be redirected through different output links. But it is not possible when we have a node capacitated network. Also, when the capacity function of either nodes or links become fuzzy, we need a fuzzy model. Most of the networks are really fuzzy in nature. We provide a systematic definition to fuzzy network in this section. To distinguish from fuzzy graphs, which are indeed models of fuzzy undirected networks, we use the term directed fuzzy network (DFN). A DFN is clearly a normalized node capacitated network.

Definition 4.4.1 Let V be a set. A **DFN** is an ordered triple $\overrightarrow{G} : (V, \sigma, \mu)$, where σ be a fuzzy set on V and μ be a fuzzy subset of $V \times V$ such that $\mu(u, v) \leq \sigma(u) \wedge \sigma(v)$. Note that μ is an unsymmetric fuzzy relation on V. Elements of σ^* are called **nodes** of $\overrightarrow{G}$ and elements of μ^* are called **arcs** of $\overrightarrow{G}$. If (u, v) is an arc, we write it as uv. We refer $\sigma(u)$ as the **normalized capacity** of the node and $\mu(uv)$ as the **normalized capacity** of the arc uv.

As in networks if uv is an arc, we call u as the tail of the arc and v as the head. As in a network, there are three different categories of arcs in a FDN, namely sources, sinks and intermediate nodes. They are defined as follows.

Definition 4.4.2 A **directed walk** or **di-walk** in DFN $\overrightarrow{G} : (V, \sigma, \mu)$ is a finite nonempty sequence $v_0, e_1, v_1, e_2, \ldots, e_k, v_k$, where elements of σ^* and μ^* alternatively such that e_i has tail v_{i-1} and head v_i for $i = 1, 2, \ldots, k$. We denote it by $(v_0, v_1, \ldots, v_k)$. A **di-path** P in $\overrightarrow{G} : (V, \sigma, \mu)$ is a di-walk $(v_0, v_1, \ldots, v_k)$ such that $v_i \neq v_j$ except for v_0 and v_k referred as the origin and terminus of P. The number of arcs in the di-path is called its **length**. A di-path is said to be a **di-cycle** if origin and terminus coincide.

A DFN is said to be **di-connected** if there exists di-paths between any two elements u and v in σ^*. Clearly 'di-connected' is an equivalence relation on σ^*. It decomposes $\overrightarrow{G} : (V, \sigma, \mu)$ into di-components $\overrightarrow{G}(V_1), \overrightarrow{G}(V_2), \ldots, \overrightarrow{G}(V_t)$.

Example 4.4.3 Consider the DFN with four nodes in Fig. 4.1.

Note that the above DFN is di-connected as there is a di-path between any two nodes. There are no sources or sinks here. All nodes are intermediate nodes. The indegree $d_f^-(a) = 0.7$ and the outdegree $d_f^+(a) = 0.4$.

Definition 4.4.4 A DFN $\overrightarrow{G'} : (V, \sigma', \mu')$ is said to be a **partial sub DFN** of $\overrightarrow{G} : (V, \sigma, \mu)$ if $\sigma' \subseteq \sigma$ and $\mu' \subseteq \mu$. $\overrightarrow{G'}$ is called a **sub DFN** of $\overrightarrow{G}$ if $\sigma'(u) = \sigma(u)$ for all $u \in \sigma'^*$ and and $\mu'(xy) = \mu(xy)$ for all $xy \in \mu'^*$.

With each DFN $\overrightarrow{G} : (V, \sigma, \mu)$, we associate a fuzzy graph $G : (V, \sigma, \mu)$ by replacing each arc uv of $\overrightarrow{G}$ with an undirected edge between u and v. We call it the underlying fuzzy graph of $\overrightarrow{G}$. Also, $\overrightarrow{G}^*$ denotes the underlying graph having no weights or directions.

Fig. 4.1 A di-connected
DFN

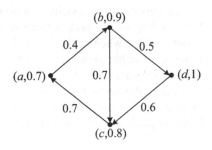

Definition 4.4.5 Let $\overrightarrow{G} : (V, \sigma, \mu)$ be a DFN and let $P : (v_0, v_1, v_2, \ldots, v_n)$ be a directed path in $\overrightarrow{G}$. The **width** of the path P, denoted as $w(P)$ is defined as $\mu(v_0v_1) \wedge \mu(v_1v_2) \wedge \cdots \wedge \mu(v_{n-1}v_n)$. $\vee\{w(P) \mid P$ is a $u - v$ di-path$\}$ is called the $u - v$ **width** in $\overrightarrow{G}$. It is denoted by $w_{\overrightarrow{G}}(u - v)$ or simply $w(u - v)$. Note that $w(u - v)$ denotes the maximum possible flow trough a single $u - v$ di-path. In communication networks, it represents the maximum bandwidth between the nodes u and v. A $u - v$ di-path P is said to be the **widest** $u - v$ **di-path** if $w(P) = w(u - v)$. It is also termed as **maximum bottleneck** $u - v$ **path** or **maximum bandwidth path**.

In Example 4.4.3, $P : (a, b, c, a)$ is a di-cycle of length 3. The width of P is 0.4. Note that there exists unique di-paths between any two nodes in this DFN. The width of any pair of nodes in this DFN is the width of the unique di-path connecting those pairs. $w(a - d) = 0.4$.

Definition 4.4.6 Let $\overrightarrow{G} : (V, \sigma, \mu)$ be a DFN. A node $u \in \sigma^*$ is said to be a **cutnode** of $\overrightarrow{G} : (V, \sigma, \mu)$ if $w(x - y) = 0$ for some x and y in $\overrightarrow{G} \setminus u$. It is said to be a **reduction node** (**r-node**) if $w_{\overrightarrow{G} \setminus u}(x - y) < w_{\overrightarrow{G}}(x - y)$. That is, the removal of u from $\overrightarrow{G}$ reduces the width of some pair of nodes in $\overrightarrow{G}$. Similarly, an arc uv in $\overrightarrow{G}$ is said to be a **di-bridge** if $w(x - y) = 0$ for some x and y in $\overrightarrow{G} \setminus uv$. An arc uv is called a **reduction arc** (**r-arc**) if $w_{\overrightarrow{G} \setminus uv}(x - y) < w_{\overrightarrow{G}}(x - y)$. That is, the width of a pair is affected by the removal of arc uv.

Proposition 4.4.7 *Let $\overrightarrow{G} : (V, \sigma, \mu)$ be a DFN. A node $u \in \sigma^*$ is a reduction node if and only if there exist two nodes x and y different from u in $\overrightarrow{G}$ such that every widest $x - y$ di-path in $\overrightarrow{G}$ passes through u.*

Proof If $u \in \sigma^*$ is a reduction node in $\overrightarrow{G}$, then $w_{\overrightarrow{G} \setminus u}(x - y) < w_{\overrightarrow{G}}(x - y)$ and hence all widest $x - y$ paths pass through u. Conversely, if all widest $x - y$ paths pass through u, then the maximum of widths of all $x - y$ di-paths in $\overrightarrow{G} \setminus u$ will be less than $w(x - y)$ and hence u becomes a reduction node. ∎

Proposition 4.4.8 *Let $\overrightarrow{G} : (V, \sigma, \mu)$ be a DFN. An arc $uv \in \mu^*$ is a reduction arc if and only if there exist two nodes x and y in $\overrightarrow{G}$ such that every widest $x - y$ di-path in $\overrightarrow{G}$ passes through uv.*

The proof of Proposition 4.4.8 is similar to proof of Proposition 4.4.7.

Example 4.4.9 Consider the DFN in Fig. 4.2. For simplicity assume that $\sigma(u) = 1$ for all $u \in \sigma^*$.

There are two reduction nodes in this DFN namely c and e (circled) and there is a cutnode namely d (in square). There are no di-bridges in this DFN, but there are four reduction arcs namely arcs ac, cd, de and eg.

Fig. 4.2 A DFN with a
cutnode and two reduction
nodes

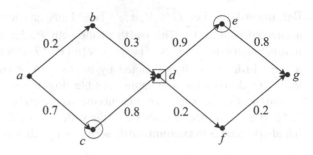

4.5 Strength Reducing Sets

We next consider the results of [8].

In network theory, a $u - v$ separating set S of nodes is a collection of nodes in a network G whose removal disconnects G and, u and v belonging to different components of $G \backslash S$ [16]. See also [6, 12, 15]. Similarly, a $u - v$ separating set of arcs is defined. Because the reduction in strength is more important and frequent in graphs and networks, we define strength reducing sets of nodes and arcs in a DFN as follows.

Definition 4.5.1 Let $\overrightarrow{G} : (V, \sigma, \mu)$ be a DFN. An arc uv is said to be α-**strong** if $\mu(uv) > w_{\overrightarrow{G} \backslash uv}(u - v)$. It is β-**strong** if $\mu(uv) = w_{\overrightarrow{G} \backslash uv}(u - v)$ and δ-**arc** if $\mu(uv) < w_{\overrightarrow{G} \backslash uv}(u - v)$. An arc xy which is either β or α is called a **strong arc**.

Definition 4.5.2 Let u and v be any two nodes of a DFN $\overrightarrow{G} : (V, \sigma, \mu)$ such that the arc uv is not strong. A set $S \subseteq \sigma^*$ of nodes is said to be a $u - v$ **strength reducing set of nodes** if $w_{\overrightarrow{G} \backslash S}(u - v) < w_{\overrightarrow{G}}(u - v)$, where $\overrightarrow{G} \backslash S$ is the sub DFN of $\overrightarrow{G}$ obtained by removing all nodes in S.

Definition 4.5.3 A set of arcs $E \subseteq \mu^*$ is said to be a $u - v$ **strength reducing set of arcs** if $w_{\overrightarrow{G} \backslash E}(u - v) < w_{\overrightarrow{G}}(u - v)$, where $\overrightarrow{G} \backslash E$ is the sub DFN of $\overrightarrow{G}$ obtained by removing all arcs in E.

Definition 4.5.4 A $u - v$ strength reducing set of nodes (arcs) with n elements is said to be a **minimum** $u - v$ **strength reducing set of nodes (arcs)** if there exist no $u - v$ strength reducing set of nodes (arcs) with less than n elements. A minimum $u - v$ strength reducing set of nodes is denoted by $S_{\overrightarrow{G}}(u, v)$ and a minimum $u - v$ strength reducing set of arcs is denoted by $E_{\overrightarrow{G}}(u, v)$.

The following provide characterizations for node and arc strength reducing sets in a DFN.

Theorem 4.5.5 *Let* $\overrightarrow{G} : (V, \sigma, \mu)$ *be a connected DFN and* u, v *be any two nodes in G such that uv is not strong. Then a set S of nodes in $\overrightarrow{G}$ is a $u - v$ strength reducing set if and only if every widest di-path from u to v contains at least one node of S.*

Proof Suppose that S is a $u - v$ strength reducing set of nodes in $\overrightarrow{G}$ and let P be a widest $u - v$ di-path in $\overrightarrow{G}$. If P contains no node of S, the removal of S keep P intact and hence $\overrightarrow{G} \backslash S$ contains P. Thus, $w_{\overrightarrow{G} \backslash S}(u - v) = w_{\overrightarrow{G}}(u - v)$, which contradicts the fact that S is a $u - v$ strength reducing set of nodes. Thus, P must contains at least one member of S. It is obvious that this result is not true when arc uv is strong. Any strong arc uv is a widest $u - v$ di-path having no node from S.

Conversely, suppose that every widest di-path from u to v contain at least one node of S, where $S \subseteq \sigma^*$ and u, v not in S. Then the removal of S destroys all widest $u - v$ di-paths in $\overrightarrow{G}$ and hence $w_{\overrightarrow{G} \backslash S}(u - v) < w_{\overrightarrow{G}}(u - v)$. Thus, it follows that S is a $u - v$ strength reducing set of nodes in $\overrightarrow{G}$.   ∎

Theorem 4.5.6 *Let* $\overrightarrow{G} : (V, \sigma, \mu)$ *be a connected DFN and* u, v *be any two nodes in* $\overrightarrow{G}$*. Then a set E of arcs in* $\overrightarrow{G}$ *is a $u - v$ strength reducing set if and only if every widest di-path from u to v contains at least one arc of E.*

Proof The proof is similar to that of Theorem 4.5.5.   ∎

Next we present a DFN version of one of the celebrated results in graph theory due to Karl Menger (1927) [10].

Theorem 4.5.7 (Node version of Menger's Theorem for DFN) *Let* $\overrightarrow{G} : (V, \sigma, \mu)$ *be a DFN. For any two nodes* $u, v \in \sigma^*$ *such that uv is not strong, the maximum number of internally disjoint widest $u - v$ di-paths in* $\overrightarrow{G}$ *is equal to the number of nodes in a minimal $u - v$ strength reducing set.*

Proof We prove the result by induction on the strong size $ss(\overrightarrow{G})$ (number of strong arcs) of G. When $ss(G) = 0$, $\overrightarrow{G} : (V, \sigma, \mu)$ is such that $\mu^* = \emptyset$ and the result is trivially true for any pair of nodes $u, v \in \sigma^*$.

Assume that the theorem is true for all DFN $\overrightarrow{G} : (V, \sigma, \mu)$ with strong size less than m, where $m \geq 1$. Let $\overrightarrow{G}$ be a DFN of strong size m. Let $u, v \in \sigma^*$ be such that uv is not strong. If u and v are in different components of $\overrightarrow{G} : (V, \sigma, \mu)$, then the theorem is obviously true. So assume that u and v belongs to the same component of $\overrightarrow{G} : (V, \sigma, \mu)$. Then either uv is not in μ^* or uv is a δ-arc. In both cases $u - v$ strength reducing sets of nodes exists in $\overrightarrow{G}$. (If uv is strong, then reduction of any number of nodes will not reduce the width between u and v and hence no strength reducing set of nodes exists.)

Now suppose that $S_{\overrightarrow{G}}(u, v)$ is a minimum strength reducing set of nodes in $\overrightarrow{G}$ with $|S_G(u, v)| = k \geq 1$. By Theorem 4.5.5, each widest $u - v$ di-path must contain at least one member from $S_{\overrightarrow{G}}(u, v)$. Hence, any $u - v$ strength reducing set should contain at least as many nodes as the number of internally disjoint widest $u - v$ di-paths. In other words, there exists at most k internally disjoint widest $u - v$ di-paths. We show that $\overrightarrow{G}$ contains exactly k internally disjoint widest $u - v$ di-paths.

If $k = 1$, then $|S_{\overrightarrow{G}}(u, v)| = 1$. Let $S_{\overrightarrow{G}}(u, v) = \{w\}$. Then $w_{\overrightarrow{G} \backslash \{w\}}(u - v) < w_{\overrightarrow{G}}(u - v)$. That is, w is a reduction node of $\overrightarrow{G}$. So every widest $u - v$ di-path must pass through w. Hence, the number of internally disjoint $u - v$ paths is one and the result is true. So assume that $k \geq 2$.

(1) $\overrightarrow{G}$ has a minimum $u - v$ strength reducing set of nodes containing a node x such that both ux and xy are α-strong arcs. Let $S_{\overrightarrow{G}}(u, v)$ be the minimum $u - v$ strength reducing set of nodes with the above mentioned property. Then $S_{\overrightarrow{G}}(u, v) \backslash \{x\}$ is a minimum $u - v$ strength reducing set in $\overrightarrow{G} \backslash \{x\}$ having $k - 1$ nodes. Because both ux and xv are α-strong arcs, they are clearly strong and hence $ss(\overrightarrow{G} \backslash \{x\}) < ss(\overrightarrow{G})$. By induction, it follows that $\overrightarrow{G} \backslash \{x\}$ contains $k - 1$ internally disjoint widest $u - v$ di-paths. Because ux and xv are α-strong, P is a widest $u - v$ di-path. Thus, we have k internally disjoint widest $u - v$ di-paths in $\overrightarrow{G}$.

(2) For every minimum $u - v$ strength reducing set $S_{\overrightarrow{G}}(u, v)$ in $\overrightarrow{G}$, either every node in $S_{\overrightarrow{G}}(u, v)$ is an α-strong neighbor of u (that is, if w is a node in $S_{\overrightarrow{G}}(u, v)$, then uw is an arc which is the unique widest $u - w$ di-path) but not of v or every node in $S_{\overrightarrow{G}}(u, v)$ is an α-strong neighbor of v but not of u.

Suppose that every node in $S_{\overrightarrow{G}}(u, v)$ is an α-strong neighbor of u but not of v. Consider a widest $u - v$ di-path P in $\overrightarrow{G}$. Let x be the first node of P which is in $S_{\overrightarrow{G}}(u, v)$. Then ux is α-strong and because xy is not α-strong, there exists at least one node say y other than u and v such that xy is β-strong. Denote the arc xy by e.

We show the following: Every $u - v$ strength reducing set in $\overrightarrow{G} \backslash e$ has exactly k nodes.

On the contrary assume that there exists a minimum $u - v$ strength reducing set in $\overrightarrow{G} \backslash e$ with $k - 1$ nodes say $Z = \{z_1, z_2, \ldots, z_{k-1}\}$. Then $Z \cup \{x\}$ is a minimum $u - v$ strength reducing set in $\overrightarrow{G}$. Note that every z_i, $i = 1, 2, .., k - 1$ and x are α-strong neighbors of u. Because $Z \cup \{y\}$ is also a minimum $u - v$ strength reducing set in $\overrightarrow{G}$, it follows that y is an α-strong neighbor of u contradicting the fact that arc xy is β-strong. (The arcs ux, uy and xy forms a triangle with arc xy as the weakest arc. The unique weakest arc of a cycle is a δ-arc). Thus, k is the minimum number of nodes in a $u - v$ strength reducing set in $\overrightarrow{G} \backslash e$. Because $ss(\overrightarrow{G} \backslash \{e\}) < ss(\overrightarrow{G})$, it follows by induction that there are k internally disjoint $u - v$ di-paths in $\overrightarrow{G} \backslash e$ and hence in $\overrightarrow{G}$.

(3) There exists a $u - v$ strength reducing set W in $\overrightarrow{G}$ such that no member of W is an α-strong neighbor of both u and v and W contains at least one node which is not an α-strong neighbor of u and at least one node which is not an α-strong neighbor of v.

Let W be a minimum $u - v$ strength reducing set with k elements having the above properties. Let $W = \{w_1, w_2, \ldots, w_k\}$. Consider all widest di-paths from u to v. Then because W is minimum, w_i, $i = 1, 2, \ldots, k$ must belong to at least one such di-path. Let G_u be the sub DFN of $\overrightarrow{G}$ consisting of all $u - w_i$ sub di-paths of all widest $u - v$ di-paths in which $w_i \in W$ is the only node of the di-path belonging to W. Note that if $w_{\overrightarrow{G}}(u, v) = t$, then $\mu(x, y) \geq t$ for all arc xy in these di-paths.

Let $\overrightarrow{G}'_u$ be the sub DFN constructed from $\overrightarrow{G}_u$ by adding a new node v' and joining v' to each node w_i for $i = 1, 2, \ldots, k$. Let $\sigma(v') = 1$ and $\mu(w_i, v') = \sigma(w_i)$ for every $i = 1, 2, \ldots, k$. The sub DFN G_v and G'_v are defined similarly.

Because W contains a node that is not an α-strong neighbor of u and a node that is not an α-strong neighbor of v (Note that all newly introduced arcs are strong), we have $ss(G'_u) < ss(G)$ and $ss(G'_v) < ss(G)$.

Clearly, $S_{\overrightarrow{G}'_u}(u, v') = k$ and $S_{\overrightarrow{G}'_v}(u', v) = k$. So by induction $\overrightarrow{G}'_u$ contains k internally disjoint $u - v'$ di-paths say $A_i, i = 1, 2, \ldots, k$, where A_i contains w_i. Also, $\overrightarrow{G}'_v$ contains k internally disjoint $u' - v$ di-paths say $B_i, i = 1, 2, \ldots, k$, where B_i contains w_i. Let A'_i be the $u - w_i$ sub di-paths of A_i and B'_i be the $w_i - v$ sub di-path of B_i for $1 \leq i \leq k$. Now k internally disjoint widest $u - v$ di-paths can be formed by joining A_i and B_i for $i = 1, 2, \ldots, k$ and the theorem is proved by induction. ∎

Next we state the arc version of Theorem 4.5.7. The proof is very similar to that of Theorem 4.5.7.

Theorem 4.5.8 (Arc version of Menger's Theorem for DFN) *Let $\overrightarrow{G}$: (V, σ, μ) be a connected DFN and let $u, v \in \sigma^*$. Then the maximum number of arc disjoint widest $u - v$ di-paths in $\overrightarrow{G}$ is equal to the number of arcs in a minimum (with respect to the number of arcs) $u - v$ strength reducing set.*

4.6 Flow Problems in DFN

In this section, we discuss flow related problems in DFNs. A DFN, being a node capacitated network, will have an extra flow condition on the capacity of nodes. We shall discuss DFNs with exactly one source and one sink for convenience.

Definition 4.6.1 Let $\overrightarrow{G}$: (V, σ, μ) be a DFN with source s and sink t. A **flow** on $\overrightarrow{G}$ is a function $\psi : \mu^* \to [0, 1]$ satisfying the following conditions.

(*i*) Arc capacity limit: For every $e \in \mu^*$, $\psi(e) < \mu(e)$.

(*ii*) Node capacity limit: For every node $u \in \sigma^*$, $\sum_{v \in \sigma^*} \psi(vu) \leq \sigma(u)$.

(*iii*) Conservation of flow: For every node $u \in \sigma^* \backslash \{s, t\}$, $\sum_{v \in \sigma^*} \psi(vu) = \sum_{w \in \sigma^*} \psi(uw)$.

The value of the flow ψ on $\overrightarrow{G}$ is defined as $\sum_{u \in \sigma^*} \psi(su)$.

In a DFN, an arc uv is called ψ-**zero** if $\psi(uv) = 0$, ψ-**positive** if $\psi(uv) > 0$, ψ-**unsaturated** if $\psi(uv) < \mu(uv)$ and ψ-**staurated** if $\psi(uv) = \mu(uv)$.

Example 4.6.2 Consider the DFN in Fig. 4.3 from [8] with a flow function.

In this DFN, the flow is 0.8. Note that the arc sb is ψ-unsaturated. There are no ψ-zero arcs in this DFN.

In network theory, an $s - t$ cut or simply a cut is a partition of V into two subsets A and B such that $s \in A$ and $t \in B$ and the cut value $c(A, B)$ is defined as the sum of capacities of all arcs going from set A to the set B. It is proved that the if f is

Fig. 4.3 A DFN with a flow
function

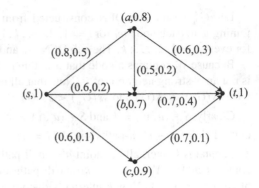

Fig. 4.4 A DFN with a
particular M-cut

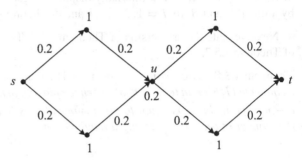

a flow and (A, B) is a cut, then value of the flow is always less than or equal to
$c(A, B)$. But in DFN, this upper bound is very loose because, the capacity of a node
also plays a crucial rule in the flow. Therefore, capacities of nodes are also taken
care of in finding the bounds. Towards this we define a mixed cut (M-cut for short)
in DFNs as follows.

Definition 4.6.3 Let $\overrightarrow{G}$: (V, σ, μ) be a connected DFN with source s and sink t. A
mixed cut (M-cut) of $\overrightarrow{G}$ is a set consisting of arcs and nodes such that whose removal
disconnects $\overrightarrow{G}$ with source in one component and sink in the other. We denote an
M-cut as $M : (E, S)$, where E is a set of arcs and S the set of nodes in the cut. The
capacity of such an M-cut is defined as $c(M : (E, S)) = \sum_{e \in E} \mu(e) + \sum_{v \in S} \sigma(v)$.

Example 4.6.4 Consider the DFN in Fig. 4.4 from [8], with a particular M-cut.
   In the DFN given in Fig. 4.4, the capacity of node u is 0.2. Being a cutnode every
path from source to sink should pass through u and the maximum possible flow from
s to t is 0.2. The set $M = \{u\}$ is a mixed cut. Note that the capacity of any cut in this
network is at least 0.4. Hence, we have the following obvious result.

Lemma 4.6.5 *Let* $\overrightarrow{G}$: (V, σ, μ) *be a connected DFN with a flow function* ψ *defined
on it. Then for any M-cut* $M : (E, S)$*, the value of the flow,* $v(\psi) \leq c(M : (E, S))$.

Proof Let $\overrightarrow{G} : (V, \sigma, \mu)$ be a connected DFN with a flow function ψ. Let $M : (E, S)$ be an M-cut in $\overrightarrow{G}$. Replace each node u in S by an arc of flow $\sigma(u)$ and capacities of other nodes in $\overrightarrow{G}$ one so that inflow and outflow at u are preserved. Let the resulting DFN be $\overrightarrow{G}'$. The newly formed arcs together with arcs in E forms a cut K in $\overrightarrow{G}'$. Let it be (N, N^c). Then $v(\psi) \leq c(K)$. Also, $c(M) = c(K)$. Thus, it follows that $v(\psi) \leq c(M : (E, S))$. ∎

As a consequence we have the following corollary.

Corollary 4.6.6 *Given a DFN* $\overrightarrow{G}$, *let* ψ^* *be the maximum flow in* $\overrightarrow{G}$ *and* M^* *be minimum capacity M-cut in* $\overrightarrow{G}$. *Then* $v(\psi^*) \leq c(M^*)$.

Lemma 4.6.7 *Given a DFN* $\overrightarrow{G}$, *and an M-cut M. Then there exists a flow* ψ *on* $\overrightarrow{G}$ *such that* $v(\psi) = c(M)$.

Proof Let $\overrightarrow{G}$ be a DFN and an M-cut M. As in the proof of Lemma 4.6.5, construct a network $\overrightarrow{G}'$. Obtain the corresponding cut X in $\overrightarrow{G}'$. By Theorem 1 in [2], there exists a flow f in $\overrightarrow{G}'$ such that $v(f) = c(X)$. Let ψ be the corresponding flow in $\overrightarrow{G}$ after reverting the M-cut. Then $v(\psi) = v(f) = c(X) = c(M)$. ∎

Next we prove an important result for DFNs.

Theorem 4.6.8 (Max-Flow Min Cut theorem for DFN) *Let* $\overrightarrow{G} : (V, \sigma, \mu)$ *be a connected DFN and let* $s, t \in \sigma^*$ *be the source and sink respectively. Then the maximum possible* $s - t$ *flow in* $\overrightarrow{G}$ *is equal to the minimum of the capacities of all M-cuts in* $\overrightarrow{G}$.

Proof Let Ψ^* be the maximum flow in $\overrightarrow{G}$ and M^* the be the minimum M-cut in $\overrightarrow{G}$. By Lemmas 4.6.5 and 4.6.7, corresponding to any flow Ψ there M-cut M of $c(M) = v(\Psi) \leq c(M^*)$. But no M-cut can have capacity less than an M-cut of minimum capacity. Thus, $v(\Psi) = c(M^*)$. By Corollary 4.6.6, $v(\Psi^*) \leq c(M^*) = v(\Psi)$. But no flow in $\overrightarrow{G}$ can have more than maximum flow. Hence, $v(\Psi^*) = c(M^*)$. ∎

Fuzzy networks provide a normalized structure for several flow networks in nature. The flow of data, humans, power, and so on, are fuzzy in nature and hence a precise network modeling is difficult. Also, most of these networks are node capacitated. The router capacity, terminal capacity, and so on, are crucial in providing uninterrupted flow between nodes. Accurate data concerning human trafficking does not exist. Some information concerning human trafficking has been developed in a linguistic manner using terms such as very low, low, medium, high, very high. It will be worth to develop max-flow min-cut theory for linguistic capacities and flows.

4.7 Network Flow

Let V be a finite set and let E denote a set of directed edges on V. Let $\gamma : E \to \{0, 0.1, 0.3, 0.5, 0.7, 0.9, 1\}$. Then $G = (V, E, \gamma)$ is called a **directed network with capacities** γ.

Definition 4.7.1 Let $G = (V, E, \gamma)$ be a directed network. Let $\mu : E \to Q \cap [0, 1]$ be such that for all $e \in E$, $\mu(e) \le \gamma(e)$. Suppose for each vertex j which is not the source or the sink that $\sum_{i \in V} \mu(i, j) = \sum_{i \in V} \mu(j, i)$. Then μ is called a **flow** in G.

Theorem 4.7.2 ([7]) *Let μ be a flow in a directed network G. If a is the source and z is the sink, then*

$$\sum_{i \in V} \mu(a, i) = \sum_{i \in V} \mu(i, z).$$

Let $\oplus$ denote the conorm, algebraic sum, i.e., for all $a, b \in [0, 1]$, $a \oplus b = a + b - ab$.

Theorem 4.7.3 *Let (V, e) be a directed network. Let $\mu : E \to Q \cap [0, 1]$. Suppose that $\sum_{i \in V}^{\oplus} \mu(i, j) = \sum_{i \in V}^{\oplus} \mu(j, i)$. If a is the source and z is the sink, then $\sum_{i \in V}^{\oplus} \mu(a, j) = \sum_{i \in V}^{\oplus} \mu(i, z)$.*

Example 4.7.4 Suppose for each vertex j, which is not the source or the sink, that $\sum_{i \in V} \mu(i, j) = \sum_{i \in V} \mu(j, i)$. Then it is not necessarily the case that $\sum_{i \in V}^{\oplus} \mu(i, j) = \sum_{i \in V}^{\oplus} \mu(j, i)$. Let $V = \{a = a_1, a_2, a_3, a_4, a_5 = z\}$ and $E = \{(a_1, a_2), (a_2, a_3)$ $(a_2, a_{45})(a_3, a_5)(a_4, a)\}$. Let $\mu : E \to Q \cap [0, 1]$ be such that

$$\gamma(a_1, a_2) = 0.6, \gamma(a_2, a_3) = 0.2, \gamma(a_2, a_4) = 0.4,$$
$$\gamma(a_3, a_5) = 0.2, \gamma(a_4, a_5) = 0.4.$$

Then for all a_j which is not the source or the sink, clearly $\sum_{a_i \in V} \mu(a_i, a_j) = \sum_{a_i \in V} \mu(a_j, a_i)$. However, $\sum_{a_i \in V}^{\oplus} \mu(a_i, a_2) = \mu(a_1, a_2) = 0.6 \ne 0.52 = 0.2 \oplus 0.4 = \mu(a_2, a_3) \oplus \mu(a_2, a_4)$. Also, note that $\sum_{a_i \in V}^{\oplus} \mu(a, a_i) = 0.6 \ne 0.52 = \sum_{a_i \in V}^{\oplus} \mu(a_i, z)$.

Example 4.7.5 Let $V = \{s, a, b, c, t\}$ and $E = \{(s, a), (a, c), (c, t), (s, b), (b, c), (c, t)\}$. For the network given by $G = (V, E)$, let μ define the capacities of the edges. Define the flows μ and ν as follows:

$$\gamma(s, a) = 0.3, \ \gamma(a, c) = 0.3, \ \gamma(s, b) = 0.2, \ \gamma(b, c) = 0.2, \ \gamma(c, t) = 0.4,$$
$$\mu(s, a) = 0.2, \ \mu(a, c) = 0.2, \ \mu(s, b) = 0.2, \ \mu(b, c) = 0.2, \ \mu(c, t) = 0.4,$$
$$\nu(s, a) = 0.3, \ \nu(a, c) = 0.3, \ \nu(s, b) = 0.1, \ \nu(b, c) = 0.1, \ \nu(c, t) = 0.4.$$

Both μ and ν give the maximal flow of 0.4. For μ, if we replace addition by the conorm algebraic sum, then $0.2 + 0.2 - (0.2)(0.2) = 0.36$ and for ν, $0.3 + 0.1 -$

$(0.3)(0.1) = 0.37$. We see that taking a maximal flow and replacing addition by a conorm does not necessarily yield the maximal flow for the conorm.

We next consider the Frank-Weber conorm $\circledast$, where for all $a, b \in [0, 1]$, $a \circledast b = 1 \wedge (a + b)$. For $a_1, \ldots, a_n \in [0, 1]$, we write $\sum_{i=1}^{\circledast n} a_i$ for $a_1 \circledast \cdots \circledast a_n$.

Proposition 4.7.6 *For* $j \in V \backslash \{a, z\}$, $\sum_{i \in V} F_{ij} = \sum_{i \in V} F_{ji}$ *implies* $\sum_{i \in V}^{\circledast} F_{ij} = \sum_{i \in V}^{\circledast} F_{ji}$.

Proof Suppose $\sum_{i \in V} F_{ij} = \sum_{i \in V} F_{ji} \geq 1$. Then $(\sum_{i \in V} F_{ij}) \wedge 1 = (\sum_{i \in V} F_{ji}) \wedge 1 = 1$ and so $\sum_{i \in V}^{\circledast} F_{ij} = \sum_{i \in V}^{\circledast} F_{ji} = 1$. Suppose $\sum_{i \in V} F_{ij} = \sum_{i \in V} F_{ji} < 1$. Then $\sum_{i \in V}^{\circledast} F_{ij} = \sum_{i \in V} F_{ij} = \sum_{i \in V} F_{ji} = \sum_{i \in V}^{\circledast} F_{ji}$. ∎

$0.7 \circledast 0.5 = 1 = 0.6 \circledast 0.4$, but $0.7 + 0.5 \neq 0.6 + 0.4$ and so the converse of Proposition 4.7.6 does not hold.

Example 4.7.7 Let $V = \{a, b, c, d, z\}$ and let $E = \{(a, b), (a, c), (b, d), (c, d), (d, z)\}$. Define $\mu : E \to [0, 1]$ as follows:

$$\mu(a, b) = 0.7, \quad \mu(a, c) = 0.5, \quad \mu(b, d) = 0.7, \quad \mu(c, d) = 0.5, \quad \mu(d, z) = 1.$$

Let $F_{xy}^{\circledast} = \mu(x, y)$ for all $x, y \in V$. Then $F_{bd}^{\circledast} + F_{cd}^{\circledast} = (0.7 + 0.5) \wedge 1 = 1 = F_{dz}^{\circledast}$. In fact, for all $j \in A \backslash \{a, z\}$, $\sum_{i \in V}^{\circledast} F_{ij} = \sum_{i \in V}^{\circledast} F_{ji}$. However, $F_{bd} + F_{cd} = 0.7 + 0.5 = 1.2$ and $F_{dz} = 1$. Thus, $\sum_{i \in V}^{\circledast} F_{ij} = \sum_{i \in V}^{\circledast} F_{ji} \not\Rightarrow \sum_{i \in V} F_{ij} = \sum_{i \in V} F_{ji}$.

Proposition 4.7.8 *Suppose for* $j \in V \backslash \{a, z\}$ *that* $\sum_{i \in V} F_{ij} = \sum_{i \in V} F_{ji}$. *Then* $\sum_{i \in V}^{\circledast} F_{ai} = \sum_{i \in V}^{\circledast} F_{iz}$.

Proof Clearly, $\sum_{j \in V}(\sum_{i \in V} F_{ij}) = \sum_{j \in V}(\sum_{i \in V} F_{ji})$. Thus,

$$0 - \sum_{j \in V}\left(\sum_{i \in V} F_{ij} - \sum_{i \in V} F_{ji}\right)$$

$$= \left(\sum_{i \in V} F_{iz} - \sum_{i \in V} F_{zi}\right) + \left(\sum_{i \in V} F_{ia} - \sum_{i \in V} F_{ai}\right) + \sum_{j \in V \backslash \{a, z\}}\left(\sum_{i \in V} F_{ij} - \sum_{i \in V} F_{ji}\right)$$

$$= \sum_{i \in V} F_{iz} - \sum_{i \in V} F_{ai}$$

because $F_{zi} = 0 = F_{ia}$ for all $i \in V$ and $\sum_{i \in V} F_{ij} = \sum_{i \in V} F_{ji}$ if $j \in V \backslash \{a, z\}$. By Proposition 4.7.6, $\sum_{i \in V}^{\circledast} F_{ij} = \sum_{i \in V}^{\circledast} F_{ji}$ for $a \neq j \neq z$. The desired result now holds. ∎

Proposition 4.7.9 *Suppose that* $\sum_{i \in V} F_{iz}$ *is the max flow for addition. Then* $\sum_{i \in V}^{\circledast} F_{iz}$ *is the max flow for the Frank-Weber conorm.*

Proof Suppose that $\sum_{i \in V} F_{iz} \geq 1$. Then $\sum_{i \in V}^{\circledast} F_{iz} = 1$ and thus $\sum_{i \in V}^{\circledast} F_{iz}$ is the max flow for $\circledast$. Suppose that $\sum_{i \in V} F_{iz} < 1$. Then $\sum_{i \in V}^{\circledast} F_{iz} = \sum_{i \in V} F_{iz}$. Suppose that $\sum_{i \in V}^{\circledast} F_{iz}$ is not the maximal flow for $\circledast$. Then there is a flow $\sum_{i \in V}^{\circledast} F_{iz}' > \sum_{i \in V}^{\circledast} F_{iz}$. Suppose $\sum_{i \in V}^{\circledast} F_{iz}' < 1$. Then $\sum_{i \in V} F_{iz}' = \sum_{i \in V}^{\circledast} F_{iz}' > \sum_{i \in V}^{\circledast} F_{iz} = \sum_{i \in V} F_{iz}$ which contradicts the maximality of $\sum_{i \in V} F_{iz}$. Suppose that $\sum_{i \in V}^{\circledast} F_{iz}' = 1$. Then $\sum_{i \in V} F_{iz}' \geq 1$, a contradiction of the assumption $\sum_{i \in V} F_{iz}' < 1$. ∎

In the following, we consider the flow of traffic from Mexico into the United States [14]. The total flow, transient plus nontransient, is rated high. The transient flow is rated medium. We wish to find the rating of the flow which is destination, but not transient. Let x represent this number and let u denote a conorm. Then we wish to solve the equation $u(a, x) = c$ for x, where a denotes the rating for transient flow and c is the rating for total flow. Let u denote the algebraic sum conorm. Then $0.6 + x - (0.6)x = 0.8$. Solving, we find $x = 0.5$. Let u denote the Dombi conorm. Then

$$\frac{0.6 + x - 2(0.6)x}{1 - 0.6x} = 0.8.$$

Solving, we find $x = \frac{5}{7} \approx 0.7$. If u is the Frank-Weber conorm, then $(0.6 + x) \wedge 1 = 0.8$. Hence, $x = 0.2$.

This suggests that the selection of an appropriate conorm will require some effort if indeed one exists.

References

1. R.K. Ahuja, T.L. Magnanti, J.B. Orlin, *Network Flows: Theory, Algorithms, and Applications* (Prentice Hall, New Jersey, 1993)
2. J.A. Bondy, U.S.R. Murty, *Graph Theory with Applications*, 5th edn. (Macmillan Press, New York, 1982)
3. H. Bustince, E. Barrenechea, J. Fernandez, M. Pagola, J. Montero, C. Guerra, Contrast of a fuzzy relation. Inf. Sci. **180**, 1326–1344 (2010)
4. M.H. Dahshan, Maximum bandwidth node disjoint paths. Int. J. Adv. Comput. Sci. Appl. 3(3), 48–56 (2012)
5. H.N. Gabow, R.E. Tarjan, Algorithms for two bottleneck optimization problems. J. Algorithms **9**, 411–417 (1988)
6. F. Harary, *Graph Theory* (Addison-Wesley, New York, 1973)
7. R. Johnsonbaugh, *Discrete Mathematics* (MacMillan, New York, 1984)
8. S. Mathew, J.N. Mordeson, Directed fuzzy networks as a model to illicit flows. New Math. Nat. Comput. **13**, 219–229 (2017)
9. S. Mathew, M.S. Sunitha, Node connectivity and arc connectivity of a fuzzy graph. Inf. Sci. **180**, 519–531 (2010)
10. K. Menger, Zur allgemeinen Kurventheorie. Fund. Math. **10**, 96–115 (1927)
11. J.N. Mordeson, P.S. Nair, *Fuzzy Graphs and Fuzzy Hypergraphs* (Physica-Verlag, New York, 2000)
12. A. Rosenfeld, Fuzzy graphs, in *Fuzzy Sets and Their Applications to Cognitive and Decision Processes*, ed. by L.A. Zadeh, K.S. Fu, K. Tanaka, M. Shimura (Academic Press, Dublin, 1975), pp. 77–95

13. A. Shapira, R. Yuster, U. Zwick, All-pairs bottleneck paths in vertex weighted graphs, Proc. SODA 978-985 (2007)
14. United Nations Office on Drugs and Crime (UNODC) Trafficking in Persons Global Patterns, April 2006
15. R.T. Yeh, S.Y. Bang, Fuzzy relations, fuzzy graphs and their applications to clustering analysis, in *Fuzzy Sets and Their Applications*, ed. by L.A. Zadeh, K.S. Fu, M. Shimura (Academic Press, Dublin, 1975), pp. 125–149
16. L.A. Zadeh, Fuzzy sets. Inf. Control **8**, 338–353 (1965)
17. L.A. Zadeh, Is there a need for fuzzy logic? Inf. Sci. **178**, 2751–2779 (2008)

References

Chapter 5
Complementary Fuzzy Incidence Graphs

5.1 (s, t)-Fuzzy Incidence Graphs

Recall that if x is a member of a set X and $t \in [0, 1]$, then we define the fuzzy subset x_t of X by $x_t(x) = t$ and $x_t(y) = 0$ for all $y \in X \setminus \{x\}$. We call x_t a **fuzzy singleton**.

The results in the first five sections are based on [14]. We develop properties concerning (s, t)-fuzzy incidence graphs and then apply the results to the problem of illegal immigration. We first review some definitions.

Let (V, E) be a graph. Then (V, E, I) is called an **incidence graph** if $I \subseteq V \times E$. If $(u, vw) \in I$, then (u, vw) is called an **incidence pair** or simply a **pair**. Two edges uv and vw are said to be **adjacent** if all four pairs (u, uv), (v, uv), (v, vw), and (w, vw) are in I.

Let (V, E) be a graph, σ a fuzzy subset of V, μ a fuzzy subset of E, and Ψ a fuzzy subset of $V \times E$. If $\Psi(u, e) \leq \sigma(u) \wedge \mu(e)$ for all $(u, e) \in I \times E$, then Ψ is called a **fuzzy incidence** of G. If Ψ is a fuzzy incidence, then $G = (\sigma, \mu, \Psi)$ is called a **fuzzy incidence graph**.

Definition 5.1.1 $G = (\sigma, \mu, \Psi)$ is called a **quasi-fuzzy incidence graph** with respect to c if for all $(x, xy) \in I$ and for all $t \in [0, 1]$, $(x, xy)_t \subseteq \Psi$ implies either $t \leq \sigma(x) \wedge \mu(xy)$ or $t > c(\sigma(x) \wedge \mu(xy))$.

We next consider some results from [13, 14].

Theorem 5.1.2 $G = (\sigma, \mu, \Psi)$ *is a quasi-fuzzy incidence graph with respect to c if and only if for all $x, y \in V$, $\Psi(x, xy) \wedge e_c \leq \sigma(x) \wedge \mu(xy)$.*

Proof Suppose G is a quasi-fuzzy incidence graph. Let $x, y \in V$. Suppose $(x, xy)_{t_0} \subseteq \Psi$. Then $\Psi(x, xy) \geq t_0$. Suppose $t_0 > \sigma(x) \wedge \mu(xy)$. Then by hypothesis, $t_0 > c(\sigma(x) \wedge \mu(xy))$. Thus, $t_0 > e_c$. Suppose $\Psi(x, xy) \wedge e_c > \sigma(x) \wedge \mu(xy)$. (We show this leads to a contradiction.) Then $e_c > \sigma(x) \wedge \mu(xy)$. Hence, $\sigma(x) \wedge \mu(xy) < e_c \leq c(\sigma(x) \wedge \mu(xy)) < t_0$ and $e_c = c(e_c) \leq c(\sigma(x) \wedge \mu(xy))$. Let t be such that

© Springer International Publishing AG 2018
J. N. Mordeson et al., *Fuzzy Graph Theory with Applications to Human Trafficking*, Studies in Fuzziness and Soft Computing 365,
https://doi.org/10.1007/978-3-319-76454-2_5

$\sigma(x) \wedge \mu(xy) < t < e_c$. Then $t < t_0$ and so $\Psi(x, xy) \geq t$. Because $t > \sigma(x) \wedge \mu(xy)$, we have by an argument as above that $t > c(\sigma(x) \wedge \mu(xy))$ and $t > e_c$, a contradiction. Thus, $\Psi(x, xy) \wedge e_c \leq \sigma(x) \wedge \mu(xy)$.

Conversely, suppose for all $x, y \in V$, $\Psi(x, xy) \wedge e_c \leq \sigma(x) \wedge \mu(xy)$. Suppose $(x, xy)_t \subseteq \Psi$, i.e., $\Psi(x, xy) \geq t$. If $t \leq \sigma(x) \wedge \mu(xy)$, we have the desired result. Suppose $t > \sigma(x) \wedge \mu(xy)$. Then $\Psi(x, xy) > \sigma(x) \wedge \mu(xy)$. Because also $\Psi(x, xy) \wedge e_c \leq \sigma(x) \wedge \mu(xy)$, $\sigma(x) \wedge \mu(xy) \geq e_c$. Hence, $c(\sigma(x) \wedge \mu(xy)) \leq c(e_c) = e_c$. Thus, $t > \sigma(x) \wedge \mu(xy) \geq e_c \geq c(\sigma(x) \wedge \mu(xy))$. ∎

Given any $t \in [0, 1]$, there exists an infinite number of functions $c : [0, 1] \to [0, 1]$ with equilibrium t. This and Theorem 5.1.2 motivate for the following definition.

Definition 5.1.3 $G = (\sigma, \mu, \Psi)$ is called a **quasi-fuzzy incidence graph** with respect to $t \in (0, 1]$ if for all $(x, y) \in E$, $\Psi(x, xy) \wedge t \leq \sigma(x) \wedge \mu(xy)$.

In fact, we consider the following definition.

Definition 5.1.4 Let $s, t \in [0, 1]$ with $s < t$. Then $G = (\sigma, \mu, \Psi)$ is called an $(s, t]$-**fuzzy incidence graph** if G^* is an incidence graph and for all $x, y \in V$, $\Psi(x, xy) \wedge t \leq (\sigma(x) \wedge \mu(xy)) \vee s$.

Let $G = (\sigma, \mu, \Psi)$ be an $(s, t]$-fuzzy graph. If $s = 0$ and $t = 1$, then G is just a fuzzy incidence graph. If $s = 0$, then G is a quasi-fuzzy incidence graph with respect to t.

Let $G^r = (\sigma^r, \mu^r, \Psi^r)$ for all $r \in [0, 1]$, where σ^r, μ^r, and Ψ^r are r-level sets. Clearly, G^r is an incidence graph.

Theorem 5.1.5 G is an $(s, t]$-*fuzzy incidence graph if and only if for all* $r \in (s, t]$, G^r *is an incidence subgraph of* (V, E, I).

Proof Suppose G is an $(s, t]$-fuzzy incidence graph. Let $r \in (s, t]$. Suppose $(x, xy) \in \Psi^r$. Then $\Psi(x, xy) \geq r$. Hence, $(\sigma(x) \wedge \mu(xy)) \vee s \geq \Psi(x, xy) \wedge t \geq r \wedge t = r$. Because $r > s$, we have $\sigma(x) \wedge \mu(xy) \geq r$. Thus, $\sigma(x) \geq r$ and $\mu(xy) \geq r$. Hence, $x \in \sigma^r$ and $xy \in \mu^r$. Thus, G^r is a subgraph of (V, E, I).

Conversely, suppose G^r is a subgraph of (V, E, I) for all $r \in (s, t]$. Let $xy \in E$ and let $\Psi(x, xy) = r_0$. Suppose $r_0 \leq s$. Then $\Psi(x, xy) \wedge t = \Psi(x, xy) = r_0 \leq s \leq (\sigma(x) \wedge \mu(xy)) \vee s$. Suppose $s < r_0 \leq t$. Then because G^{r_0} is an incidence subgraph of (V, E, I), we have $x \in \sigma^{r_0}$, $xy \in \mu^{r_0}$. Hence, $\sigma(x) \wedge \mu(xy) \geq r_0 = \Psi(x, xy)$. Thus, $(\sigma(x) \wedge \mu(xy)) \vee s \geq \Psi(x, xy) \wedge t$. Suppose $r_0 > t$. Then $\Psi(x, xy) \wedge t = t$. Because $\Psi(x, xy) > t$, we have $(x, xy) \in \Psi^t$. Because G^t is an incidence subgraph of (V, E, I), we have $x \in \sigma^t$, $xy \in \mu^t$. Thus, $\sigma(x) \wedge \mu(x, xy) \geq t$. Hence, $\Psi(x, xy) \wedge t \leq (\sigma(x) \wedge \mu(xy)) \vee s$. Thus, G is an $(s, t]$-fuzzy incidence graph. ∎

Let $r, s, t \in [0, 1]$. Suppose $s < t$. Let X be a set and λ a fuzzy subset of X. Let $\lambda_{st}^r = \{x \in X \mid \lambda(x) \vee s \geq r \wedge t\}$. Then λ_{st}^r is called an r_{st}-level set (or cut) of λ. Let $G_{st}^r = (\sigma_{st}^r, \mu_{st}^r, \Psi_{st}^r)$.

Theorem 5.1.6 G is an (s, t)-fuzzy incidence graph if and only if for all $r \in [0, 1]$, G_{st}^r is an incidence subgraph of (V, E, I).

Proof Suppose G is an (s, t)-fuzzy incidence graph. Let $r \in [0, 1]$. Suppose $(x, xy) \in \Psi_{st}^r$. Then $\Psi(x, xy) \vee s \geq r \wedge t$. Now $(\sigma(x) \wedge \mu(xy)) \vee s \geq \Psi(x, xy) \wedge t$ and so $(\sigma(x) \wedge \mu(xy)) \vee s \geq (\Psi(x, xy) \wedge t) \vee s = (\Psi(x, xcy) \vee s) \wedge (t \vee s) \geq (r \wedge t) \wedge t = r \wedge t$. Thus, $\sigma(x) \vee s \geq r \wedge t$ and $\mu(xy) \vee s \geq r \wedge t$. Hence, $x \in \sigma_{st}^r$, $xy \in \mu^r$. Thus, $(\sigma_{st}^r, \mu_{st}^r, \Psi_{st}^r)$ is an incidence subgraph of (V, E, I).

Conversely, suppose $(\sigma_{st}^r, \mu_{st}^r, \Psi_{st}^r)$ is an incidence subgraph of (V, E, I) for all $r \in [0, 1]$. Let $(x, xy) \in I$ and let $\Psi(x, xy) = r_0$. Then $\Psi(x, xy) \vee s \geq r_0 \wedge t$. Hence, $(x, xy) \in \Psi_{st}^{r_0}$. Thus, $x \in \sigma_{st}^{r_0}, xy \in \mu_{st}^{r_0}$. Hence, $\sigma(x) \vee s \geq r_0 \wedge t$ and $\mu(xy) \vee s \geq r_0 \wedge t$. Thus, $(\sigma(x) \wedge \mu(xy)) \vee s \geq r_0 \wedge t = \Psi(x, xy) \wedge t$. Hence, G is an (s, t)-fuzzy incidence graph. ∎

5.2 (s, t)-Fuzzy Incidence Cut Vertices and (s, t)-Fuzzy Incidence End Vertices

Let $G = (\sigma, \mu)$ be a fuzzy graph. Let $x, y \in V$. A path from x to y in G is a sequence $\pi : x = x_0, x_1, \ldots, x_n = y$ of points x_i in V such that $\sigma(x_i) > 0$ for $i = 0, 1, \ldots, n$ and $\mu(x_{i-1}, x_i) > 0$ for $i = 1, \ldots, n$. The **strength** of the path π is $\wedge \{ \mu(x_{i-1}, x_i) \mid i = 1, \ldots, n \}$. Define $\text{CONN}_G(x, y)$ to be the maximum of the strengths of all paths from x to y. If $x \in V$, we use the notation $G \backslash x$ to denote the fuzzy graph determined by setting $\sigma(x) = 0$ and $\mu(xy) = 0$ for all $y \in V$ such that $xy \in E$. Properties of fuzzy cut vertices and fuzzy end vertices can be found in [4–6].

Definition 5.2.1 Let $G = (\sigma, \mu, \Psi)$ be a fuzzy incidence graph.

(*i*) Let $x \in V$. Then x is called a **fuzzy incidence cut vertex** if there exists $w, z \in V \cup E \backslash \{x\}$ such that $\text{ICONN}_G(w, z) > \text{ICONN}_{G \backslash x}(w, z)$.

(*ii*) Let $xy \in E$. Then xy is called a **fuzzy incidence cut edge** if there exists $w, z \in V \cup E \backslash \{xy\}$ such that $\text{ICONN}_G(w, z) > \text{ICONN}_{G \backslash xy}(w, z)$.

Definition 5.2.2 Let $G = (\sigma, \mu, \Psi)$ be a fuzzy incidence graph.

(*i*) Let $x \in V$. Then x is called a **fuzzy incidence end vertex** if there exists $y \in V \backslash \{x\}$ such that $\Psi(x, xy) > \vee \{\Psi(x, xz) \mid z \in V, z \neq y\}$.

(*ii*) Let $xy \in E$. Then xy is called a **fuzzy incidence end edge** if $\Psi(x, xy) \neq \Psi(y, xy)$.

Definition 5.2.3 Let $s, t \in [0, 1]$ be such that $s < t$. Let $G = (\sigma, \mu, \Psi)$ be a fuzzy incidence graph.

(*i*) Let $x \in V$. Then x is called an (s, t)-**fuzzy incidence cut vertex** if there exists $w, z \in V \backslash \{x\}$ such that $\text{ICONN}_G(w, z) \wedge t > \text{ICONN}_{G \backslash x}(w, z) \vee s$.

(*ii*) Let $xy \in E$. Then xy is called an (s, t)-**fuzzy incidence cut edge** if there exists $w, z \in V \backslash \{x\}$ such that $\text{ICONN}_G(w, z) \wedge t > \text{ICONN}_{G \backslash xy}(w, z) \vee s$.

Definition 5.2.4 Let $s, t \in [0, 1]$ be such that $s < t$. Let $G = (\sigma, \mu, \Psi)$ be a fuzzy incidence graph

(1) Let $x \in V$. Then x is called an $(s, t]$-**fuzzy incidence end vertex** if there exists $y \in V \backslash \{x\}$ such that $\Psi(x, xy) \wedge t > \vee \{\Psi(x, xz) \mid z \in V, z \neq y\} \vee s$.

If $s = 0$ and $t = 1$, then $(0, 1]$-fuzzy incidence end vertices and $(0, 1]$-fuzzy incidence cut vertices are simply fuzzy incidence end vertices and fuzzy incidence cut vertices of G, respectively.

Suppose that $x \in V$ is an $(s, t]$-fuzzy incidence end vertex. Then the $y \in V$ such that $\Psi(x, xy) \wedge t > \vee \{\Psi(x, xz) \mid z \in V, z \neq y\} \vee s$ is unique. This follows because for all $u \in V \backslash \{y\}$, $\Psi(x, xy) \geq \Psi(x, xy) \wedge t > \mu(x, ux)$. Also, if x is an $(s, t]$-fuzzy incidence end vertex, then x is also an $(r, w]$-fuzzy incidence vertex for all $r \in [0, s]$ and $w \in [t, 1]$.

Example 5.2.5 Let $V = \{x, y, z\}$ and $E = \{(x, y), (x, z), (y, z)\}$. Let $s = 0.05$ and $t = 0.95$. Define the fuzzy subset σ of V and the fuzzy subset μ of E as follows: $\sigma(x) = \sigma(y) = \sigma(z) = 1$ and $\mu(x, y) = \mu(y, z) = \mu(x, z) = 1$ and $\Psi(x, xy) = \Psi(y, yz) = \Psi(y, xy) = \Psi(z, xz) = \Psi(z, yz) = 1$ and $\Psi(x, xz) = 0.99$. Then x is a fuzzy incidence end vertex, but not an $(s, t]$-fuzzy incidence end vertex. We also have that y is a fuzzy incidence cut vertex, but not an $(s, t]$ -fuzzy incidence cut vertex.

Example 5.2.6 Let $V = \{x, y, z\}$ and $E = \{(x, y), (x, z), (y, z)\}$. Let $s = 0.05$ and $t = 0.95$. Define the fuzzy subset σ of V and the fuzzy subset μ of E as follows: $\sigma(x) = \sigma(y) = \sigma(z) = 1$ and $\mu(x, y) = \mu(y, z) = \mu(x, z) = 1$. Define Ψ as follows: $\Psi(x, xz) = 0.04$ and $\Psi(x, xy) = 0.01$ and $\Psi(u, uv) = 1$ elsewhere. Then x is a fuzzy incidence end vertex, but not an $(s, t]$-fuzzy incidence end vertex. We also have that y is a fuzzy incidence cut vertex, but not an $(s, t]$-fuzzy incidence cut vertex because $\text{ICONN}_G(x, z) = 0.04 > 0.01 = \text{ICONN}_{G \backslash y}(x, z)$ and $\text{ICONN}_G(x, z) \wedge 0.05 = 0.04 \not> 0.05 = \text{ICONN}_{G \backslash y}(x, z) \vee 0.05$.

Definition 5.2.7 Let $x, y \in V$. Then xy is called an **incidence neighbor** of x if $\Psi(x, xy) > 0$. xy is called a t-**strong incidence neighbor** of x if xy is a neighbor of x and $\mu(x, xy) \wedge t \geq \text{CONN}_G(x, xy) \wedge t$.

Theorem 5.2.8 *Let $z \in V$ be an $(s, t]$-fuzzy incidence cut vertex of G. Then z has at least two t-strong incidence neighbors.*

Proof Because z is an $(s, t]$-fuzzy cut vertex of G, there exists $x, y \in V \cup E \backslash \{z\}$ such that $\text{CONN}_G(x, y) \wedge t > \text{CONN}_{G \backslash z}(x, y) \vee s$. Hence, there exists a path P from x to y and passing through z such that $s(P) \wedge t = \text{CONN}_G(x, y) \wedge t$. Because $x \neq z \neq y$, z has two neighbors in P, say zu and zv. Because (z, zu) and (z, zv) are on P, $\Psi(u, uz) \wedge t \geq s(P) \wedge t$ and $\Psi(z, vz) \wedge t \geq s(P) \wedge t$. Suppose $\Psi(u, uz) \wedge t < \text{CONN}_G(u, uz) \wedge t$. Then there exists a path Q from u to z such that $s(Q) \wedge t > \Psi(u, uz) \wedge t$. Let $u'z$ in Q be a neighbor of z. We show $u' \neq v$. Suppose $u' = v$. Let Q' be the path $x \ldots u \ldots u' \ldots v \ldots y$. Then $s(Q') \wedge t \geq s(P) \wedge t$, but Q' does not pass through z. However, this is impossible because $\text{CONN}_{G \backslash z}(x, y) \wedge t <$

$\text{CONN}_G(x, y) \wedge t$. Let P' denote the path $x \ldots u \ldots u'zv \ldots y$. Then $s(P') \wedge t \geq s(P) \wedge t = \text{CONN}_G(x, y) \wedge t$. Suppose $\Psi(z, zu') \wedge t < \text{CONN}_G(z, zu) \wedge t$. Then we can apply the above process again. Now length$(P') >$ length(P). Hence, this process must end in a finite number of steps yielding a neighbor u^*z of z such that $\mu(z, zu^*) \wedge t \geq \text{CONN}_G(z, zu) \wedge t$. As above $u^* \neq v$. Hence, a similar argument can be used to yield a neighbor v^*z of z such that $u^* \neq v^*$ and $\Psi(z, zv^*) \wedge t \geq \text{CONN}_G(z, zu) \wedge t$. Thus, z has two t-strong neighbors, namely u^*z and v^*z. ∎

Definition 5.2.9 Let $x \in V$. Then x is called a **weak fuzzy incidence end node** if there exists $t \in (0, h(\Psi)]$ such that x is an end node in G^t, where $h(\Psi)$ is the height of Ψ, i.e., $h(\Psi) = \vee\{\Psi(x, xy) \mid x, y \in V\}$.

Definition 5.2.10 Let $x \in V$. Then x is called a **partial fuzzy incidence end node** if x is an end node in G^t for all $t \in (d(\Psi), h(\Psi)] \cup \{h(\Psi)\}$, where $d(\Psi)$ is the depth of Ψ, i.e., $d(\Psi) = \wedge\{\Psi(x, xy) \mid x, y \in V\}$.

Proposition 5.2.11 *Let $x \in V$. Then x is a $(0, 1]$-fuzzy incidence end vertex if and only if x is a weak fuzzy incidence end vertex.*

Proof Suppose x is a $(0, 1]$-fuzzy incidence end vertex. Then there exists $y \in V$ such that $\Psi(x, y) = \Psi(x, xy) \wedge 1 > \vee\{\Psi(x, xz) \mid z \in V, z \neq y\}$. Thus, x is an end vertex in G^a, where $a = \Psi(x, xy)$. Hence, x is a weak fuzzy incidence end vertex. Conversely, suppose x is a weak fuzzy incidence end vertex. Then x is an end node in G^b for some $b \in (0, h(\Psi)]$. Hence, there exists a unique $y \in V$ such that $\Psi(x, xy) > b$ and for all other $z \in V$, $\mu(x, xz) < b$. Thus, x is an $(0, 1]$-fuzzy end incidence vertex. ∎

Corollary 5.2.12 *Let $x \in V$. Then x is a partial fuzzy incidence end vertex if and only if x is a $(d(\mu), 1]$-fuzzy incidence end vertex.*

Corollary 5.2.13 *Let $x \in V$. Then x is an end vertex of G^* if and only if x is a $(0, d(\Psi)]$-fuzzy end vertex.*

5.3   Complementary Fuzzy Incidence Graphs

Recall that **complementary fuzzy graph** $\widehat{G} = (\tau, \nu)$ is a pair such that τ is a fuzzy subset of V and ν is a fuzzy subset of E such that for all $x, y \in V$, $\nu(x, y) \geq \tau(x) \vee \tau(y)$. Let $\tau^{\#}$ denote the cosupport of τ and $\nu^{\#}$ the cosupport of ν. Let $\widehat{G}^{\#} = (\tau^{\#}, \nu^{\#})$.

The notions of a fuzzy graph and a complementary fuzzy graph can be combined in a natural way to form an intuitionistic fuzzy graph. Results on intuitionistic fuzzy graphs can be found in [1–3].

Let $G = (\sigma, \mu, \Psi)$ be a fuzzy incidence graph and (τ, ν) be a complementary fuzzy graph. Let Ω be a fuzzy subset of I. Then $\widehat{G} = (\tau, \nu, \Omega)$ is called a **complementary fuzzy incidence graph** if for all $(x, xy) \in I$, $\Omega(x, xy) \geq \tau(x) \vee \nu(xy)$.

5.4 Complementary Quasi-Fuzzy Incidence Graphs

In this section, we provide the basic results needed for the remainder of the section. The proofs of parts (i) and (ii) of the following results are duals of each other. Let c denote an involutive fuzzy complement. Then it is easily shown that for all $a, b \in [0, 1]$, $a \wedge b = c(c(a) \vee c(b))$ and $a \vee b = c(c(a) \wedge c(b))$. From this observation, it can be seen part (i) is obtainable from part (ii) and vice versa. Thus, a result concerning (σ, μ) has a complementary result for (σ^c, μ^c). Assume a general result R holds for (σ, μ). Given (τ, ν), the result R holds for (τ^c, ν^c). Hence, its dual holds for (τ, ν). Consequently, we prove only part (ii) in the following results because the case for part (i) has been considered previously.

Definition 5.4.1 Let c and $\widehat{c}$ be involutive fuzzy complements. Let (σ, μ) and (τ, ν) be fuzzy graphs. Let Ψ and Ω be fuzzy subsets of I.

(i) $G = (\sigma, \mu, \Psi)$ is called a **quasi-fuzzy incidence graph with respect to** c if for all $x, y \in V$ and for all $t \in [0, 1]$, $\Psi(x, xy) \geq t \Rightarrow t \leq \sigma(x) \wedge \mu(xy)$ or $t > c(\sigma(x) \wedge \mu(xy))$.

(ii) $\widehat{G} = (\tau, \nu, \Omega)$ is called a **complementary quasi-fuzzy incidence graph with respect to** $\widehat{c}$ if for all $x, y \in V$ and for all $t \in [0, 1]$, $\Omega(x, xy) \leq t \Rightarrow t \geq \tau(x) \vee \nu(xy)$ or $t < \widehat{c}(\tau(x) \vee \nu(xy))$.

(iii) $(G, \widehat{G})$ is called a **quasi-fuzzy graph** with respect to $(c, \widehat{c})$ if G is a quasi-fuzzy incidence graph with respect to c and $\widehat{G}$ is a complementary quasi-fuzzy incidence graph with respect to $\widehat{c}$.

Theorem 5.4.2 (i) $G = (\sigma, \mu, \Psi)$ is a quasi-fuzzy incidence graph with respect to c if and only if for all $x, y \in V$, $\Psi(x, xy) \wedge e_c \leq \sigma(x) \wedge \mu(xy)$.

(ii) Let $\widehat{G} = (\tau, \nu, \Omega)$. Then $\widehat{G}$ is a complementary quasi-fuzzy incidence graph with respect to $\widehat{c}$ if and only if for all $x, y \in V$, $\Omega(x, xy) \vee e_{\widehat{c}} \geq \tau(x) \vee \nu(xy)$.

Proof (ii) Let $x, y \in V$. Suppose $\Omega(x, xy) \leq t$. Suppose $t < \tau(x) \vee \nu(xy)$. Then by hypothesis, $t < \widehat{c}(\tau(x) \vee \nu(xy))$. Thus, $t < e_{\widehat{c}}$. Suppose $\Omega(x, xy) \vee e_{\widehat{c}} < \tau(x) \vee \nu(xy)$. We obtain a contradiction. Then $e_{\widehat{c}} < \tau(x) \vee \nu(xy)$. Hence, $\tau(x) \vee \nu(xy) > e_{\widehat{c}} \geq c(\tau(x) \vee \nu(xy)) > t$ and $e_{\widehat{c}} = c(e_{\widehat{c}}) \geq \widehat{c}(\tau(x) \vee \nu(xy))$. Let t' be such that $\tau(x) \vee \nu(xy) > t' > e_{\widehat{c}}$. Then $t' > t$ and so $\Omega(x, xy) \leq t'$. Because $t' < \tau(x) \vee \nu(xy)$, we have by an argument as above that $t' < \widehat{c}(\tau(x) \vee \nu(xy))$ and $t' < e_{\widehat{c}}$, a contradiction. Thus, $\Omega(x, y) \vee e_{\widehat{c}} \geq \tau(x) \vee \mu(xy)$.

Conversely, suppose for all $x, y \in V$, $\Omega(x, xy) \vee e_{\widehat{c}} \geq \tau(x) \vee \nu(xy)$. Suppose $\Omega(x, xy) \leq t$. If $t \geq \tau(x) \vee \nu(xy)$, we have the desired result. Suppose $t < \tau(x) \vee \nu(xy)$. Then $\Omega(x, xy) < \tau(x) \vee \nu(xy)$. Because also $\Omega(x, xy) \vee e_{\widehat{c}} \geq \tau(x) \vee \nu(xy)$, $\tau(x) \vee \nu(xy) \leq e_{\widehat{c}}$. Hence, $\widehat{c}(\tau(x) \vee \nu(xy)) \geq c(e_{\widehat{c}}) = e_{\widehat{c}}$. Thus, $t < \tau(x) \vee \nu(xy) \leq e_{\widehat{c}} \leq \widehat{c}(\tau(x) \vee \nu(xy))$. ∎

Definition 5.4.3 (i) $G = (\sigma, \mu, \Psi)$ is called a **quasi-fuzzy incidence graph** with respect to $t \in (0, 1]$ if for all $x, y \in V$, $\Psi(x, xy) \wedge t \leq \sigma(x) \wedge \mu(y)$.

(ii) $\widehat{G} = (\tau, \nu, \Omega)$ is called a **complementary quasi-fuzzy incidence graph** with respect to $\widehat{t} \in [0, 1)$ if for all $x, y \in V$, $\Omega(x, xy) \vee \widehat{t} \geq \tau(x) \vee \nu(xy)$.

In fact, we consider the following definition.

Definition 5.4.4 (i) Let $s, t \in [0, 1]$ with $s < t$. Then $G = (\sigma, \mu, \Psi)$ is called an $(s, t]$-**fuzzy incidence graph** if G^* is a graph and for all $x, y \in V$, $\Psi(x, xy) \wedge t \leq (\sigma(x) \wedge \mu(xy)) \vee s$.

(ii) Let $\widehat{s}, \widehat{t} \in [0, 1)$ with $\widehat{s} > \widehat{t}$. Then $\widehat{G} = (\tau, \nu, \Omega)$ is called a **complementary** $[\widehat{t}, \widehat{s})$-**fuzzy incidence graph** if $\widehat{G}^{\#}$ is a graph and for all $x, y \in V$, $\Omega(x, xy) \vee \widehat{t} \geq (\tau(x) \vee \nu(y)) \wedge \widehat{s}$.

Let $\widehat{r}, \widehat{s}, \widehat{t} \in [0, 1]$. Suppose $\widehat{t} < \widehat{s}$. Let X be a set and λ be a fuzzy subset of X. Let $\lambda_{\widehat{ts}}^{\widehat{r}} = \{x \in X \mid \lambda(x) \wedge \widehat{s} \leq \widehat{r} \vee \widehat{t}\}$. Then $\lambda_{\widehat{ts}}^{\widehat{r}}$ is called an $\widehat{r_{ts}}$-**level set** (or cut) of λ. Let $G_{\widehat{ts}}^{\widehat{r}} = (\tau_{\widehat{ts}}^{\widehat{r}}, \nu_{\widehat{ts}}^{\widehat{r}})$.

Theorem 5.4.5 (i) G is an $(s, t]$-fuzzy incidence graph if and only if for all $r \in (s, t]$, G^r is an incidence subgraph of (V, E, I).

(ii) $\widehat{G}$ is a $[\widehat{t}, \widehat{s})$-fuzzy complementary incidence graph if and only if for all $\widehat{r} \in [0, 1]$, $\widehat{G}^r$ is an complementary incidence subgraph of (V, E).

Proof (ii) Suppose $\widehat{G}$ is an $[\widehat{t}, \widehat{s})$-fuzzy complementary incidence graph. Let $\widehat{r} \in [0, 1]$. Suppose $(x, xy) \in \Omega_{\widehat{ts}}^{\widehat{r}}$. Then $\Omega(x, xy) \wedge \widehat{s} \leq \widehat{r} \vee \widehat{t}$. Now $(\tau(x) \vee \nu(xy)) \wedge \widehat{s} \leq \Omega(x, xy) \vee \widehat{t}$ and so $(\tau(x) \vee \nu(xy)) \wedge \widehat{s} \leq (\Omega(x, xy) \vee \widehat{t}) \wedge \widehat{s} = (\Omega(x, xy) \wedge \widehat{s}) \vee (\widehat{t} \wedge \widehat{s}) \leq (\widehat{r} \vee \widehat{t}) \vee \widehat{t} = (\widehat{r} \vee \widehat{t})$. Thus, $\tau(x) \wedge \widehat{s} \leq (\widehat{r} \vee \widehat{t})$ and $\nu(xy) \wedge \widehat{s} \leq (\widehat{r} \vee \widehat{t})$. Hence, $x \in \tau_{\widehat{ts}}^{\widehat{r}}$ and $xy \in \nu_{\widehat{ts}}^{\widehat{r}}$. Thus, $\widehat{G}_{\widehat{ts}}^{\widehat{r}}$ is a subgraph of (V, E).

Conversely, suppose for all $\widehat{r} \in [0, 1]$, $\widehat{G}_{\widehat{ts}}^{\widehat{r}}$ is an incidence subgraph of (V, E, I). Let $(x, xy) \in I$ and let $\Omega(x, xy) = \widehat{r_0}$. Then $\Omega(x, xy) \wedge \widehat{s} \leq \widehat{r_0} \vee \widehat{t}$. Hence, $(x, xy) \in \Omega_{\widehat{ts}}^{\widehat{r_0}}$. Thus, $x \in \tau_{\widehat{ts}}^{\widehat{r_0}}$ and $xy \in \nu_{\widehat{ts}}^{\widehat{r_0}}$. Hence, $\tau(x) \wedge \widehat{s} \leq \widehat{r_0} \vee \widehat{t}$ and $\nu(xy) \wedge \widehat{s} \leq \widehat{r_0} \vee \widehat{t}$. Thus, $(\tau(x) \vee \nu(xy)) \wedge \widehat{s} \leq \widehat{r_0} \vee \widehat{t} = \Omega(x, xy) \vee \widehat{t}$. Hence, $\widehat{G}$ is a $[\widehat{t}, \widehat{s})$-fuzzy complementary incidence graph. ∎

Theorem 5.4.6 (i) G is an $(s, t]$-fuzzy incidence graph if and only if for all $r \in [0, 1]$, G_{st}^r is an incidence subgraph of (V, E, I).

(ii) $\widehat{G}$ is a $[\widehat{t}, \widehat{s})$-fuzzy complementary incidence graph if and only if for all $\widehat{r} \in [0, 1]$, $\widehat{G}_{\widehat{ts}}^{\widehat{r}}$ is an incidence subgraph of (V, E, I).

Proof (ii) Suppose $\widehat{G}$ is a $[\widehat{t}, \widehat{s})$-fuzzy complementary incidence graph. Let $\widehat{r} \in [0, 1]$. Suppose $(x, xy) \in \Omega_{\widehat{ts}}^{\widehat{r}}$. Then $\Omega(x, xy) \wedge \widehat{s} \leq \widehat{r} \vee \widehat{t}$. Now $(\tau(x) \vee \nu(xy)) \wedge \widehat{s} \leq \Omega(x, xy) \vee \widehat{t}$ and so $(\tau(x) \vee \nu(xy)) \wedge \widehat{s} \leq (\Omega(x, xy) \vee \widehat{t}) \wedge \widehat{s} = (\Omega(x, xy) \wedge \widehat{s}) \vee (\widehat{t} \wedge \widehat{s}) \leq (\widehat{r} \vee \widehat{t}) \vee \widehat{t} = \widehat{r} \vee \widehat{t}$. Thus, $\tau(x) \wedge \widehat{s} \leq \widehat{r} \vee \widehat{t}$ and $\nu(xy) \wedge \widehat{s} \leq \widehat{r} \vee \widehat{t}$. Hence, $x \in \tau_{\widehat{ts}}^{\widehat{r}}$ and $xy \in \nu_{\widehat{ts}}^{\widehat{r}}$. Thus, $(\tau_{\widehat{ts}}^{\widehat{r}}, \nu_{\widehat{ts}}^{\widehat{r}}, \Omega_{\widehat{ts}}^{\widehat{r}})$ is an incidence subgraph of (V, E).

Conversely, suppose $(\tau_{\widehat{ts}}^{\widehat{r}}, \nu_{\widehat{ts}}^{\widehat{r}}, \Omega_{\widehat{ts}}^{\widehat{r}})$ is an incidence subgraph of (V, E) for all $\widehat{r} \in [0, 1]$. Let $(x, xy) \in I$ and let $\Omega(x, xy) = \widehat{r_0}$. Then $\Omega(x, xy) \wedge \widehat{s} \leq \widehat{r_0} \vee t$. Hence, $(x, xy) \in \Omega_{\widehat{ts}}^{\widehat{r}}$. Thus, $\tau(x) \wedge \widehat{s} \leq \widehat{r_0} \vee t$. Hence, $(\tau(x) \vee \nu(xy)) \wedge \widehat{s} \leq \widehat{r_0} \vee t = \Omega(x, xy) \vee \widehat{t}$. Thus, $\widehat{G}$ is $[\widehat{t}, \widehat{s})$-fuzzy complementary incidence graph. ∎

5.5 Complementary Fuzzy Cut Vertices and Fuzzy End Vertices

In this section, we present results concerning fuzzy cut vertices and fuzzy end vertices. We provide an interpretation of a fuzzy graph, where vertices can be considered to be countries and edges connections between countries that allow trafficking.

If $s = 0$ and $t = 1$, then $(0, 1]$-fuzzy end vertices and $(0, 1]$-fuzzy cut vertices are simply fuzzy end vertices and fuzzy cut vertices of G, respectively.

Suppose that $x \in V$ is an $(s, t]$-fuzzy end vertex. Then the $y \in V$ such that $\mu(x, y) \wedge t > \vee\{\mu(x, z) \mid z \in V, z \neq y\} \vee s$ is unique. This follows because for all $u \in V\backslash\{y\}$, $\mu(x, y) \geq \mu(x, y) \wedge t > \mu(x, u)$. Also, if x is an $(s, t]$-fuzzy end vertex, then x is also an $(r, w]$-fuzzy vertex for all $r \in [0, s]$ and $w \in [t, 1]$.

The next example shows that a fuzzy end vertex may not be a an $(s, t]$-fuzzy end vertex for certain s and t. An interpretation of this may be as follows. The values s and t may represent a benefit of supporting the fuzzy end node x. The support of a fuzzy end node is not of sufficient value because if edge $\{x, y\}$ were deleted, $\mu(x, z) \not\leq t = 0.95$.

Example 5.5.1 Let $V = \{x, y, z\}$ and $E = \{\{x, y\}, \{x, z\}, \{y, z\}\}$. Let $s = 0.05$ and $t = 0.95$. Define the fuzzy subset σ of V and the fuzzy subset μ of E as follows: $\sigma(x) = \sigma(y) = \sigma(z) = 1$ and $\mu(x, y) = \mu(y, z) = 1$ and $\mu(x, z) = 0.99$. Then x is a fuzzy end vertex, but not an $(s, t]$-fuzzy end vertex. We also have that y is a fuzzy cut vertex, but not an $(s, t]$-fuzzy cut vertex. This can be interpreted as saying that attacking or deleting the fuzzy cut vertex is not of sufficient worth because the connectedness of the graph is not sufficiently reduced, where sufficiently is defined by $t = 0.95$.

Example 5.5.2 Let $V = \{x, y, z\}$ and $E = \{\{x, y\}, \{x, z\}, \{y, z\}\}$. Let $s = 0.05$ and $t = 0.95$. Define the fuzzy subset σ of V and the fuzzy subset μ of E as follows: $\sigma(x) = \sigma(y) = \sigma(z) = 1$ and $\mu(x, y) = \mu(y, z) = 0.04$ and $\mu(x, z) = 0.01$. Then x is a fuzzy end vertex, but not an $(s, t]$-fuzzy end vertex. We also have that y is a fuzzy cut vertex, but not an $(s, t]$-fuzzy cut vertex.

Remark 5.5.3 *Let $x \in V$. If x is an end vertex in (V, E), then x is a fuzzy end vertex and an $(s, t]$-fuzzy end vertex for all $t \in (0, 1]$ and for all $s < t \leq \mu(x, y)$, where y is the unique neighbor of x in G^*.*

Definition 5.5.4 Let $\widehat{G} = (\tau, \nu, \Omega)$ be a complementary fuzzy incidence graph with respect to $\widehat{c}$. Let $x, y \in V \cup E$. A **complementary incidence path** from x to y is an incidence path sequence P such that $\tau(x) < 1$, $\nu(xy) < 1$, and $\Omega(x, xy) < 1$ for all $x, xy, (x, xy)$ in P. Define the **strength** of the P to be $s(P) = \vee\{\Omega(x, xy) \mid xy$ in $P\}$. Define ICONN$^c_{\widehat{G}}(x, y)$ to be the minimum of the strengths of all complementary paths from x to y.

Definition 5.5.5 Let $\widehat{G} = (\tau, \nu, \Omega)$ be a complementary fuzzy incidence graph. Let $x \in V$. Then x is called a **complementary fuzzy incidence cut vertex** if there exist $y, z \in V \cup E \setminus \{x\}$ such that $\text{ICONN}_{\widehat{G}}^c(y, z) < \text{ICONN}_{\widehat{G}\setminus x}^c(y, z)$.

Definition 5.5.6 Let $\widehat{G} = (\tau, \nu, \Omega)$ be a complementary fuzzy incidence graph and let $x \in V$. Then x is called a **complementary fuzzy incidence end vertex** if there exists $y \in V \setminus \{x\}$ such that $\Omega(x, xy) < \wedge\{\Omega(x, xz) \mid z \in V, z \neq y\}$.

Definition 5.5.7 Let $\widehat{t}, \widehat{s} \in [0, 1)$ be such that $\widehat{t} < \widehat{s}$. Let $\widehat{G} = (\tau, \nu, \Omega)$ be a complementary fuzzy incidence graph.

Let $x \in V$. Then x is called a **complementary $[\widehat{t}, \widehat{s})$-fuzzy incidence cut vertex** if there exists $y, z \in V \cup E \setminus \{x\}$ such that $\text{ICONN}_{\widehat{G}}^c(y, z) \vee \widehat{t} < \text{ICONN}_{\widehat{G}\setminus x}^c(y, z) \wedge \widehat{s}$.

Let $xy \in E$. Then xy is called a **complementary $[\widehat{t}, \widehat{s})$-fuzzy incidence cut edge** if there exists $y, z \in V \cup E \setminus \{xy\}$ such that $\text{ICONN}_{\widehat{G}}^c(y, z) \vee \widehat{t} < \text{ICONN}_{\widehat{G}\setminus xy}^c (y, z) \wedge \widehat{s}$.

Definition 5.5.8 Let $\widehat{t}, \widehat{s} \in [0, 1]$ be such that $\widehat{t} < \widehat{s}$. Let $\widehat{G} = (\tau, \nu, \Omega)$ be a complementary fuzzy incidence graph and let $x \in V$. Then x is called a **complementary $[\widehat{t}, \widehat{s})$-fuzzy incidence end vertex** if there exists $y \in V \setminus \{x\}$ such that $\Omega(x, xy) \vee \widehat{t} < \wedge\{\Omega(x, xz) \mid z \in V, z \neq y\} \wedge \widehat{s}$.

If $\widehat{t} = 0$ and $\widehat{s} = 1$, then complementary $[0, 1)$-fuzzy incidence end vertices and complementary $[1, 0)$-fuzzy incidence cut vertices are simply fuzzy incidence end vertices and fuzzy incidence cut vertices of G, respectively.

Suppose that $x \in V$ is a complementary $[\widehat{t}, s)$-fuzzy incidence end vertex. Then the $y \in V$ such that $\Omega(x, xy) \vee \widehat{t} < \wedge\{\Omega(xx, z) \mid z \in V, z \neq y\} \wedge \widehat{s}$ is unique. This follows because for all $u \in V \setminus \{y\}$, $\Omega(x, xy) \leq \Omega(x, xy) \vee \widehat{t} < \Psi(x, xu)$. Also, if x is a complementary $[\widehat{t}, \widehat{s})$-fuzzy incidence end vertex, then x is also an $[w, r)$-fuzzy incidence end vertex for all $r \in [\widehat{s}, 1]$ and $w \in [0, \widehat{t}]$.

Example 5.5.9 Let $V = \{x, y, z\}$ and $E = \{xy, xz, yz\}$. Let $\widehat{s} = 0.95$ and $\widehat{t} = 0.05$. Define the fuzzy subset τ of V and the fuzzy subset ν of E as follows: $\tau(x) = \tau(y) = \tau(z) = 0$, $\nu(x, y) = \nu(y, z) = \nu(x, z) = 0$, $\Omega(x, xz) = 0.01$ and $\Omega(u, uv) = 0$ elsewhere. Then x is a complementary fuzzy incidence end vertex, but not a complementary $[\widehat{t}, s)$-fuzzy incidence end vertex. We also have that y is a complementary fuzzy incidence cut vertex, but not a complementary $[\widehat{t}, \widehat{s})$-fuzzy incidence cut vertex.

Example 5.5.10 Let $V = \{x, y, z\}$ and $E = \{xy, xz, yz\}$. Let $\widehat{s} = 0.95$ and $\widehat{t} = 0.05$. Define the fuzzy subset τ of V and the fuzzy subset ν of E as follows: $\tau(x) = \tau(y) = \tau(z) = 0$ and $\nu(xy) = \nu(yz) = \nu(xz) = 0$, $\Omega(x, xy) = \Omega(y, yz) = 0.96$, $\Omega(x, xz) = 0.99$, and $\Omega(u, uv) = 0$ elsewhere. Then x is a complementary fuzzy incidence end vertex, but not an $[\widehat{t}, s)$-fuzzy end vertex. We also have that y is a complementary fuzzy incidence cut vertex, but not a complementary $[\widehat{t}, s)$-fuzzy incidence cut vertex.

Remark 5.5.11 *Let $x \in V$. If x is a complementary incidence end vertex in (V, E, I), then x is a complementary fuzzy incidence end vertex and an $[\widehat{t}, s)$-fuzzy* incidence end vertex for all $\hat{t} \in [0, 1)$ and for all $\hat{s} > \hat{t} \geq \Omega(x, y)$, where y is the unique complementary incidence neighbor of x in $G^{\#}$.

Definition 5.5.12 Let $x, y \in V$. Then xy is called a **complementary incidence neighbor** of x if $\Omega(x, xy) < 1$. xy is called a $\hat{t}$-**strong complementary incidence neighbor** of x if xy is a complementary incidence neighbor of x and $\Omega(x, xy) \vee \hat{t} \leq$ $\mathrm{CONN}_{\widehat{G}}^{c}(x, xy) \vee \hat{t}$.

Theorem 5.5.13 (i) *Let* $z \in V$ *be an* $(s, t]$-*fuzzy incidence cut vertex of* G. *Then* z *has at least two t-strong incidence neighbors.*

(ii) *Let* $z \in V$ *be a complementary* $[\hat{t}, s)$-*fuzzy incidence cut vertex of* $\widehat{G} = (\tau, \nu, \Omega)$. *Then* z *has at least two complementary* $\hat{t}$-*strong incidence neighbors.*

Proof (ii) Because z is a complementary $[\hat{t}, s)$-fuzzy incidence cut vertex of $\widehat{G}$, there exists $x, y \in V \cup E \backslash \{z\}$ such that $\mathrm{ICONN}_{\widehat{G}}^{c}(x, y) \vee \hat{t} < \mathrm{ICONN}_{\widehat{G} \backslash z}^{c}(x, y) \wedge \hat{s}$. Hence, there exists a complementary path P from x to y and passing through z such that $s(P) \vee \hat{t} = \mathrm{ICONN}_{\widehat{G}}^{c}(x, y) \vee \hat{t}$. Because $x \neq z \neq y$, z has two complementary neighbors in P, say uz and vz. Because (u, uz) and (z, vz) are on P, $\Omega(u, uz) \vee \hat{t} \leq s(P) \vee \hat{t}$ and $\Omega(v, vz) \vee \hat{t} \leq s(P) \vee \hat{t}$. Suppose $\Omega(u, uz) \vee \hat{t} >$ $\mathrm{ICONN}_{\widehat{G}}^{c}(u, uz) \vee \hat{t}$. Then there exists a complementary path Q from u to uz such that $s(Q) \vee \hat{t} < \nu(u, uz) \vee \hat{t}$. Let u' in Q be a complementary neighbor of z. We show $u' \neq v$. Suppose $u' = v$. Let Q' be the path $x \ldots u \ldots u' \ldots v \ldots y$. Then $s(Q') \vee \hat{t} \leq s(P) \vee \hat{t}$, but Q' does not pass through z. However, this is impossible because $\mathrm{ICONN}_{\widehat{G} \backslash z}^{c}(x, y) \vee \hat{t} > \mathrm{ICONN}_{\widehat{G}}^{c}(x, y) \vee \hat{t}$. Let P' denote the path $x \ldots u \ldots u'zv \ldots y$. Then $s(P') \vee \hat{t} \leq s(P) \vee \hat{t} = \mathrm{ICONN}_{\widehat{G}}^{c}(x, y) \vee \hat{t}$. Suppose $\Omega(u', u'z) \vee \hat{t} > \mathrm{CONN}_{\widehat{G}}^{c}(u, uz) \vee \hat{t}$. Then we can apply the above process again. Now length$(P') >$ length(P). Hence, this process must end in a finite number of steps yielding a complementary neighbor $u^{*}z$ of z such that $\nu(u^{*}, u^{*}z) \vee \hat{t} \leq$ $\mathrm{CONN}_{\widehat{G}}^{c}(u, uz) \vee \hat{t}$. As above $u^{*} \neq v$. Hence, a similar argument can be used to yield a complementary neighbor $v^{*}z$ of z such that $u^{*} \neq v^{*}$ and $\mu(v^{*}, v^{*}z) \vee \hat{t} \leq$ $\mathrm{ICONN}_{\widehat{G}}^{c}(u, uz) \vee \hat{t}$. Thus, z has two complementary $\hat{t}$-strong neighbors, namely $u^{*}z$ and $v^{*}z$. ∎

Example 5.5.14 Let $V = \{x, y, z\}$ and $E = \{\{x, y\}, \{x, z\}, \{y, z\}\}$. Define $\mu : E \rightarrow$ $[0, 1]$ as follows: $\mu(x, y) = 0.7$, $\mu(x, z) = 0.9$, and $\mu(y, z) = 0.8$. Then x is not a fuzzy cut vertex. Now $\mathrm{CONN}_{G}(x, y) = 0.8$ and $\mathrm{CONN}_{G}(x, z) = 0.9$. Now z is a 1-strong neighbor of x because $\mu(x, z) \wedge 1 \geq \mathrm{CONN}_{G}(x, z) \wedge 1$, but y is not a 1-strong neighbor of x because $\mu(x, y) \not\geq \mathrm{CONN}_{G}(x, y)$. However, z is a fuzzy cut vertex of G because $\mathrm{CONN}_{G}(x, y) = 0.8 > 0.7 = \mathrm{CONN}_{G \backslash z}(x, y)$. Now z has two 1-strong neighbors, namely x and y, because $\mu(z, x) = \mathrm{CONN}_{G}(z, x)$ and $\mu(z, y) =$ $\mathrm{CONN}_{G}(z, y)$.

Definition 5.5.15 Let $x \in V$. Then x is called a **complementary weak fuzzy incidence end node** if there exists $\hat{t} \in [0, h^{c}(\Omega))$ such that x is an end node in $\widehat{G}^{\hat{t}}$, where $h^{c}(\Omega)$ is the complementary height of Ω, i.e., $h^{c}(\Omega) = \wedge \{\Omega(x, xy) \mid (x, xy) \in I\}$.

Definition 5.5.16 Let $x \in V$. Then x is called a **complementary partial fuzzy incidence end node** if x is a complementary incidence end node in $\widehat{G}^{\widehat{t}}$ for all $\widehat{t} \in [h^c(\Omega), d^c(\Omega)) \cup \{d^c(\Omega)\}$, where $d^c(\Omega)$ is the complementary depth of Ω, i.e., $d^c(\Omega) = \vee\{\Omega(x, xy) \mid (x, xy \in I\}$.

Definition 5.5.17 Let $x \in V$. Then x is called a **partial fuzzy incidence end node** if x is an incidence end node in G^t for all $t \in (d(\Psi), h(\Psi)] \cup \{h(\Psi)\}$, where $d(\Psi)$ is the depth of Ψ, i.e., $d(\Psi) = \wedge\{\Psi(x, xy) \mid (x, xy) \in I\}$.

Proposition 5.5.18 (*i*) *Let* $x \in V$. *Then* x *is an* $(0, 1]$ *-fuzzy incidence end vertex if and only if* x *is a weak fuzzy incidence end vertex.*

(*ii*) *Let* $x \in V$. *Then* x *is a complementary* $[0, 1)$*-fuzzy incidence end vertex if and only if* x *is a complementary weak fuzzy incidence end vertex.*

Proof (*ii*) Suppose x is a complementary $[0, 1)$-fuzzy end vertex. Then there exists $y \in V$ such that $\Omega(x, y) = \Omega(x, y) \vee 0 < \wedge\{\Omega(x, xz) \mid z \in V, z \neq y\}$. Thus, x is a complementary incidence end vertex in $\widehat{G}^a$, where $a = \Omega(x, xy)$. Hence, x is a complementary weak fuzzy incidence end vertex.

Conversely, suppose x is a complementary weak fuzzy incidence end vertex. Then x is a complementary incidence end node in $\widehat{G}^b$ for some $b \in [0, h^c(\Omega))$. Hence, there exists a unique $y \in V$ such that $\Omega(x, xy) < b$ and for all other $z \in V$, $\Omega(x, xz) > b$. Thus, x is a complementary $[0, 1)$-fuzzy incidence end vertex. ∎

Example 5.5.19 Let $V = \{x, y, z\}$. Define the fuzzy relation Ω on I as follows:

$$\Omega(x, xy) = 0.7, \ \Omega(y, yz) = 0.8, \text{ and } \Omega(x, xz) = 0.9,$$
$$\Omega(u, uv) = 1 \text{ otherwise.}$$

Then $0.7 = \Omega(x, xy) = \Omega(x, xy) \vee 0 < \wedge\{\Omega(x, xu) \mid u \in V, u \neq y\} = \Omega(x, xz)$ $= 0.9$. Also, x is a complementary weak fuzzy end vertex for $\widehat{G}^a = (\tau^a, \nu^a)$, where $a = \nu(x, y) = 0.7$ because $\Omega^{.7} = \{(x, xy)\}$. Now $h^c(\Omega) = 0.7$. Let $b = 0.7$. Then there exists an unique $y \in V$ such that $\Omega(x, xy) \leq b$ and for all other $z \in V$, $\Omega(x, xz) > b$. Thus, x is a complementary $[0, 1)$ -fuzzy incidence end vertex.

Corollary 5.5.20 (*i*) *Let* $x \in V$. *Then* x *is a partial fuzzy incidence end vertex if and only if* x *is a* $(d(\Psi), 1]$*-fuzzy end vertex.*

(*ii*) *Let* $x \in V$. *Then* x *is a complementary partial fuzzy incidence end vertex if and only if* x *is a* $[d^c(\Omega), 1)$*-fuzzy end vertex.*

Corollary 5.5.21 (*i*) *Let* $x \in V$. *Then* x *is an incidence end vertex of* G^* *if and only if* x *is a* $(0, d(\Psi)]$*-fuzzy end vertex.*

(*ii*) *Let* $x \in V$. *Then* x *is a complementary incidence end vertex of* $\widehat{G}^{\#}$ *if and only if* x *is a complementary* $[0, d^c(\Omega))$*-fuzzy incidence end vertex.*

5.6 Illegal Immigration and Vague Fuzzy Incidence Graphs

We introduce the notion of a vague incidence graph and its eccentricity. We apply the results to problems involving human trafficking and illegal immigration. We are particularly interested in the roll played by countries' vulnerabilities and their government's response to human trafficking. There are four main vulnerabilities of illegal immigration and five main categories of government responses [16]. The categories are explained in detail in [16] and Chap. 6. We use measurements of these categories for countries given in [16] and Chap. 6. We show of the leading illegal immigration routes through Mexico to the United States that Somalia has the highest eccentricity.

In [8] and [9], Dinesh introduced the notion of the degree of incidence of a vertex and an edge in fuzzy graph theory. The results were expanded upon in [11] and applied to the problem of illegal immigration in [10].

In [7], Darabian and Borzooei defined the notion of eccentricity for vague graphs and applied the concept to the problem of illegal immigration. We extend the notion of eccentricity to vague incidence graphs and apply our results to the problem of illegal immigration. The eccentricity of a route is a measure of the route's susceptibility to illegal immigration. Intuitionistic fuzzy graphs, [1–3], are essentially vague graphs with an additional requirement.

Let $G = (V, E, I)$ be an incidence graph. Let $n \in \mathbb{N}$ and $u, v_0, \ldots, v_n \in V$. Recall that the sequences

$$v_0, (v_0, v_0v_1), v_0v_1, (v_1, v_0v_1), v_1, \ldots, v_{n-1}, (v_{n-1}, v_{n-1}v_n),$$
$$v_{n-1}v_n, (v_n, v_{n-1}v_n), v_n;$$
$$v_0, (v_0, v_0v_1), v_0v_1, (v_1, v_0v_1), v_1, \ldots, v_{n-1}, (v_{n-1}, v_{n-1}v_n), v_{n-1}v_n;$$
$$uv_0, (v_0, uv_0), v_0, (v_0, v_0v_1), v_0v_1, (v_1, v_0v_1), v_1, \ldots, v_{n-1},$$
$$(v_{n-1}, v_{n-1}v_n), v_{n-1}v_n, (v_n, v_{n-1}v_n), v_n;$$
$$uv_0, (v_0, uv_0), v_0, (v_0, v_0v_1), v_0v_1, (v_1, v_0v_1), v_1, \ldots, v_{n-1},$$
$$(v_{n-1}, v_{n-1}v_n), v_{n-1}v_n$$

are called **walks**. If the vertices are distinct, the sequences are called **paths**.

Definition 5.6.1 Let (V, E) be a graph.

(i) Let σ be a fuzzy subset of V and μ be a fuzzy subset of E. Let Ψ be a fuzzy subset of $V \times E$ such that for all $(v, e) \in V \times E$, $\Psi(v, e) \leq \sigma(v) \wedge \mu(e)$. Then Ψ is called a **fuzzy incidence** of (V, E).

(ii) Let τ be a fuzzy subset of V and ν a fuzzy subset of E. Let Ω be a fuzzy subset of $V \times E$ such that for all $(v, e) \in V \times E$, $\Omega(v, e) \geq \tau(v) \vee \nu(e)$. Then Ω is called a **complementary fuzzy incidence** of (V, E).

Definition 5.6.2 Let (V, E) be a graph.

(i) Let (σ, μ) be a fuzzy subgraph of (V, E). If Ψ is a fuzzy incidence of (V, E), then (σ, μ, Ψ) is called a **fuzzy incidence graph**.

(*ii*) Let (τ, ν) be a complementary fuzzy subgraph of (V, E). If Ω is a complementary fuzzy incidence of (V, E), then (τ, ν, Ω) is called a **complementary fuzzy incidence graph.**

Let k be a positive integer. Define $\Psi^k (u, uv) = \vee\{\Psi(u, uu_1) \wedge \cdots \wedge \mu(u_{n-1},$ $u_{n-1}v) \mid P : u = u_0, u_1, \ldots, u_{k-1}, u_k = v$ is a path of vertices of length k from u to $v\}$. Let $\Psi^\infty(u, uv) = \vee\{\Psi^k(u, uv) \mid k \in \mathbb{N}\}$. Define $\Omega^k (uv) = \wedge\{\Omega(u, uu_1) \vee \cdots \vee \Omega(u_{n-1}, u_{n-1}v) \mid P : u = u_0, u_1, \ldots, u_{k-1}, u_k = v$ is a path of length k from u to $v\}$. Let $\Omega^\infty(u, uv) = \wedge\{\Psi^k(u, uv) \mid k \in \mathbb{N}\}$.

A path of vertices $P : u = u_0, u_1, \ldots, u_{k-1}, u_k = v$ is said to be Ψ-**strong for** (u, uv) if $\Psi^k(u, uv) = \Psi^\infty(u, uv)$ and Ω-**strong for** (u, uv) if $\Omega^k(u, uv) = \Omega^\infty(u, uv)$.

Definition 5.6.3 Let $F = (\sigma, \mu, \Psi)$ be a fuzzy incidence graph on (V, E) and $C = (\tau, \nu, \Omega)$ be a complementary fuzzy incidence graph on (V, E). Then the pair (F, C) is called a **vague fuzzy incidence graph** on (V, E).

5.7 Eccentricity of Vague Fuzzy Incidence Graphs

The results in this section are based on [12].

We introduce the notion eccentricity in a vague fuzzy incidence graph and develop the results necessary for our applications.

Definition 5.7.1 Define the fuzzy subset δ of $V \times V$ by for all $(u, v) \in V \times V$,

$$\delta(u, v) = \wedge\left\{ \sum_{i=1}^{k_j} \left(\frac{1}{\Psi(x_{i-1}, x_{i-1}x_i)} + \frac{1}{1 - \Omega(x_{i-1}, x_{i-1}x_i)} \right) \right.$$
$$\left. \mid P_j : u = x_{j0}, x_{j1}, \ldots, x_{jk-1}, x_{jk} = v; \ k_j \in \mathbb{N} \right\}.$$

We see that it is not necessarily the case that $\Psi(u, uv) = \Psi(v, vu)$. However, in our applications it will follow that $\Psi(u, uv) = \Psi(v, vu)$.

Definition 5.7.2 Define the function $e : V \to \mathbb{R}$ by for all $u \in V$, $e(u) = \vee\{\delta(u, v) \mid v \in V\}$. The function e is called the **eccentricity** of u for all u in V.

In the previous definition, u is an origin country and v is a destination country of u.

We say that the pair (u, uv) is a **strong pair** if $\Psi(u, uv) \geq \Psi^\infty(u, uv)$ and $\Omega(u, v) \leq \Omega^\infty(u, uv)$.

Theorem 5.7.3 *Let (F, C) be a vague fuzzy incidence graph. Suppose (u, uv) is a strong pair. Then $\delta(u, v) = \frac{1}{\Psi(u,uv)} + \frac{1}{1-\Omega(u,uv)}$.*

Proof Let $P : u = x_0, x_1, \ldots, x_{k-1}, x_k = v$ be a path of nodes from u to v. Then $\Psi(u, uv) \geq \Psi(x_0, x_0 x_1) \wedge \cdots \wedge \Psi(x_{k-1}, x_{k-1} x_k) = \Psi(x_{m-1}, x_{m-1} x_m)$ for some m, $1 \leq m \leq k$. Thus,

$$\frac{1}{\Psi(u, uv)} \leq \frac{1}{\Psi(x_{m-1}, x_{m-1} x_m)}$$

$$\leq \frac{1}{\Psi(x_0, x_0 x_1)} + \cdots + \frac{1}{\Psi(x_{k-1}, x_{k-1} x_k)}.$$

Also, $\Omega(u, uv) \leq \Omega(x_0, x_0 x_1) \vee \cdots \vee \Omega(x_{k-1}, x_{k-1} x_k) = \Omega(x_{j-1}, x_{j-1} x_j)$ for some $j, 1 \leq j \leq k$. Thus, $\Omega(u, uv) \leq \Omega(x_{j-1}, x_{j-1} x_j)$ and so $1 - \Omega(u, uv) \geq 1 - \Omega(x_{j-1}, x_{j-1} x_j)$. Hence,

$$\frac{1}{1 - \Omega(u, uv)} \leq \frac{1}{1 - \Omega(x_{j-1}, x_{j-1} x_j)}$$

$$\leq \frac{1}{1 - \Omega(x_0, x_{j0} x_1)} + \cdots + \frac{1}{1 - \Omega(x_{k-1}, x_{k-1} x_k)}.$$

Because P is arbitrary, the desired result holds.   ∎

Let (u, uv) be a pair. Then (u, uv) is called Ψ-**effective** if $\Psi(u, uv) = \mu(uv) \wedge \sigma(u)$ and Ω-**effective** if $\Omega(u, uv) = \nu(uv) \vee \tau(u)$. The pair is called μ-**effective** if $\mu(uv) = \sigma(u) \wedge \sigma(v)$ and ν-**effective** if $\nu(uv) = \tau(u) \vee \tau(v)$. Clearly, (u, uv) is Ψ-effective if and only if $\Psi(u, uv) = \mu(uv)$ and Ω-effective if and only if $\Omega(u, uv) = \nu(uv)$.

A fuzzy incidence graph is said to be Ψ-**complete** if for all $u, v \in V$, (u, uv) is Ψ-effective. A complementary fuzzy incidence graph is said to be Ω-**complete** if for all $u, v \in V$, (u, uv) is Ω-effective. A vague fuzzy incidence graph is said to be (Ψ, Ω)-**complete** if it is both Ψ-complete and Ω-complete.

A fuzzy incidence graph is said to be μ-**complete** if for all $u, v \in V$, (u, uv) is μ-effective. A complementary fuzzy incidence graph is said to be ν-**complete** if for all $u, v \in V$, (u, uv) is ν-effective. A vague fuzzy incidence graph is said to be (μ, ν)-**complete** if it is both μ-complete and ν-complete.

Lemma 5.7.4 *Let (F, C) be a vague fuzzy incidence graph. Let (u, uv) be a pair.*
  (i) If (u, uv) is Ψ-effective and μ-effective, then (u, uv) is Ψ-strong.
  (ii) If (u, uv) is Ω-effective and ν-effective, then (u, uv) is Ω-strong.

Proof (i) Let $P : u = x_{k0}, \ldots, x_{k,k-1}, x_{kk} = v$ be a Ψ-strong path of length k for (u, v). Then

$$\Psi^k(u, uv) = \Psi(u, x_{k1}u) \wedge \cdots \wedge \Psi(x_{k,k-1}, x_{k,k-1}v)$$
$$\leq \mu(ux_{k1}) \wedge \cdots \wedge \mu(x_{k,k-1}v)$$
$$\leq \sigma(u) \wedge \sigma(x_{k1}) \wedge \cdots \wedge \sigma(x_{k,k-1}) \wedge \sigma(v)$$
$$\leq \sigma(u) \wedge \sigma(v)$$
$$= \mu(uv)$$
$$= \Psi(u, uv).$$

There exists a positive integer n such that $\Psi^\infty(u, uv) = \vee_{i=1}^n \Psi^i(u, uv)$. Thus, $\Psi^\infty(u, uv) = \vee_{i=1}^n \Psi^i(u, uv) \leq \Psi(u, uv)$.

(ii) Let $P : u = x_{k0}, \ldots, x_{k,k-1}, x_{kk} = v$ be a Ω-strong path of length k for (u, v). Then

$$\Omega^k(u, uv) = \Omega(u, x_{k1}u) \vee \cdots \vee \Omega(x_{k,k-1}, x_{k,k-1}v)$$
$$\geq \nu(u, k_{k1}) \vee \cdots \vee \nu(x_{k,k-1}v)$$
$$\geq \tau(u) \vee \cdots \vee \tau(x_{k,k-1}) \vee \tau(v)$$
$$\geq \tau(u) \vee \tau(v)$$
$$= \nu(uv)$$
$$= \Omega(u, uv).$$

There exists a positive integer n such that $\Omega^\infty(u, uv) = \wedge_{i=1}^n \Omega^i(u, uv)$. Thus, $\Omega^\infty(u, uv) = \wedge_{i=1}^n \Omega^i(u, uv) \geq \Omega(u, uv)$. ∎

Example 5.7.5 Let $V = \{u, v, x\}$ and $E = \{uv, ux, vx\}$. Define the fuzzy subset σ of V as follows: $\sigma(u) = \sigma(v) = \sigma(x) = 0.7$. Define the fuzzy subset μ of E as follows: $\mu(uv) = 0.7$, $\mu(ux) = 0.6$, $\mu(vx) = 0.7$. Define the fuzzy subset Ψ of $V \times E$ as follows:

$$\Psi(u, uv) = 0.7 = \Psi(v, uv),$$
$$\Psi(u, ux) = 0.6 = \Psi(x, ux),$$
$$\Psi(v, vx) = 0.7 = \Psi(x, vx).$$

Then all pairs are Ψ-effective. However, ux is not μ-effective because $\mu(ux) = 0.6 \neq 0.7 = \sigma(u) \wedge \sigma(x)$. Also, (u, uv) is not Ψ-strong because $\Psi(u, ux) = 0.6 \not\geq 0.7 = \Psi^\infty(u, ux)$.

Example 5.7.6 Let $V = \{u, v, x\}$ and $E = \{uv, ux, vx\}$. Define the fuzzy subset σ of V as follows: $\sigma(u) = \sigma(v) = \sigma(x) = 0.7$. Define the fuzzy subset μ of E as follows: $\mu(uv) = \mu(ux) = \mu(vx) = 0.7$. Define the fuzzy subset Ψ of $V \times E$ as follows :

$$\Psi(u, uv) = 0.7 = \Psi(v, uv),$$
$$\Psi(u, ux) = 0.6 = \Psi(x, ux),$$
$$\Psi(v, vx) = 0.7 = \Psi(x, vx).$$

Then all edges are μ-effective, but the pair (u, ux) is not Ψ-effective.

Lemma 5.7.7 *Let (F, C) be a vague fuzzy incidence graph. Let c be an involutive complement.*

(i) If the path $P : u = x_0, x_1, \ldots, x_k = v$ is strong for Ω^k, then P is strong for $(c\Omega)^k$.

(ii) If the path $P : u = x_0, x_1, \ldots, x_k = v$ is strong for Ψ^k, then P is strong for $(c\Psi)^k$.

Proof *(i)* Because P is strong for Ω^k, $\Omega^k(u, uv) = \Omega(u, ux_1) \vee \cdots \vee \Omega(x_{k-1}, x_{k-1}v)$. Suppose P is not strong for $(c\Omega)^k$. Then there exists a path $Q : u = y_0, y_1, \ldots, y_{j-1}, y_k = v$ such that

$$(c\Omega)^k(u, uv) = (c\Omega)(u, uy_1) \wedge \cdots \wedge (c\Omega)(y_{k-1}, y_{k-1}v)$$
$$> (c\Omega)(u, ux_1) \wedge \cdots \wedge (c\Omega)(x_{k-1}, x_{k-1}v).$$

Thus,

$$c[(c\Omega)(u, uy_1) \wedge \cdots \wedge (c\Omega)(y_{k-1}, y_{k-1}v)]$$
$$< c[(c\Omega)(u, ux_1) \wedge \cdots \wedge (c\Omega)(x_{k-1}, x_{k-1}v)].$$

Hence,

$$\Omega(u, ux_1) \vee \cdots \vee \Omega(x_{k-1}, x_{k-1}v)$$
$$< \Omega(u, uy_1) \vee \cdots \vee \Omega(y_{k-1}, y_{k-1}v),$$

a contradiction.

(ii) The proof is similar to that in *(i)* so we omit it. ∎

Example 5.7.8 Let $V = \{x, y, z\}$ and $E = \{xy, xz, yz\}$. Define the fuzzy subsets σ and τ of V as follows:

$$\sigma(x) = \frac{1}{2}, \ \sigma(y) = \frac{1}{2}, \ \sigma(z) = \frac{1}{4};$$
$$\tau(x) = \frac{1}{4}, \ \tau(y) = \frac{3}{8}, \ \tau(z) = \frac{1}{4}.$$

Define the fuzzy subsets μ and ν of E as follows:

$$\mu(xy) = \frac{1}{2}, \ \mu(xz) = \frac{1}{4}, \ \mu(yz) = \frac{1}{4};$$

$$v(xy) = \frac{3}{8}, \ v(xz) = \frac{1}{4}, \ v(yz) = \frac{3}{8}.$$

Then (σ, μ) is a complete fuzzy subgraph of (V, E) and (τ, v) is a complete complementary fuzzy subgraph of (V, E). Now $\wedge\{\mu(xv) \mid v \in V\} = \frac{1}{2} \wedge \frac{1}{4} = \frac{1}{4} = \mu(xz)$ and $\vee\{v(xv) \mid v \in V\} = \frac{3}{8} \vee \frac{1}{4} = \frac{3}{8} = v(xy)$. Now

$$\vee\left\{\frac{1}{\mu(xv)} + \frac{1}{1 - v(xv)} \ \middle| \ v \in V\right\}$$

$$= \vee\left\{\frac{1}{\frac{1}{2}} + \frac{1}{1 - \frac{3}{8}}, \frac{1}{\frac{1}{4}} + \frac{1}{1 - \frac{1}{4}}\right\}$$

$$= \left(2 + \frac{8}{5}\right) \vee \left(4 + \frac{4}{3}\right)$$

$$< 4 + \frac{8}{5}$$

$$= \frac{1}{\frac{1}{4}} + \frac{1}{1 - \frac{3}{8}} = \frac{1}{\wedge\{\mu(xv) \mid v \in V\}} + \frac{1}{1 - \vee\{v(xv) \mid v \in V\}}.$$

Let $\Psi(u, uv) = \sigma(u) \wedge \sigma(v)$ and $\Omega(u, uv) = \tau(u) \vee \tau(v)$ for all $u, v \in V$. Then $\Psi(u, uv) = \mu(uv)$ and $\Omega(u, uv) = \tau(uv)$ for all $u, v \in V$.

From Example 5.7.8, we see that the last equality in the proof of (i) of Proposition 5.7.9 is false without the additional hypothesis.

Proposition 5.7.9 Let (F, C) be an vague fuzzy incidence graph. Suppose for all $x, y, u, v \in V$ that $\Psi(x, xy) \leq \Psi(u, uv)$ if and only if $\Omega(x, xy) \geq \Omega(u, uv)$.
(i) If (F, C) is (Ψ, Ω)-complete, then for all $u \in V$,

$$e(u) = \frac{1}{\wedge\{\mu(uv) \mid v \in V\}} + \frac{1}{1 - \vee\{v(uv) \mid v \in V\}}.$$

(ii) If (F, C) is (μ, v)-complete and (Ψ, Ω)-complete, then

$$e(u) = \frac{1}{\wedge\{\sigma(\mu) \mid v \in V\}} + \frac{1}{1 - \vee\{\tau(u) \mid v \in V\}}.$$

Proof (i) Because (F, C) is (Ψ, Ω)-complete, every pair (u, uv) is effective and so by Lemma 5.7.4 is strong. Hence, by Theorem 5.7.3, we have

$$\delta(u, v) = \frac{1}{\Psi(u, uv)} + \frac{1}{1 - \Omega(u, uv)}.$$

Because each node is adjacent to the other nodes, we have for every $u \in V$ that

$$e(u) = \vee\{\delta(u, v) \mid v \in V\}$$

$$= \vee\left\{\frac{1}{\Psi(u, uv)} + \frac{1}{1 - \Omega(u, uv)} \mid v \in V\right\}$$

$$= \vee\left\{\frac{1}{\mu(uv)} + \frac{1}{1 - \nu(uv)} \mid v \in V\right\}$$

$$= \frac{1}{\wedge\{\mu(uv) \mid v \in V\}} + \frac{1}{1 - \vee\{\nu(uv) \mid v \in V\}}$$

(ii) The result here follows from (i) because $\wedge\{\mu(uv) \mid v \in V\} = \wedge\{\sigma(u) \wedge \sigma(v) \mid v \in V\} = \wedge\{\sigma(v) \mid v \in V\}$ and $\vee\{\nu(uv) \mid v \in V\} = \vee\{\tau(u) \vee \tau(v) \mid v \in V\} = \vee\{\vee\tau(v) \mid v \in V\}$. ∎

Illegal Immigration to the United States Through Mexico

Increasing numbers of people from Asia and Africa are seeking to enter the U.S. illegally over the Mexican border. The vast majority of migrants detained were from the Americas, however a significant number were from Asian and African countries. Our results are based on [10, 12].

We use the average of the four vulnerabilities and the average of the four government responses in what follows. We let σ denote government response and τ denote vulnerability for the countries listed in the paths from the origin country to the U.S. Then the eccentricity of the origin country is determined. The eccentricity provides us with a number which measures the susceptibility of human trafficking in the paths of the origin country to the U.S.

For intuitionistic fuzzy incidence graphs, $\Psi(x, xy) \leq \sigma(x) \wedge \mu(xy)$ and $\Omega(x, xy) \geq \tau(x) \vee \nu(xy)$. Note that the sum of the government response values and the vulnerability values for Mexico, U.S., Ethiopia, and Nigeria actually add to a number larger than 1. No data is available for the government response of Somalia. We have made adjustments for these countries in the following tables. The paths to the U.S. presented below are the ones which have the highest flow of illegal immigration, [17].

We let σ denote government response and τ denote vulnerability for the countries listed in the paths from the origin country to the U.S. Then the eccentricity of the origin country is determined. We define Ψ and Ω in terms of the t-norm multiplication and the conorm algebraic sum, respectively.

In the following, we let $\Psi(x, xy) = \sigma(x)\sigma(y)$ and $\Omega(x, xy) = \tau(x) + \tau(y) - \tau(x)\tau(y)$.

	China		Columbia		Guatemala		Mexico		U.S.
σ	0.36		0.53		0.56		0.57		0.82
Ψ		0.19		0.30		0.32		0.47	
τ	0.45		0.42		0.42		0.43		0.18
Ω		0.68		0.66		0.67		0.53	

Let P denote the path: China, Columbia, Guatemala, Mexico, U.S. Because there is only one path from China to U.S.,

$$\delta(\text{China, U.S.}) = \left(\frac{1}{0.19} + \frac{1}{1 - 0.68}\right) + \left(\frac{1}{0.30} + \frac{1}{1 - 0.66}\right) + \left(\frac{1}{0.32} + \frac{1}{1 - 0.67}\right)$$
$$+ \left(\frac{1}{0.47} + \frac{1}{1 - 0.53}\right)$$
$$= (5.26 + 3.12) + (3.33 + 2.94) + (3.12 + 3.03) + (2.13 + 2.13)$$
$$= 25.06 = e(\text{China}).$$

	India		Guatemala		Mexico		U.S.
σ	0.46		0.56		0.57		0.82
Ψ		0.26		0.32		0.47	
τ	0.53		0.42		0,43		0.18
Ω		0.73		0.67		0.53	

$$\delta(\text{India, U.S.}) = \left(\frac{1}{0.26} + \frac{1}{1 - 0.73}\right) + \left(\frac{1}{0.32} + \frac{1}{1 - 0.67}\right) + \left(\frac{1}{0.47} + \frac{1}{1 - 0.53}\right)$$
$$= (3.85 + 3.70) + (3.12 + 3.03) + (2.13 + 2.13)$$
$$= 17.96 = e(\text{India}).$$

	Ethiopia		S. Africa		Brazil		Ecuador		Mexico		U.S.
σ	0.42		0.49		0.66		0.51		0.57		0.82
Ψ		0.21		0.32		0.34		0.29		0.47	
τ	0.58		0.49		0.31		0.35		0.43		0.18
Ω		0.79		0.65		0.55		0.63		0.53	

$$\delta(\text{Ethiopia, U.S.}) = \left(\frac{1}{0.21} + \frac{1}{1 - 0.79}\right) + \left(\frac{1}{0.32} + \frac{1}{1 - 0.65}\right)$$
$$+ \left(\frac{1}{0.34} + \frac{1}{1 - 0.55}\right) + \left(\frac{1}{0.29} + \frac{1}{1 - 0.63}\right)$$
$$+ \left(\frac{1}{0.47} + \frac{1}{1 - 0.53}\right)$$
$$= (4.76 + 4.76) + (3.12 + 2.86) + (2.94 + 2.22)$$
$$+ (3.45 + 2.70) + (2.13 + 2.13)$$
$$= 31.07 = e(\text{Ethiopia}).$$

	Somalia	EAU	Russia	Cuba	Columbia	Mexico	U.S.
σ	0.28	0.56	0.30	0.21	0.53	0.57	0.82
Ψ		0.16	0.17	0.06	0.11	0.30	0.47
τ	0.72	0.26	0.42	0.32	0.42	0.43	0.18
Ω		0.79	0.57	0.61	0.61	0.67	0.53

$$\delta(\text{Somalia, U.S.}) = \left(\frac{1}{0.16} + \frac{1}{1-0.79}\right) + \left(\frac{1}{0.17} + \frac{1}{1-0.57}\right)$$

$$+ \left(\frac{1}{0.06} + \frac{1}{1-0.61}\right) + \left(\frac{1}{0.11} + \frac{1}{1-0.61}\right)$$

$$+ \left(\frac{1}{0.30} + \frac{1}{1-0.67}\right) + \left(\frac{1}{0.47} + \frac{1}{1-0.53}\right)$$

$$= (6.25 + 4.76) + (5.89 + 2.33)$$

$$+ (16.67 + 2.56) + (9.09 + 2.56)$$

$$+ (3.33 + 3.03) + (2.13 + 2.13)$$

$$= 60.73 = e(\text{Somalia}).$$

	Nigeria	Spain	Cuba	Columbia	Mexico	U.S.
σ	0.44	0.71	0.21	0.53	0.57	0.82
Ψ		0.31	0.15	0.07	0.30	0.47
τ	0.56	0.20	0.32	0.42	0.43	0.18
Ω		0.65	0.46	0.61	0.67	0.53

$$\left(\frac{1}{0.31} + \frac{1}{1-0.65}\right) + \left(\frac{1}{0.15} + \frac{1}{1-0.46}\right) + \left(\frac{1}{0.07} + \frac{1}{1-0.61}\right)$$

$$+ \left(\frac{1}{0.30} + \frac{1}{1-0.67}\right) + \left(\frac{1}{0.47} + \frac{1}{1-0.53}\right)$$

$$= (3.23 + 2.86) + (6.67 + 1.85) + (14.29 + 2.56)$$

$$+ (3.33 + 3.03) + (2.13 + 2.13)$$

$$= 42.08.$$

There are two paths from Nigeria to U.S., but one is a subpath of the other. Thus,

$$\delta(\text{Nigeria, U.S.}) = 42.08 \wedge 21.90 = 21.90,$$
$$e(\text{Nigeria}) = 42.08 \vee 21.90 = 42.08.$$

We see that of the leading illegal immigration routes through Mexico to the United States, Somalia has the highest eccentricity.

In the following, we let $\mu(xy) = \sigma(x) \wedge \sigma(y)$ and $\nu(xy) = \tau(x) \vee \tau(y)$.

For intuitionistic fuzzy graphs, $\mu(xy) \leq \sigma(x) \wedge \sigma(y)$ and $\nu(xy) \geq \tau(x) \vee \tau(y)$. In the following, we assume, $\Psi(x, xy) = \sigma(x) \wedge \sigma(y)$ and $\Omega(x, xy) = \tau(x) \vee \tau(y)$. Thus, $\mu(xy) = \sigma(x) \wedge \sigma(y)$ and $\nu(xy) = \tau(x) \vee \tau(y)$. Note that the sum of the government response values and the vulnerability values for Mexico, U.S., Ethiopia, and Nigeria actually add to a number larger than 1. No data is available for the government response of Somalia. We have made adjustments for these countries in the following tables. The paths to the U.S. presented below are the ones which have the highest flow of illegal immigration, [17].

	China		Columbia		Guatemala		Mexico		U.S.
σ	0.36		0.53		0.56		0.57		0.82
Ψ		0.36		0.53		0.56		0.57	
τ	0.45		0.42		0.42		0.43		0.18
Ω		0.45		0.42		0.43		0.43	

Let P denote the path: China, Columbia, Guatemala, Mexico, U.S. Because there is only one path from China to U.S., we have

$$\delta(\text{China, U.S.}) = \left(\frac{1}{0.36} + \frac{1}{1 - 0.45}\right) + \left(\frac{1}{0.53} + \frac{1}{1 - 0.42}\right)$$
$$+ \left(\frac{1}{0.56} + \frac{1}{1 - 0.43}\right) + \left(\frac{1}{0.57} + \frac{1}{1 - 0.43}\right)$$
$$= 2.78 + 1.82 + 1.89 + 1.72 + 1.79$$
$$+ 1.75 + 1.75 + 1.75$$
$$= 15.25 = e(\text{China}).$$

	India		Guatemala		Mexico		U.S.
σ	0.46		0.56		0.57		0.82
μ		0.46		0.56		0.57	
τ	0.53		0.42		0.43		0.18
ν		0.53		0.43		0.43	

$$\delta(\text{India, U.S.}) = \left(\frac{1}{0.46} + \frac{1}{1 - 0.53}\right) + \left(\frac{1}{0.56} + \frac{1}{1 - 0.43}\right)$$
$$+ \left(\frac{1}{0.57} + \frac{1}{1 - 0.43}\right)$$
$$= 2.17 + 2.13 + 1.79 + 1.75 + 1.75 + 1.75$$
$$= 11.34 = e(\text{India}).$$

	Ethiopia		S. Africa		Brazil		Ecuador		Mexico		U.S.
σ	0.42		0.49		0.66		0.51		0.57		0.82
Ψ		0.42		0.49		0.51		0.51		0.57	
τ	0.58		0.49		0.31		0.35		0.43		0.18
Ω		0.58		0.49		0.35		0.43		0.43	

$$\delta(\text{Ethiopia,U.S.}) = \left(\frac{1}{0.42} + \frac{1}{1-0.58}\right) + \left(\frac{1}{0.49} + \frac{1}{1-0.49}\right)$$
$$+ \left(\frac{1}{0.51} + \frac{1}{1-0.35}\right) + \left(\frac{1}{0.51} + \frac{1}{1-0.43}\right)$$
$$+ \left(\frac{1}{0.57} + \frac{1}{1-0.43}\right)$$
$$= 2.38 + 2.38 + 2.04 + 1.96 + 1.96$$
$$+ 1.54 + 1.96 + 1.75 + 1.75 + 1.75$$
$$= 19.47 = e(\text{Ethiopia}).$$

	Somalia		EAU		Russia		Cuba		Columbia		Mexico		U.S.
σ	0.28		0.56		0.30		0.21		0.53		0.57		0.82
Ψ		0.28		0.30		0.21		0.21		0.53		0.57	
τ	0.72		0.26		0.42		0.32		0.42		0.43		0.18
Ω		0.72		0.42		0.42		0.42		0.42		0.43	

$$\delta(\text{Somalia, U.S.}) = \left(\frac{1}{0.28} + \frac{1}{1-0.72}\right) + \left(\frac{1}{0.30} + \frac{1}{1-0.42}\right)$$
$$+ \left(\frac{1}{0.21} + \frac{1}{1-0.42}\right) + \left(\frac{1}{0.21} + \frac{1}{1-0.42}\right)$$
$$+ \left(\frac{1}{0.53} + \frac{1}{1-0.42}\right) + \left(\frac{1}{0.57} + \frac{1}{1-43}\right)$$
$$= 3.57 + 3.57 + 3.33 + 1.72 + 4.76 + 1.72$$
$$+ 4.76 + 1.72 + 1.89 + 1.72 + 1.75 + 1.75$$
$$= 32.26 = e(\text{Somalia}).$$

	Nigeria		Spain		Cuba		Columbia		Mexico		U.S.
σ	0.44		0.71		0.21		0.53		0.57		0.82
Ψ		0.44		0.21		0.21		0.53		0.57	
τ	0.56		0.20		0.14		0.42		0.43		0.18
Ω		0.56		0.20		0.42		0.43		0.43	

$$\left(\frac{1}{0.44} + \frac{1}{1-0.56}\right) + \left(\frac{1}{0.21} + \frac{1}{1-0.20}\right) + \left(\frac{1}{0.21} + \frac{1}{1-0.42}\right)$$

$$+ \left(\frac{1}{0.53} + \frac{1}{1-0.43}\right) + \left(\frac{1}{0.57} + \frac{1}{1-0.43}\right)$$

$$= 2.27 + 2.27 + 4.76 + 1.25 + 4.76 + 1.72$$

$$+ 1.89 + 1.75 + 1.75 + 1.75$$

$$= 24.17.$$

	Nigeria		Spain		Columbia		Mexico		U.S.
σ	0.44		0.71		0.53		0.57		0.82
Ψ		0.44		0.53		0.53		0.57	
τ	0.56		0.20		0.42		0.43		0.18
Ω		0.56		0.42		0.43		0.43	

$$\left(\frac{1}{0.44} + \frac{1}{1-0.56}\right) + \left(\frac{1}{0.53} + \frac{1}{1-0.42}\right)$$

$$+ \left(\frac{1}{0.53} + \frac{1}{1-0.43}\right) + \left(\frac{1}{0.57} + \frac{1}{1-0.43}\right)$$

$$= 2.27 + 2.27 + 1.89 + 1.72 + 1.89$$

$$+ 1.75 + 1.75 + 1.75$$

$$= 15.29.$$

There are two paths from Nigeria to U.S., but one is a subpath of the other. Thus

$$\delta(\text{Nigeria, U.S.}) = 23.38 \wedge 15.29 = 15.29,$$

$$e(\text{Nigeria}) = 23.38 \vee 15.29 = 23.38.$$

We see that the eccentricity of Somalia is the largest. Consequently, Somalia would be at the top of the list to influence the government to respond to trafficking in order to reduce the over all susceptibility to trafficking in humans to the U.S.

We considered the main routes of trafficking to the United States. The eccentricity of the origin country was determined. The eccentricity provides us with a number which measures the susceptibility of human trafficking in the paths of the origin country to the U.S. We found that Somalia has the highest eccentricity. We also considered the flow of human trafficking from countries to the United States with respect to the size of the flow. The ideal situation is that the flow be less than or equal to the minimum of the vulnerabilities (or of the government responses) of the two countries involved. This is not always the case, so a quasi-fuzzy incidence graph is obtained. Measures on how much the flow should be reduced to obtain a fuzzy incidence graph were obtained. In future research, we plan to pursue these ideas by replacing the minimum t-norm with an arbitrary t-norm and the conorm maximum with an arbitrary conorm.

References

1. K.T. Atanassov, Intuitionistic fuzzy sets. Fuzzy Sets Syst. **20**, 87–96 (1986)
2. K.T. Atanassov, On intuitionistic fuzzy graphs and intuitionistic fuzzy relations, in *Proceedings of the 6th IFSA World Congress*, vol 1 (San Paulo, Brazil, 1995), pp. 551–554
3. K.T. Atanassov, A. Shannon, On a generalization of intuitionistic fuzzy graphs. Notes Intuit. Fuzzy Sets **12**, 24–29 (2006)
4. K.R. Bhutani, J.N. Mordeson, A. Rosenfeld, On degrees of end nodes and cut nodes in fuzzy graphs. Iran. J. Fuzzy Syst. **1**, 57–64 (2004)
5. K.R. Bhutani, J.N. Mordeson, P.K. Saha, $(s, t]$ -fuzzy graphs, in *JCIS, Proceedings* (2005), pp. 37–40
6. K.R. Bhutani, A. Rosenfeld, Fuzzy end nodes in fuzzy graphs. Inf. Sci. **152**, 323–326 (2003)
7. E.Darabian, R.A. Borzooei, Results on vague graphs with applications in human trafficking. New Math. Nat. Comput. to appear
8. T. Dinesh, A study on graph structures, incidence algebras and their fuzzy analogues, Ph.D. Thesis, Kummar University, (Kerala, India, 2012)
9. T. Dinesh, Fuzzy incidence graph - an introduction. Adv. Fuzzy Sets Syst. **21**(1), 33–48 (2016)
10. D.S. Malik, S. Mathew, J.N. Mordeson, Fuzzy Incidence Graphs: Applications to Human Trafficking, submitted
11. J.N. Mordeson, Fuzzy incidence graphs. Adv. Fuzzy Sets Syst. **21**(2), 1–13 (2016)
12. J.N. Mordeson, S. Mathew, A. Borzooei, Vulnerability and Government Response to Human Trafficking: Vague Fuzzy Incidence Graphs, Submitted
13. J.N. Mordeson, D.S. Malik, C.S. Richards, J.A. Trebbian, M.A. Boyce, M.P. Byrne, B.J. Cousino, Fuzzy graphs and complentary fuzzy graphs. J. Fuzzy Math. **24**, 271–288 (2016)
14. J.N. Mordeson, S. Mathew, $(s, t]$-fuzzy incidence graphs. J. Fuzzy Math. **25**, 723–734 (2017)
15. Central Intelligence Agency, Field Listing: Trafficking in Persons (2014). http://www.cia.gov/library/publications/the-world-factbook/fields/2196.html
16. Global Slavery Index (2016). http://www.globalslaveryindex.org/findings
17. Routes to the US: Mapping Human Smuggling Networks. http://www.insightcrime.org/news-analysis/routes-to-us-mapping
18. Trafficking in Persons, *Global Patterns* (Appendices-United Nations Office on Drugs and Crime, Citation Index, 2006)
19. U.S. Department of State, Diplomacy in Action, Trafficking in Persons Report 2013: Tier Placements. http://www.state.gov/j/rls/tiprpt/2013/210548.htm

Chapter 6
Human Trafficking: Source, Transit, Destination Designations

6.1 Introduction

The following is stated in [10].

"Trafficking in persons has been defined as the recruitment, transportation, transfer, harboring or receipt of persons by means of threat or use of force or other forms of coercion, of abduction, of fraud, of deception, of the abuse of power or of a position of vulnerability or the giving or receiving of payments or benefits to achieve the consent of a person having control over another person, for the purpose of exploitation, [6].

Although governments, international organizations, and civil society are devoting efforts to combat sex trafficking, there still exists a lack of accurate information concerning trafficking. There exist three main challenges in combating sex trafficking: reduction of demand, target the traffickers, and protect the trafficking victims, [5]. These three challenges are often referred to as the 3 Ps: Prevention Protection and Prosecution. The 3 Ps have been the main focus in many studies, [4, 5, 11]. Because of the lack of accurate information on global sex trafficking, we take a linguistic approach in this chapter to determine an upper bound in linguistic terms on the trafficking from one region to another. We use the terms very low, low, medium, high, and very high to describe the source, transit and destination designations of countries. The designations are taken from [13]. The designations were determined in the following manner. Data was provided for 161 countries and special administrative territories by 113 source institutions. If a country was mentioned once in a document generated by one source institution as either an origin, transit or destination country, it appears in the Trafficking Database. For any one of these three designations, the citation index takes the number of sources reporting that variable to a country and indicates the appropriate citation category, very low, low, medium, high, very high, in comparison to all other countries, [13]. We then determine source, transit and destination designations for regions by using a weighted average based on population of the countries in their respective region. Different studies define regions differently.

© Springer International Publishing AG 2018
J. N. Mordeson et al., *Fuzzy Graph Theory with Applications to Human Trafficking*, Studies in Fuzziness and Soft Computing 365,
https://doi.org/10.1007/978-3-319-76454-2_6

We use the definitions in [13]."

In [3], a study of how governments are tackling modern slavery was undertaken. We conclude the chapter by providing measures that relate flow to the United States with the vulnerability and government responses of other countries.

6.2 Source Transit and Destination

The size of flow from country to country is taken from [13]. It is reported in linguistic terms. Information is provided with respect to the reported human trafficking in terms of origin, transit, and/or destination according to the citation index. The data is provided in two columns. Information in the left column as to whether a country ranks (very) low, medium, (very) high depends upon the total number of sources which made reference to this country as one of origin, transit, or destination. Information provided in the right column provides further detail to the information provided in the left column. If a country is reported as one of origin, information in the right column will be provided on the countries to which victims are reportedly trafficked. Likewise, if the country is reported to be a destination country, information in the right column will indicate the countries of origin from which victims are reportedly being trafficked or where the victims are eventually destined. A different scale was used to determine whether the related countries in the right column are ranked high, medium or low. If a country in the right column was mentioned by one or two sources, the related country was ranked low. If the linkage between the countries in the two columns was reported by 3–5 sources, the related country was ranked medium. If 5 or more sources linked two countries, the country in the right was ranked high. This method of combining linguistic data provides an ideal reason for the use of mathematics of uncertainty to study the problem of trafficking in persons. For example, by assigning numbers in the interval [0, 1] to the linguistic data, the data can be combined in a mathematical way. We use the notions of a t-norm and conorm.

The results of this section are based on [10]. We assign the number 0.1 for very low, 0.3 for low, 0.5 for medium, 0.7 for high, and 0.9 for very high. This allows for the use of fuzzy logic to study the problem. We define the flow from a country to another as a combination of the origin flow and the transit flow through the country. We define the flow into a country as a combination of the transit flow through it and the destination flow into it. We accomplish this combination by using a particular conorm. We use the symbol ∘ to denote the algebraic conorm, i.e., for all a, b in the closed interval $[0, 1], a \circ b = a + b - ab$. The population figures in the tables to follow are in the thousands.

Africa

Of the countries in Africa, Nigeria is the only country ranked very high as an origin country. The countries Benin, Ghana, and Morocco are ranked high. Morocco is the only country rated as a very high destination country.

Eastern Africa	Source	Transit	Destination	Population	Weight	$s \circ t$	$t \circ d$
Burundi	0.3	0	0.1	7548	0.03	0.3	0.1
Djibouti	0.3	0	0.1	793	0	0.3	0.1
Eritrea	0.3	0	0	4401	0.02	0.3	0
Ethiopia	0.5	0	0.1	77431	0.27	0.5	0.1
Kenya	0.5	0	0.5	34256	0.12	0.5	0.5
Madagascar	0.3	0	0	18606	0.07	0.3	0
Malawi	0.5	0.1	0.1	12884	0.05	0.55	0.19
Mozambique	0.5	0.1	0.1	19792	0.07	0.55	0.19
Rwanda	0.3	0	0	9038	0.03	0.3	0
Somalia	0.3	0	0	8228	0.03	0.3	0
Uganda	0.5	0	0.3	28816	0.10	0.5	0.3
United Republic of Tanzania	0.5	0.1	0.3	38329	0.13	0.55	0.37
Zambia	0.5	0.1	0.1	11668	0.04	0.55	0.19
Zimbabwe	0.3	0.1	0.1	13010	0.05	0.55	0.19

Total Population 284,800

Middle Africa	Source	Transit	Destination	Population	Weight	$s \circ t$	$t \circ d$
Angola	0.5	0	0	15941	0.15	0.5	0
Cameroon	0.5	0.3	0.5	16322	0.15	0.65	0.65
Chad	0.1	0.1	0.1	9749	0.09	0.19	0.19
Congo, Democratic Republic of	0.3	0	0.1	57549	0.55	0.3	0.1
Congo, Republic of	0.5	0	0.1	3999	0.04	0.5	0.1
Equatorial Guinea	0.3	0.1	0.5	504	0	0.37	0.55
Gabon	0.3	0.5	0.5	1385	0.01	0.65	0.75

Total Population 105,409

Northern Africa	Source	Transit	Destination	Population	Weight	$s \circ t$	$t \circ d$
Algeria	0.5	0.3	0.1	32854	0.17	0.65	0.37
Egypt	0.1	0.5	0.3	74033	0.39	0.55	0.65
Libyan Arab Jamahiriya	0	0	0.3	5853	0.03	0	0.3
Morocco	0.7	0.3	0.1	31478	0.17	0.79	0.37
Sudan	0.3	0	0.1	36233	0.19	0.3	0.1
Tunisia	0.3	0	0	10102	0.05	0.3	0

Total Population 190,553

Southern Africa	Source	Transit	Destination	Population	Weight	$s \circ t$	$t \circ d$
Botswana	0.3	0.3	0	1765	0.03	0.51	0.3
Lesotho	0.3	0.1	0	1795	003	0.37	0.1
South Africa	0.5	0.5	0.5	47432	0.91	0.75	0.75
Swaziland	0.3	0	0	1032	0.02	0.3	0

Total Population 52, 024

Western Africa	Source	Transit	Destination	Population	Weight	$s \circ t$	$t \circ d$
Benin	0.7	0.5	0.5	8439	0.03	0.85	0.75
Burkina Faso	0.5	0.5	0.5	1322	0.01	0.75	0.75
Cape Verde	0.3	0	0	507	0	0.3	0
Cote d'voire	0.5	0.5	0.5	18154	0.07	0.75	0.75
Gambia	0.3	0	0.1	1517	0.01	0.3	0.1
Ghana	0.7	0.3	0.5	22113	0.09	0.89	0.65
Guinea	0.3	0	0	9402	0.04	0.3	0
Liberia	0.5	0	0.1	3283	0.01	0.5	0.1
Mali	0.5	0.1	0.3	13518	0.05	0.55	0.37
Mauritania				4068	0.02	0	0
Niger	0.5	0.1	0.3	13957	0.06	0.55	0.37
Nigeria	0.9	0.3	0.5	131530	0.52	0.93	0.65
Senegal	0.5	0.1	0.1	11658	0.05	0.55	0.19
Sierra Leone	0.5	0	0.1	5525	0.02	0.5	0.1
Togo	0.5	0.5	0.5	6145	0.02	0.75	0.75

Total Population 251,138

Asia

Much of the trafficking in Asia is intra-regional. The countries ranked very high as destination countries are Thailand, Japan, India, Taiwan, and Pakistan. China and Thailand are ranked very high as origin countries.

Eastern Asia	Source	Transit	Destination	Population	Weight	$s \circ t$	$t \circ d$
China	0.9	0.1	0.7	1315844	0.86	0.91	0.73
Hong Kong	0.5	0.5	0.7	7041	0	0.75	0.85
China, Macao SAR	0.1	0	0.5		0	0.1	0.5
Taiwan, Province of China	0.5	0	0.7		0	0.5	0.7
Dem. People's Rep. of Korea	0.5	0	0	22488	0.01	0.5	0
Japan	0	0.1	0.9	128085	0.08	0.1	0.91
Republic of Korea	0.3	0.1	0.5	47817	0.03	0.37	0.55

Total Population 1,528,918

South-Central Asia	Source	Transit	Destination	Population	Weight	$s \circ t$	$t \circ d$
Afghanistan	0.5	0	0	29863	0.02	0.5	0
Bangladesh	0.7	0.1	0.3	141822	0.09	0.73	0.37
Bhutan	0.3	0	0.1	2163	0	0.3	0.1
India	0.7	0.5	0.7	1103371	0.71	0.85	0.85
Iran	0.3	0	0.5	69515	0.04	0.3	0.5
Maldives	0.3	0	0.1	329	0	0.3	0.1
Nepal	0.7	0.1	0	27133	0.02	0.73	0.1
Pakistan	0.7	0.1	0.7	157935	0.10	0.73	0.73
Sri Lanka	0.5	0	0.3	20743	0.01	0.5	0.3

Total Population 1,552,874

South-Eastern Asia	Source	Transit	Destination	Population	Weight	$s \circ t$	$t \circ s$
Brunei Darussalam	0.1	0.3	0.3	374	0	0.37	0.51
Cambodia	0.7	0.1	0.7	16322	0.03	0.73	0.73
Indonesia	0.5	0.3	0.3	222781	0.40	0.65	0.51
Laos	0.7	0.3	0.3	5924	0.01	0.79	0.51
Malaysia	0.5	0.5	0.5	25347	0.05	0.75	0.75
Myanmar	0.7	0.7	0.5	50519	0.09	0.91	0.85
Philippines	0.7	0.1	0.5	83054	0.15	0.73	0.55
Singapore	0.5	0.5	0.5	4326	0.01	0.75	0.75
Thailand	0.9	0.9	0.9	64233	0.12	0.99	0.99
Vietnam	0.7	0.1	0.5	84238	0.15	0.73	0.55

Total Population 557,118

Western Asia & Turkey	Source	Transit	Destination	Population	Weight	$s \circ t$	$t \circ d$
Bahrain	0	0.1	0.5	727	0	0.1	0.55
Cyprus	0	0.5	0.7	835	0	0.5	0.85
Iraq	0.3	0	0.3	28807	0.15	0.3	0.3
Israel	0	0	0.9	6725	0.03	0	0.9
Jordan	0.3	0.1	0	5703	0..03	0.37	0.1
Kuwait	0	0	0.3	2687	0.02	0	0.3
Lebanon	0.3	0.1	0.5	3577	0.02	0.37	0.55
Oman	0	0	0.3	2567	0.01	0	0.3
Qatar	0	0	0.3	813	0	0	0.3
Saudi Arabia	0	0.1	0.7	24573	0.13	0.1	0.73
Syrian Arab Republic	0.1	0	0.5	19043	0.10	0.1	0.5
Turkey	0.5	0.7	0.9	73193	0.38	0.85	0.97
United Arab Emirates	0	0	0.7	4496	0.02	0	0.7
Yemen	0.1	0	0.3	20975	0.11	0.1	0.3

Total Population 194,721

Americas

The region of Latin America and the Caribbean is primarily a origin country, while North America is almost exclusively a destination region.

Caribbean	Source	Transit	Destination	Population	Weight	$s \circ t$	$t \circ d$
Aruba	0	0	0.3	99	0	0	0.3
Cuba	0.5	0	0	11269	0.34	0.5	0
Curacao	0	0	0.5		0	0	0.5
Dominica	0	0.1	0.1	79	0	0.1	0.19
Dominican Republic	0.7	0	0.5	8895	0.27	0.7	0.5
Haiti	0.5	0	0.3	8528	0.26	0.5	0.3
Jamaica	0.1	0.1	0.1	2651	0.08	0.19	0.19
Trinidad & Tobago	0	0	0.1	1305	0.04	0	0.1

Total Population 32,984

Central America	Source	Transit	Destination	Population	Weight	$s \circ t$	$t \circ d$
Belize	0	0.1	0.3	270	0	0.1	0.37
Costa Rica	0.1	0.3	0.3	4327	0.03	0.37	0.51
El Salvador	0.5	0.1	0.5	6881	0.05	0.55	0.55
Guatemala	0.7	0.1	0.5	12599	0.09	0.63	0.55
Honduras	0.5	0	0.1	7205	0.05	0.5	0.1
Mexico	0.7	0.5	0.5	107029	0.73	0.85	0.75
Nicaragua	0.3	0	0	5483	0.04	0.3	0
Panama	0.3	0.1	0.5	3232	0.02	0.37	0.55

Total Population 147,026

South America	Source	Transit	Destination	Population	Weight	$s \circ t$	$t \circ d$
Argentina	0.3	0	0.5	38747	0.11	0.3	0.5
Brazil	0.7	0	0.1	186405	0.51	0.7	0.1
Chile	0.1	0	0.1	16295	0.04	0.1	0.1
Columbia	0.7	0.1	0	45600	0.13	0.73	0.1
Ecuador	0.5	0	0.3	13228	0.04	0.5	0.3
Paraguay	0.1	0	0.3	6158	0.02	0.1	0.3
Peru	0.5	0	0	27968	0.08	0.5	0
Uruguay	0.1	0.1	0	3463	0.01	0.19	0.1
Venezuela	0.5	0	0.5	26749	0.07	0.5	0.5

Total Population 364,604

North America	Source	Transit	Destination	Population	Weight	$s \circ t$	$t \circ d$
Canada	0.3	0.5	0.7	32268	0.10	0.65	0.85
United States	0.3	0	0.9	298213	0.90	0.3	0.9

Total Population 330,481

Europe

Central and South Eastern Europe is reported as an origin, transit, and destination region, while Eastern Europe is mostly a destination region.

C&S Eastern Europe	Source	Transit	Destination	Population	Weight	$s \circ t$	$t \circ d$
Albania	0.9	0.9	0.5	3130	0.02	0.99	0.95
Bosnia & Herzegovina	0.5	0.7	0.7	3907	0.03	0.85	0.91
Bulgaria	0.9	0.9	0.5	7726	0.06	0.99	0.95
Croatia	0.5	0.5	0.5	4551	0.04	0.75	0.75
Czech Republic	0.7	0.7	0.7	10220	0.08	0.91	0.91
Estonia	0.7	0.1	0.5	1330	0.01	0.73	0.55
Hungary	0.7	0.9	0.5	10098	0.08	0.97	0.95
Latvia	0.7	0.3	0.5	2307	0.02	0.79	0.65
Lithuania	0.9	0.3	0.5	3431	0.03	0.93	0.65
Poland	0.7	0.9	0.7	38530	0.30	0.97	0.97
Romania	0.9	0.7	0.3	21711	0.17	0.97	0.79
Serbia & Montenegro	0.5	0.7	0.5	10503	0.08	0.85	0.85
Kosovo	0.5	0.7	0.7	1871	0.01	0.85	0.91
Slovakia	0.7	0.7	0.1	5401	0.04	0.91	0.73
Slovenia	0.5	0.3	0.3	1967	0.02	0.65	0.51
Yugoslavia	0.5	0.7	0.5	2034	0.02	0.85	0.85

Total Population 128,717

Western Europe	Source	Transit	Destination	Population	Weight	s o t	t o d
Austria	0	0.3	0.7	8189	0.02	0.3	0.79
Belgium	0	0.7	0.9	10419	0.03	0.7	0.97
Denmark	0	0	0.7	5431	0.01	0	0.7
Finland	0	0.1	0.5	5249	0.01	0.1	0.55
France	0	0.7	0.7	60496	0.15	0.7	0.91
Germany	0	0.7	0.9	82689	0.21	0.7	0.97
Greece	0	0.7	0.9	11120	0.03	0.7	0.97
Iceland	0	0	0.5	295	0	0	0.5
Ireland	0	0.1	0.3	4148	0.01	0.1	0.37
Italy	0	0.9	0.9	58093	0.15	0.9	0.99
Luxembourg	0	0	0.3	465	0	0	0.3
Malta							
Netherlands	0.1	0.5	0.9	16299	0.04	0.55	0.95
Norway	0	0.1	0.5	4620	0.01	0.1	0.55
Portugal	0	0	0.5	10495	0.03	0	0.5
Spain	0	0.3	0.7	43064	0.11	0.3	0.79
Sweden	0	0.1	0.5	9041	0.02	0.1	0.55
Switzerland	0	0.1	0.5	7252	0.02	0.1	0.55
United Kingdom	0	0.5	0.7	59668	0.15	0.5	0.85

Total Population 397,003

Commonwealth of Independent States

C I States	Source	Transit	Destination	Population	Weight	s o t	t o d
Armenia	0.7	0	0	3016	0.01	0.7	0
Azerbaijan	0.5	0.3	0	8411	0.03	0.65	0.3
Belarus	0.9	0.5	0	9755	0.04	0.95	0.5
Georgia	0.7	0.5	0.1	4474	0.02	0.85	0.55
Kazakhstan	0.7	0.5	0.5	14825	0.05	0.85	0.75
Kyrgyzstan	0.5	0.1	0.3	5264	0.02	0.55	0.37
Rep. of Moldova	0.9	0.3	0.1	4206	0.02	0.93	0.37
Russian Federation	0.9	0.5	0.5	143202	0.52	0.95	0.75
Tajikistan	0.5	0	0.1	6507	0.02	0.5	0.1
Turkmenistan	0.5	0	0	4833	0.02	0.5	0
Ukraine	0.9	0.7	0.5	46481	0.17	0.97	0.85
Uzbekistan	0.7	0	0.3	26593	0.10	0.7	0.3

Total Population 277,567

Oceania

	Source	Transit	Destination	Population	Weight	s o t	t o d
Australia	0	0	0.7	20155	0.83	0	0.7
New Zealand	0	0.3	0.5	4028	0.17	0.3	0.65

Total Population 24,183

6.3 Regional Flow Upper Bounds

In the table below, we present the regions and their degree of involvement as a source, transit, and destination region. The designations were determined by a weighted average of the source, transit, and destination values of the countries making up the region. The weights were determined by the countries population divided by the total population of the countries in the region.

The following table gives the source, transit, and designation rankings for regions.

	Source	Transit	Destination	$s \circ t$	$t \circ d$
Eastern Africa	0.46	0.03	0.18	0.47	0.20
Middle Africa	0.35	0.06	0.15	0.39	0.20
Northern Africa	0.31	0.30	0.18	0.52	0.43
Southern Africa	0.48	0.47	0.46	0.72	0.71
Western Africa	0.73	0.27	0.42	0.80	0.78
Eastern Asia	0.79	0.10	0.70	0.81	0.73
South-Central Asia	0.67	0.38	0.62	0.80	0.76
South-Eastern Asia	0.63	0.35	0.47	0.78	0.66
Western Asia & Turkey	0.27	0.28	0.62	0.47	0.73
Caribbean	0.50	0.01	0.23	0.51	0.24
Central America	0.64	0.39	0.46	0.78	0.67
South America	0.58	0.01	0.16	0.53	0.17
North America	0.30	0.05	0.88	0.34	0.89
C& S Eastern Europe	0.72	0.76	0.53	0.93	0.89
Western Europe	0.00	0.56	0.77	0.56	0.94
C of Independent States	0.83	0.44	0.41	0.90	0.67
Oceania	0.00	0.05	0.67	0.05	0.69

A natural way to define an upper bound of the flow from region R_1 to another R_2 is to first apply a conorm to the s, t and the t, d values. We felt that there were conorms more suitable that the maximum conorm. We decided to use the algebraic sum conorm $\circ$, i.e., for all a, b in the closed interval $[0, 1]$, $a \circ b = a + b - ab$. Let $\mathcal{R}$ denote the set of regions and let β be a fuzzy subset of $\mathcal{R} \times \mathcal{R}$ defined by for all $R_1, R_2 \in \mathcal{R} \times \mathcal{R}$, $\beta(R_1, R_2) = (s_1 \circ t_1) \wedge (t_2 \circ d_2)$, where $\wedge$ denotes minimum and s_1, t_1 are R_1's source and transit values and t_2, d_2 are R_2's transit and destination values, respectively. This gives a linguistic upper bound on the flow from region R_1 to region R_2. For example, β(Northern Africa, Eastern Asia) $= 0.52 \wedge 0.73 = 0.52$. The following table lists the upper bounds from one region to a different region.

	E Af	M Af	N Af	S Af	W Af	E Asia	S-C Asia	S-E Asia	W Asia & T	Car	C Am	S Am	N Am	C&S E Eur	W Eur	C I S	Oc
E Af		0.20	0.43	0.47	0.47	0.47	0.47	0.47	0.47	0.24	0.47	0.17	0.47	0.47	0.47	0.47	0.47
M Af	0.20		0.39	0.39	0.39	0.39	0.39	0.39	0.39	0.24	0.39	0.17	0.39	0.39	0.39	0.39	0.39
N Af	0.20	0.20		0.52	0.52	0.52	0.52	0.52	0.52	0.24	0.52	0.17	0.52	0.52	0.52	0.52	0.52
S Af	0.20	0.20	0.43		0.72	0.72	0.72	0.66	0.72	0.24	0.67	0.17	0.72	0.72	0.72	0.67	0.69
W Af	0.20	0.20	0.43	0.71		0.73	0.76	0.66	0.73	0.24	0.67	0.17	0.80	0.80	0.80	0.67	0.69
E Asia	0.20	0.20	0.43	0.71	0.78		0.76	0.66	0.73	0.24	0.67	0.17	0.81	0.81	0.81	0.67	0.69
S-C Asia	0.20	0.20	0.43	0.71	0.78	0.73		0.66	0.73	0.24	0.67	0.17	0.80	0.80	0.80	0.67	0.69
S-E Asia	0.20	0.20	0.43	0.71	0.78	0.73	0.76		0.73	0.24	0.67	0.17	0.78	0.78	0.78	0.67	0.69
W Asia & T	0.20	0.20	0.43	0.47	0.47	0.47	0.47	0.47		0.24	0.47	0.17	0.47	0.47	0.47	0.47	0.47
Car	0.20	0.20	0.43	0.51	0.51	0.51	0.51	0.51	0.51		0.51	0.17	0.51	0.51	0.51	0.51	0.51
C Am	0.20	0.20	0.43	0.71	0.78	0.73	0.76	0.66	0.73	0.24		0.17	0.78	0.78	0.78	0.67	0.69
S Am	0.20	0.20	0.43	0.53	0.53	0.53	0.53	0.53	0.53	0.24	0.53		0.53	0.53	0.53	0.53	0.53
N Am	0.20	0.20	0.34	0.34	0.34	0.34	0.34	0.34	0.34	0.24	0.34	0.17		0.34	0.34	0.34	0.34
C&S E Eur	0.20	0.20	0.43	0.71	0.78	0.73	0.76	0.66	0.73	0.24	0.67	0.17	0.89		0.93	0.67	0.69
W Eur	0.20	0.20	0.43	0.56	0.56	0.56	0.56	0.56	0.56	0.24	0.56	0.17	0.56	0.56		0.56	0.56
C I S	0.20	0.20	0.43	0.71	0.78	0.73	0.76	0.66	0.73	0.24	0.67	0.17	0.89	0.89	0.90		0.69
Oc	0.05	0.05	0.05	0.05	0.05	0.05	0.05	0.05	0.05	0.05	0.05	0.05	0.05	0.05	0.05	0.5	

Future Research

"Consider the region Oceania. Then the linguistic upper bound of the flow from Oceania to South America is given by $0.064 \wedge 0.261 = 0.064$. However, Oceania is made up of three countries, Australia, Fiji, and New Zealand. According to the United Nations report, Trafficking in Persons: Global Patterns, [13], no flow from Australia to any country exists, there is a very low flow from Fiji to Australia (none outside the region), and there exists low flow for New Zealand as transit country for flow from Thailand to Australia (none outside the region). Thus, μ(Oceania, South America) $= 0$. In fact, for any region R $\neq$ Oceania, μ(Oceania, R) $= 0$. Hence, the flow of traffic in Oceania stays within the region. It is reasonable to define μ(Oceania, Oceania) to be somewhere between low and very low. A future research problem would be to determine linguistically, the flow between any two regions [10]."

We begin this project by examining the flow from countries of all regions into the countries of the Americas. The Americas consist of four regions, the Caribbean, Central America, South America (which make up what is called in [13] Latin America and the Caribbean), and North America.

North America

We present the flow from countries in the region Central and South Eastern Europe into the countries of North America. There is no flow from Western Europe into North America.

Central & South Eastern Europe	United States	Canada
Albania		
Bosnia & Herzegovina		
Bulgaria	Low	Low
Croatia		
Czech Republic	Medium	Low
Estonia	Low	
Hungary	Medium	Low
Latvia	Medium	
Lithuania	Medium	Low
Poland	Medium	
Romania	Low	Low
Serbia & Montenegro	Low	
Kosovo		
Slovakia		
Slovenia	Low	
Former Yugoslav Rep. of Macedonia		

Western Europe	United States	Canada
All countries		

We next present the flow from countries in the region Commonwealth of Independent States into the countries of North America.

Commonwealth of Independent States	United States	Canada
Armenia	Medium	Medium
Azerbaijan	Medium	Medium
Belarus	Medium	Medium
Georgia	Medium	Medium
Kazakhstan	Medium	Medium
Kyrgyzstan	Medium	Medium
Republic of Moldova	Medium	Medium
Russian Federation	High	Medium
Tajikistan	Medium	Medium
Turkmenistan	Medium	Medium
Ukraine	High	High
Uzbekistan	Medium	Medium

The remaining tables give the flow from countries of the given regions into North America.

Caribbean	United States	Canada
Aruba		
Cuba	Low	Low
Curacao		
Dominica		
Dominican Republic	Low	Low
Haiti	Low	Low
Jamaica		
Trinidad and Tobago		

Central America	United States	Canada
Belize		
Costa Rica	Low	
El Salvador	Medium	
Guatemala	Medium	
Honduras	Low	Low
Mexico	High	Medium
Nicaragua	Low	Low
Panama	Low	Low

South America	Unites States	Canada
Argentina		
Brazil	Medium	Low
Chili		
Columbia	Medium	Medium
Ecuador	Medium	Low
Paraguay		
Peru	Low	Low
Uruguay		
Venezuela	Low	

South Central Asia	United States	Canada
Afghanistan	Low	Low
Bangladesh	Low	
Bhutan		
India	Medium	Low
Iran		
Maldives	Low	
Nepal	Low	
Pakistan	Low	
Sri Lanka	Medium	

South Eastern Asia	United States	Canada
Brunei Darussalem		
Cambodia	Low	
Indonesia	Low	
Lao P. Dem Rep.	Low	
Malaysia		Medium
Philippines	Medium	Medium
Singapore	Low	
Thailand		Medium
Vietnam	Medium	Low

Eastern Asia	United States	Canada
China	High	
Rep. Korean	Medium	
Other		

Western Asia and Turkey	United States	Canada
Cyprus	Low	
Jordan	Low	
Lebanon	Low	
Saudi Arabia	Low	
Other		

Northern Africa	United States	Canada
All countries		

Southern Africa	United States	Canada
All countries		

Western Africa	United States	Canada
Ghana	Low	
Niger	Low	
Nigeria	Medium	Low
Other		

Eastern Africa	United States	Canada
Eritea	Low	Low
U. R. Tanzania	Low	Low
Other		

Middle Africa	United States	Canada
Cameron	Low	
Congo	Low	

Oceania Australia & New Zealand	United States	Canada
Australia		
New Zealand		

Melanesia	United States	Canada
Fiji		

The term intraregional is used to denote flow within the region.

Intraregional

The intraregional flow for North America is given in the following table.

North America	Canada	United States
Canada		Medium
United States	Low	

The next table gives the destination ranking in [11] for the countries in North America.

North America	Country of Destination Ranking
Canada	High
United States	Very High

Latin America and Caribbean

We next present the flow from counties of all regions into the countries making up the three regions of Latin America and the Caribbean.

Caribbean	Aruba	Cuba	Curacao	Dominica	Dominican Republic	Haiti	Jamaica	Trinidad & Tobago
Mexico					Low			
Other								

Intraregional

The intraregional flow of the Caribbean is given in the following table.

	Aruba	Curacao	Dominica	Haiti
Dominican Republic	Medium	Medium	Low	Medium

Haiti is ranked low as a destination country, but medium as a destination country for Dominican Republic.

Trinidad and Tobago is ranked very low as a destination country, but low as a destination country for the region Caribbean.

Central America	Belize	Costa Rica	El Salvador	Guatemala	Honduras	Mexico	Nicaragua	Panama
Bulgaria		Low						
Columbia		Low						Medium
Dominican Rep		Low				Low		Medium
Philippines		Low						
Russian Fed		Low						
Ecuador				Low	Low	Low		
Brazil						Low		
Haiti						Low		
Hungary						Low		
Slovenia						Low		

Intraregional

Central America	Belize	Costa Rica	El Salvador	Guatemala	Honduras	Mexico	Nicaragua	Panama
Belize								
Costa Rica			Low	Low				
El Salvador				Medium		Medium		
Guatemala	Low		Low			Medium		
Honduras	Low		Medium	Medium		Medium		
Mexico								
Nicaragua		Medium	Low	Medium				
Panama		Low						

The next table gives the destination ranking in [12] for the countries in Central America.

Central America	Country of Destination Ranking
Belize	Low
Costa Rica	Low
El Salvador	Medium
Guatemala	Medium
Honduras	Very Low
Mexico	Medium
Nicaragua	
Panama	Medium

South America	Argentina	Brazil	Chile	Colombia	Ecuador	Paraguay	Peru	Uruguay	Venezuela
Dominican Republic	Medium		Low						Medium
China	Low								
Latin America		Low	Low						

Intraregional

South America	Argentina	Brazil	Chile	Colombia	Ecuador	Paraguay	Peru	Uruguay	Venezuela
Argentina		Low							
Brazil									
Chile					Low				
Colombia									
Ecuador				Medium					
Paraguay		Low							
Peru									
Uruguay									
Venezuela				Medium	Low				

The next table gives the destination ranking in [11] for the countries in South America.

South America	Country of Destination Ranking
Argentina	Medium
Brazil	Very Low
Chile	Very Low
Colombia	
Ecuador	Low
Paraguay	Low
Peru	
Uruguay	
Venezuela	Medium

Arithmetic operations applied to the terms very low, low, medium, high, very high need to be determined so they agree with the way these terms are combined in [13].

In the following, we consider the flow of traffic from Mexico into the United States. The total flow, transient plus nontransient, is rated high. The transient is rated medium. We wish to find the rating of the flow which is destination, but not transient. Let x represent this number and let u denote a conorm. Then we wish to solve the equation $u(a, x) = c$ for x, where a denotes the rating for transient flow and c is the rating for total flow. Let u denote the algebraic sum conorm. Then $0.6 + x - (0.6)x = 0.8$. Solving, we find $x = 0.5$. Let u denote the Dombi conorm. Then $[0.6 + x - 2(0.6)x]/[1 - 0.6x] = 0.8$. Solving, we find $x = \dfrac{5}{7} \approx 0.7$. If u is the Frank-Weber conorm, $(0.6 + x) \wedge 1 = 0.8$. Thus, $x = 0.2$.

Now consider the flow of Canada into the U.S. The rating of the transit flow is medium while the rating of the total flow is medium. In this case, we obtain $x = 0$ for any of the above three conorms.

6.4 Fuzzy Incidence Graphs: Applications to Human Trafficking

A study of how governments are tackling modern slavery was undertaken in [3]. The assessment of government responses included 161 countries. Of these countries, 124 have criminalized human trafficking in line with the United Nations Trafficking Protocol. 91 have National Action Plans to evaluate government responses, and 150 countries provide some sort of service for victims of modern slavery.

Vulnerability to human trafficking is affected by an interaction of factors related to the presence or absence of protection and respect for rights, physical safety and security, access to the necessities of life such as, food water and health care, and patterns of migration, displacement and conflict. Statistical testing grouped 24 measures of vulnerability into four dimensions covering: (1) civil and political protections, (2) social health and economic rights, (3) personal security, and (4) refugee populations and conflict. In [3], tables are provided giving measures of vulnerability to modern slavery by country with respect to these four dimensions.

Government response to human trafficking involves the following categories: (1) survivors supported, (2) criminal justice, (3) coordination and accountability, (4) addressing risk, and (5) government and business. We ignore the fifth government response because hardly any of the countries were assigned a positive value. The Walk Free Foundation included measure on state-sanctioned forced labor to the government response rating in 2016. Government response rating to human trafficking can be found in [3].

We normalize the data in the vulnerability and government response ratings of the countries for each of the four vulnerabilities and each of the five government responses. For example, let U_j denote the 161-dimensional column vector containing the vulnerability ratings of each country, $j = 1, 2, 3, 4$. Let u_{ij} denote the i-th entry for $U_j, i = 1, 2, \ldots, 161$. For a fixed j, let m_j denote the minimum value of the u_{ij} and M_j the maximum value. Then the entries of V_j are normalized using the formula $v_{ij} = (u_{ij} - m_j)/(M_j - m_j)$. The government response ratings are normalized in the same manner. This allows us to apply fuzzy logic techniques in our study. We let

V_j denote the column vector containing the normalized ratings of U_j, $j = 1, 2, 3, 4$. Similarly, we let G_j denote the column vector containing the government response normalized ratings, $j = 1, 2, 3, 4, 5$. Because only 4 countries were given positive ratings, we delete G_5 from our approach.

Let C_{VL} denote the set of all countries which have government response ratings lower than 0.2, [3]. Using the ratings given in [3], the countries in this set are ranked as very low with respect to government response. We next consider the destination countries of the countries in C_{VL}. For example, consider Sierra Leone. Then $\mu(SierraLeone, Angola) = 0.3$, i.e., the flow from Sierra Leone to Angola is low, [3]. Now $\sigma(SierraLeone) = 0.1$ and $\sigma(Angola) = 0.24$. Thus,

$$\mu(SierraLeone, Angola) > \sigma(SierraLeone) \wedge \sigma(Angola),$$
$$0.3 > 0.1 \wedge 0.24 = 0.1.$$

Hence, $t_{xy} = 0.24$, where $x = $ SierraLeone and $y = $ Angola.

Also, $\mu(SierraLeone, Angola) = $ low and $\sigma(SierraLeone) = $ very low, $\sigma(Angola) = $ low. Thus, $t_{xy} = $ very low.

Another approach is to let $\Psi(x, xy)$ denote the flow from x to y and $\sigma(x) = $ either the vulnerability or the government response of x. Note $\Psi(x, xy) \leq \mu(xy) \wedge \sigma(x) = \sigma(x) \wedge \sigma(y)$ if $\mu(xy) = \sigma(x) \wedge \sigma(y)$. This is essentially the same as the previous approach.

The ideal situation is that the flow be less than or equal to the minimum of the vulnerabilities (or government responses) of the two countries involved. (This would yield a fuzzy incidence graph or a fuzzy graph rather than a quasi-fuzzy graph.) This is not always the case so when a quasi-fuzzy incidence graph is obtained, the t gives a measure of how far one is from a fuzzy graph.

Keeping y fixed, a measure of the amount of decrease in flow needed for a set of countries R is given by $t_R = (\vee\{(\Psi(x, xy) - \sigma(x) \wedge \sigma(y)) \mid x \in R\}) \vee 0$ or $t_R^* = \sum_{x \in R}(\Psi(x, xy) - \sigma(x) \wedge \sigma(y)) \vee 0)$.

6.5 Application

We let V_j denote the column vector containing the normalized ratings of U_j, $j = 1, 2, 3, 4$. Similarly, we let G_j denote the column vector containing the government response normalized ratings, $j = 1, 2, 3, 4, 5$. Because only 4 countries were given positive ratings, we delete G_5 from our approach.

We measure the similarity between the averages of the vulnerabilities and the averages of the government responses as well as the complements of the vulnerability averages and the government response. The reasoning to include the complements is because the data is represented in such a way that a high vulnerability is represented by large number and high government response is represented by large number.

We mainly wish to measure the similarity between a low vulnerability and a high government response. The similarity of a high vulnerability and a low government response is also of interest.

We first use the similarity measure given by $S(E, F) = (\sum_{i=1}^{n} \overline{E}(e_i) \wedge \overline{F}(f_i))/(\sum_{i=1}^{n} \overline{E}(e_i) \vee \overline{F}(f_i))$, where $\overline{E}(e_i)$ is the i-th value of the n dimensional vector $(e_1, \ldots, e_n)$ and $\overline{F}(f_i)$ is the i-th value of the n-th dimensional vector $(f_1, \ldots, f_n)$ and where the $\overline{E}(e_i), \overline{F}(f_i)) \in [0, 1]$, [7]. We let V denote the normalized values of the vulnerabilities of the averages of the 17 regions and G denote the normalized values of the vulnerabilities of government responses of the averages of the 17 regions. Then V^c denotes the complement of the values of V and G^c denotes the complement of the values of G, where we use the standard complement. The following table provides the similarity measures of the following four similarities:

$$S(V, G) = \left(\sum_{i=1}^{n} v_i \wedge g_i \right) / \left(\sum_{i=1}^{n} e_i \vee g_i \right),$$

$$S(V^c, G) = \left(\sum_{i=1}^{n} v_i^c \wedge g_i \right) / \left(\sum_{i=1}^{n} v_i^c \vee g_i \right),$$

$$S(V, G^c) = \left(\sum_{i=1}^{n} v_i \wedge g_i^c \right) / \left(\sum_{i=1}^{n} v_i \vee g_i^c \right),$$

$$S(V^c, G^c) = \left(\sum_{i=1}^{n} v_i^c \wedge g_i^c \right) / \left(\sum_{i=1}^{n} v_i^c \vee g_i^c \right),$$

	$S(V, G)$	$S(V^c, G)$	$S(V, G^c)$	$S(V^c, G^c)$
Eastern Africa	0.58	0.66	0.76	0.67
Middle Africa	0.42	0.61	0.79	0.54
Northern Africa	0.62	0.59	0.69	0.74
Southern Africa	0.81	0.71	0.75	0.86
Western Africa	0.65	0.59	0.68	0.78
Eastern Asia	0.73	0.47	0.43	0.87
South-Central Asia	0.75	0.87	0.86	0.75
South-Eastern Asia	0.67	0.75	0.79	0.72
Western Asia & Turkey	0.70	0.67	0.64	0.79
Caribbean	0.63	0.69	0.63	0.73
Central America	0.71	0.79	0.73	0.76
South America	0.60	0.74	0.66	0.70
North America	0.26	0.86	0.55	0.28
Central & South Eastern Europe	0.47	0.79	0.64	0.58
Western Europe	0, 28	0.78	0.49	0.44
Commonwealth of Independent States	0.75	0.67	0.65	0.83
Oceania	0.20	0.78	0.42	0.38

We next use the similarity measure given by $S(E, F) = (\sum_{i=1}^{n} \overline{E}(e_i) \bullet \overline{F}(f_i))/$ $(\sum_{i=1}^{n} \overline{E}(e_i) \vee \overline{F}(f_i))$, where $\bullet$ denotes dot product, [7].

	$S(V, G)$	$S(V^c, G)$	$S(V, G^c)$	$S(V^c, G^c)$
Eastern Africa	0.56	0.63	0.76	0.65
Middle Africa	0.37	0.59	0.80	0.52
Northern Africa	0.55	0.55	0.69	0.72
Southern Africa	0.81	0.69	0.73	0.85
Western Africa	0.65	0.59	0.67	0.74
Eastern Asia	0.72	0.46	0.43	0.86
South-Central Asia	0.75	0.87	0.86	0.75
South-Eastern Asia	0.62	0.75	0.79	0.69
Western Asia & Turkey	0.69	0.67	0.63	0.79
Caribbean	0.61	0.68	0.59	0.74
Central America	0.70	0.79	0.72	0.75
South America	0.56	0.75	0.65	0, 69
North America	0.26	0.86	0.56	0.28
Central & South Eastern Europe	0.46	0.78	0.62	0.57
Western Europe	0.27	0.78	0.47	0.43
Commonwealth of Independent States	0.72	0.68	0.63	0.82
Oceania	0.21	0.78	0.40	0.38

We next use the soft correlation coefficient developed in [1] to the correlation between V^c and G at various levels.

	0.5	0.6	0.7	0.8	0.9
Eastern Africa	0	0	0	0	0
Middle Africa	0	0	0	0	0
Northern Africa	0	0	0	0	0
Southern Africa	0	0	0	0	0
Western Africa	0	0	0	0	0
Eastern Asia	0	0	0	0	0
South-Central Asia	0.63	0	0	0	0
South-Eastern Asia	0.32	0	0	0	0
Western Asia&Turkey	0.60	0	0	0	0
Caribbean	0.58	0.58	0	0	0
Central America	0.76	0	0	0	0
South America	0.82	0.41	0.29	0	0
North America	1.00	1.00	0.71	0	0
Central & South Eastern Europe	0.91	0.64	0.24	0	0
Western Europe	0.91	0.82	0.62	0.25	0
Commonwealth of Independent States	0.50	0.41	0	0	0
Oceania	1.0	1.00	0.71	0	0

The data also shows that the countries with a government response greater than 0.5 are such that their vulnerabilities are lower than 0.5 and that the countries with government responses less than 0.5 have vulnerabilities higher than those countries with government responses greater than 0.5. This suggests that our fuzzy quasi-incidence graphs use a complement with equilibrium 0.5. This means that will be interested in fuzzy quasi-incidence graphs where $\Psi(x, xy) \wedge 0.5 \leq \mu(xy) \wedge \sigma(x)$.

Flow of Human Trafficking from Countries to the United States with Respect to Size of Flow

Some of the results to follow are from [8, 9].

We let very low be assigned the number 0.1, low 0.3, medium 0.5, high 0.7, and very high 0.9.

High Flow	China	Malaysia	Mexico	Russian Fed.	Thailand	Ukraine
Gov. Res.	0.36	0.40	0.62	0.30	0.49	0.47
Vulnerability	0.45	0.38	0.47	0.20	0.48	0.40

σ denotes vulnerability: $\Psi(x, U.S.) = 0.7$ and $\sigma(U.S.) = 0.18$. Now $\sigma(x)$ is given in the table. Hence, $\sigma(U.S.) \wedge \sigma(x) = \sigma(U.S.) = 0.18$ for every x in the table. Thus, $t_{xy} = 0.18$, where $y = U.S.$ Hence, $t = \wedge\{t_{xy} \mid x \text{ is in the table}\} = 0.18$. This yields a fuzzy 0.18-quasi incidence graph for this set of countries.

We have $t_R = 0.7 - 0.18 = 0.52$ and $t_R^* = 6(0.7 - 0.18) = 3.12$.

σ denotes government response: $\Psi(x, U.S.) = 0.7$ and $\sigma(U.S.) = 0.82$. Now $\sigma(x)$ is given in the table. Hence, $\sigma(U.S.) \wedge \sigma(x) = \sigma(x)$ for every x in the table. Thus, $t_{xy} = \sigma(x)$, where $y = U.S.$ Hence, $t = \wedge(t_{xy} \mid x \text{ is in the table}\} = 0.30$. This yields a fuzzy 0.30-quasi incidence graph for this set of countries.

We have $t_R = 0.7 - 0.3 = 0.4$ and $t_R^* = 0.34 + 0.30 + 0.08 + 0.40 + 0.21 + 0.23 = 1.56$.

Med. Flow	Armenia	Azerbaijan	Belarus	Brazil	Canada	Columbia	Czech Rep.
Gov. Res.	0.54	0.42	0.40	0.66	0.68	0.53	0.61
Vulnerability	0.37	0.39	0.33	0.31	0.18	0.42	0.22

Med. Flow	Ecuador	El Salvador	Georgia	Guatemala	Hungary	India	Kazakhstan
Gov. Res.	0.51	0.43	0.64	0.56	0.66	0.46	0.34
Vulnerability	0.35	0.36	0.35	0.42	0.21	0.53	0.22

Med. Flow	Kyrgyzstan	Latvia	Lithuania	Nigeria	Philippines	Poland	Rep. Korea
Gov. Res.	0.38	0.59	0.29	0.55	0.63	0.64	0.27
Vulnerability	0.35	0.24	0.23	0.69	0.50	0.22	0.29

Med Flow	Rep. Moldova	Sri Lanka	Tajikistan	Turkmenistan	Uzbekistan	Vietnam
Gov. Res.	0.60	0.48	0.39	0.29	0.29	0.55
Vulnerability	0.31	0.37	0.46	0.39	0.39	0.29

σ denotes vulnerability: $\Psi(x, U.S.) = 0.5$ and $\sigma(U.S.) = 0.18$. Now $\sigma(x)$ is given in the table. Hence, $\sigma(U.S.) \wedge \sigma(x) = 0.18$ for every x in the table. Thus, $t_{xy} = \sigma(x)$, where $y = U.S.$ Hence, $t = \wedge\{t_{xy} \mid x$ is in the table$\} = 0.18$. This yields a fuzzy 0.18-quasi incidence graph for this set of countries.

We have $t_R = 0.5 - 0.18 = 0.32$ and $t_R^* = 27(0.5 - 0.18) = 8.66$.

σ denotes government response: $\Psi(x, U.S.) = 0.5$ and $\sigma(U.S.) = 0.82$. Now $\sigma(x)$ is given in the table. Hence, $\sigma(U.S.) \wedge \sigma(x) = \sigma(x)$ for every x in the table. Thus, $t_{xy} = \sigma(x)$, where $y = U.S.$ Hence, $t = \wedge(t_{xy} \mid x$ is in the table$\} = 0.27$. This yields a fuzzy 0.27-quasi incidence graph for this set of countries.

We have $t_R = 0.5 - 0.27 = 0.23$ and $t_R^* = 0.08 + 0.10 + 0.07 + 0.04 + 0.16 + 0.12 + 0.21 + 0.23 + 0.02 + 0.11 + 0.21 + 0.21 = 1.56$.

Low Flow	Bangladesh	Bulgaria	Cambodia	Cameroon	Congo	Costa Rica	Cuba	Dom. Rep.
Gov. Res.	0.61	0.51	0.41	0.38	0.25	0.55	0.21	0.63
Vulnerability	0.48	0.28	0.46	0.56	0.54	0.27	0.32	0.39

Low Flow	Estonia	Ethiopia	Ghana	Haiti	Honduras	Indonesia	Jordan
Gov. Res.	0.42	0.44	0.30	0.25	0.39	0.48	0.49
Vulnerability	0.21	0.60	0.44	0.51	0.43	0.45	0.42

Low Flow	Laos	Lebanon	Mali	Nepal	Nicaragua	Niger	Panama
Gov. Res.	0.34	0.38	0.25	0.58	0.59	0.26	0.48
Vulnerability	0.38	0.50	0.50	0.43	0.35	0.53	0.38

Low Flow	Peru	Romania	Serbia	Singapore	Slovenia	Pakistan	Venezuela
Gov. Res.	0.52	0.47	0.59	0.29	0.63	0.42	0.29
Vulnerability	0.38	0.27	0.29	0.15	0.16	0.68	0.46

σ represents vulnerability: $\Psi(x, U.S.) = 0.3$ and $\sigma(U.S.) = 0.18$. Now $\sigma(x)$ is given in the table. Hence, $\sigma(U.S.) \wedge \sigma(x) = 0.18$ for every x in the table except Singapore and Slovenia. Thus, $t_{xy} = \sigma(x)$, where $y = U.S$ and $x \neq$ Singapore and Slovenia. Hence, $t = \wedge(t_{xy} \mid x$ is in the table, $x \neq$ Singapore and Slovenia$\} \wedge 0.15 \wedge 0.16 = 0.15$. This yields a fuzzy 0.15-quasi incidence graph for this set of countries.

We have $t_R = 0.3 - 0.15 = 0.15$ and $t_R^* = 27(0.3 - 0.18) + (0.3 - 0.15) + (0.3 - 0.16) = 3.53$.

σ represents government response: $\Psi(x, U.S.) = 0.3$ and $\sigma(U.S.) = 0.82$. Now $\sigma(x)$ is given in the table. Hence, $\sigma(U.S.) \wedge \sigma(x) = \sigma(x)$ for every x in the table. Thus, $t_{xy} = \sigma(x)$, where $y = U.S.$ Hence, $t = \wedge(t_{xy} \mid x$ is in the table$\} = 0.21$. This yields a fuzzy 0.21-quasi incidence graph for this set of countries.

We have $t_R = 0.3 - 0.21 = 0.09$ and $t_R^* = 0.05 + 0.09 + 0.05 + 0.05 + 0.04 + 0.01 + 0.01 = 0.30$.

In linguistic terms, we have a low-quasi incidence graph for the union of the three sets of countries, i.e., all the counties with flow to the U.S.

Let (σ, μ) be a fuzzy subgraph of the graph (V, E) and Ψ an incidence on (V, E). If it is not the case that $\Psi(u, uv) \leq \mu(uv) \wedge \sigma(v)$ for all $u, v \in TV$, then there exists a largest $t \in [0, 1)$ such that $\Psi(u, uv) \leq \mu(u, uv) \wedge \sigma(v)$ for all $u, v \in V$. In this case, (σ, μ, Ψ) is called a t -**quasi fuzzy incidence graph.** Quasi fuzzy incidence graphs have been studied and applied to human trafficking in [8, 9].

In all tables that follow, $\sigma(U.S.) = 0.88$. Now $\sigma(x)$ is given in the table. Hence, $\sigma(U.S.) \wedge \sigma(x) = \sigma(x)$ for every x in the table. Thus, $t_{xy} = \sigma(x)$, where $y = U.S.$ Recall that $\Psi(x, U.S.)$ is the flow from country x into the U. S.

Keeping y fixed, another measure of the increase in government response (or decrease in flow when σ denotes vulnerability) needed for a region R to be represented by fuzzy incidence graph is given by $t_R = (\vee\{(\Psi(x, xy) - \sigma(x) \wedge \sigma(y)) \mid x \in R\}) \vee 0$ or $t_R^* = \sum_{x \in R}(\Psi(x, xy) - \sigma(x) \wedge \sigma(y)) \vee 0)$.

Flow from a Region to the U. S.

In [11], details of the reported trafficking in persons situation of the country or territory under analysis was provided. information is provided with respect to the reported human trafficking in terms of origin, transit, and/or destination according to a citation index. Whether a country ranks (very) low, medium or (very) high depends upon the total number of sources which made reference to this country as one of origin, transit or destination. The results in [12] are also of interest.

Once again, we let very low be assigned the number 0.1, low 0.3, medium 0.5, high 0.7, and very high 0.9.

Eastern Africa	Ethiopia	Somalia	Tanzania
Gov. Res.	0.44	NR	0.31
Vulnerability	0.60	NR	0.55
Flow	Low	Low	Low

σ represents vulnerability: For Eastern Africa, we have $\Psi(x, U.S.) = 0.30 \not\leq \sigma(x) \wedge 0.18 = 0.18$ for x either Ethiopia or Tanzania. Hence, (σ, μ, Ψ) is a 0.18-quasi fuzzy incidence fuzzy graph. Also, $t_R = 0.12$ and $t_R^* = (0.30 - 0.18) + (0.30 - 0.18) = 0.24$.

σ represents government response: For Eastern Africa, we have $\Psi(x, U.S.) = 0.3 \leq \sigma(x)$ for x either Ethiopia or Tanzania. Hence, (σ, μ, Ψ) is a fuzzy incidence graph. Also, $t_R = 0 = t_R^*$.

Middle Africa	Cameroon	Congo
Gov. Res.	0.38	0.25
Vulnerability	0.56	0.79
Flow	Low	Low

σ represents vulnerability: For Middle Africa, we have $\Psi(x, U.S.) = 0.3 \nleq 0.18$. Hence, (σ, μ, Ψ) is a 0.18-quasi fuzzy incidence graph. Also, $t_R = 0.12$ and $t_R^* = (0.30 - 0.18) + (0.30 - 0.18) = 0.12 + 0.12 = 0.24$.

σ represents government response: For Middle Africa, we see that the above table yields a 0.25-quasi fuzzy incidence graph (σ, μ, Ψ). Also, $t_R = 0.3 - 0.25 = 0.05 = t_R^*$.

Western Africa	Ghana	Mali	Niger	Nigeria
Gov. Res.	0.30	0.25	0.26	0.55
Vulnerability	0.44	0.50	0.53	0.69
Flow	Low	Low	Low	Medium

σ represents vulnerability: For Western Africa, we have that (σ, μ, Ψ) is a 0.18-quasi fuzzy incidence graph. Here $t_R = 0.50 - 0.18 = 0.32$ and $t_R^* = (0.30 - 0.18) + (0.30 - 0.18) + (0.30 - 0.18) + (0.50 - 0.18) = 0.12 + 0.12 + 0.12 + 0.32 = 0.68$.

σ represents government response: For Western Africa, we see that the above table yields a 0.25-quasi fuzzy incidence graph (σ, μ, Ψ). Here $t_R = 0.05$ and $t_R^* = (0.3 - 0.25) + (0.3 - 0.26) = 0.09$.

East Asia	China	Korea
Gov. Res.	0.36	0.27
Vulnerability	0.45	0.29
Flow	High	Medium

σ represents vulnerability: For East Asia, we see that the above table yields a 0.18-quasi fuzzy incidence graph (σ, μ, Ψ). Here $t_R = 0.7 - 0.18 = 0.52$ and $t_R^* = (0.7 - 0.18) + (0.5 - 0.18) = 0.52 + 0.32 = 0.84$.

σ represents government response: For East Asia, we see that the above table yields a 0.27-quasi fuzzy incidence graph (σ, μ, Ψ). Here $t_R = 0.5 - 0.27 = 0.23$ and $t_R^* = (0.5 - 0.27) + (0.7 - 0.36) = 0.23 + 0.34 = 0.57$.

S Cen Asia	Bangladesh	India	Nepal	Pakistan	Sri Lanka	Maldives	Afghanistan
Gov. Res.	0.61	0.46	0.58	0.42	0.48	NR	NR
Vulnerability	0.48	0.53	0.43	0.68	0.37	NR	NR
Flow	Low	Medium	Low	0	Medium	Low	Low

σ represents vulnerability: For South Central Asia, we see that the table yields a 0.18-quasi fuzzy incidence graph. Here $t_R = 0.32$ and $t_R^* = (0.30 - 0.18) + (0.50 - 0.18) + (0.30 - 0.18) + (0.50 - 0.18) + (0.30 - 0.18) + (0.30 - 0.18) = 1.12$.

σ represents government response: For South Central Asia, we see that the above table yields a 0.46-quasi fuzzy incidence graph (σ, μ, Ψ). Here $t_R = 0.04$ and $t_R^* = 0.04 + 0.02 = 0.06$.

S East Asia	Cambodia	Laos	Indonesia	Philippines	Thailand	Singapore	Malaysia	Vietnam
Gov. Res.	0.41	0.34	0.48	0.63	0.49	0.29	0.40	0.58
Vulnerability	0.46	0.38	0.45	0.50	0.48	0.10	0.38	0.29
Flow	Low	Low	Low	Medium	High	Low	0	Medium

σ represents vulnerability: For South East Asia, we see that the above table yields a 0.10-quasi fuzzy incidence graph (σ, μ, Ψ). Here $t_R = 0.70 - 0.18 = 0.52$ and

$$t_R^* = (0.3 - 0.18) + (0.30 - 0.18) + (0.3 - 0.18) + (0.5 - 0.18)$$
$$+(0.7 - 0.18) + (0.3 - 0.18) + (0.3 - 0.10) + (0.5 - 0.18)$$
$$= 1.84.$$

σ represents government response: For South East Asia, we see that the above table yields a 0.29-quasi fuzzy incidence graph (σ, μ, Ψ). Here $t_R = 0.21$ and $t_R^* = 0.01 + 0.21 = 0.22$.

Western Asia & Turkey	Iraq	Jordan	Lebanon
Gov. Res.	NR	0.49	0.38
Vulnerability	NR	0.42	0.50
Flow	Low	Low	Low

σ represents vulnerability: For Western Asia and Turkey, we see that the above table yields a 0.18-quasi fuzzy incidence graph (σ, μ, Ψ). Here $t_R = 0.12$ and $t_R^* = (0.30 - 0.18) + (0.30 - 0.18) = 0.24$.

σ represents government response: For Western Asia and Turkey, we see that the above table yields a quasi fuzzy incidence graph (σ, μ, Ψ). Here $t_R = 0 = t_R^*$.

Caribbean	Cuba	Dominican Republic
Gov. Res.	0.21	0.63
Vulnerability	0.32	0.39
Flow	Low	Low

σ represents vulnerability: For the Caribbean, we see that the above table yields a 0.18-quasi fuzzy incidence graph (σ, μ, Ψ). Here $t_R = 0.12$ and $t_R^* = (0.30 - 0.18) + (0.30 - 0.18) = 0.24$.

σ represents government response: For the Caribbean, we see that the above table yields a 0.21-quasi fuzzy incidence graph (σ, μ, Ψ). Here $t_R = 0.09 = t_R^*$.

Cent America	Costa Rica	El Salvador	Guatemala	Honduras	Mexico	Nicaragua	Panama
Gov. Res.	0.55	0.43	0.56	0.39	0.62	0.59	0.48
Vulnerability	0.27	0.36	0.42	0.43	0.47	0.35	0.33
Flow	Low	Medium	Medium	Low	High	Low	Low

σ represents vulnerability: For Central America, we see that the above table yields a 0.18-quasi fuzzy incidence graph (σ, μ, Ψ). Here $t_R = 0.5 - 0.18 = 0.32$ and

$$t_R^* = (0.30 - 0.18) + (0.50 - 0.18) + 0.50 - 0.18)$$
$$+(0.30 - 0.18) + (0.70 - 0.18) + (0.30 - 0.18) + (0.30 - 0.18)$$
$$= 1.84$$

σ represents government response: For Central America, we see that the above table yields a 0.43-quasi fuzzy incidence graph (σ, μ, Ψ). Also, $t_R = 0.08$ and $t_R^* = 0.07 + 0.08 = 0.15$.

South America	Brazil	Columbia	Ecuador	Peru	Venezuela
Gov. Res.	0.66	0.53	0.51	0.52	0.29
Vulnerability	0.31	0.42	0.35	0.38	0.46
Flow	Medium	Medium	Medium	Low	Low

σ represents vulnerability: For Central America, we see that the above table yields a 0.18-quasi fuzzy incidence graph (σ, μ, Ψ). Here $t_R = 0.32$ and $t_R^* = (0.50 - 0.18) + (0.50 - 0.18) + (0.50 - 0.18) + (0.3 - 0.18) + (0.30 - 0.18) = 1.20$.

σ represents government response: For South America, we see that the above table yields a 0.29-quasi fuzzy incidence graph (σ, μ, Ψ). Also, $t_R = 0.01 = t_R^*$.

North America	Canada
Gov. Res.	0.68
Vulnerability	0.18
Flow	Medium

σ represents vulnerability: For North America, we see that the above table yields a 0.18-quasi fuzzy incidence graph (σ, μ, Ψ). Here $t_R = 0.32 = t_R^*$.

σ represents government response: For North America, we see that the above table yields a quasi fuzzy incidence graph (σ, μ, Ψ). Also, $t_R = 0 = t_R^*$.

Cent SE Europe	Bulgaria	Czech Republic	Estonia	Latvia
Gov. Res.	0.51	0.61	0.42	0.49
Vulnerability	0.28	0.22	0.21	0.24
Flow	Low	Medium	Low	Medium

Cent SE Europe	Lithuania	Poland	Romania	Serbia	Slovenia
Gov. Res.	0.55	0.64	0.47	0.59	0.63
Vulnerability	0.23	0.22	0.27	0.29	0.16
Flow	Medium	Medium	Low	Low	Low

σ represents vulnerability: For Central South Eastern Europe, we see that the above table yields a 0.18-quasi fuzzy incidence graph (σ, μ, Ψ). Here $t_R = 0.32$ and $t_R^* = 1.88$.

σ represents government response: For Central South Eastern Europe, we see that the above table yields a 0.49-quasi fuzzy incidence graph (σ, μ, Ψ). Also, $t_R = 0.01 = t_R^*$.

CIStates	Armenia	Azerbaijan	Belarus	Kazakhstan	Kyrgyzstan	Moldova
Gov. Res.	0.54	0.42	0.40	0.34	0.34	0.60
Vulnerability	0.37	0.39	0.33	0.32	0.35	0.31
Flow	Medium	Medium	Medium	Medium	Medium	Medium

CIStates	Russia	Tajikistan	Turkmenistan	Ukraine	Uzbekistan	Georgia
Gov. Res.	0.30	0.29	0.29	0.47	0.29	0.64
Vulnerability	0.42	0.46	0.39	0.40	0.39	0.35
Flow	High	Medium	Medium	High	Medium	Medium

σ represents vulnerability: For Commonwealth of Independent States, we see that the above table yields a 0.18-quasi fuzzy incidence graph (σ, μ, Ψ). Here $t_R = 0.52$ and $t_R^* = 4.24$.

σ represents government response: For the Commonwealth of Individual States, we see that the above table yields a 0.29-quasi fuzzy incidence graph (σ, μ, Ψ). Also, $t_R = 0.40$ and $t_R^* = 0.08 + 0.10 + 0.16 + 0.16 + 0.40 + 0.21 + 0.21 + 0.23 + 0.21 = 1.76$.

We see that in order to be represented by a fuzzy incidence graph, the Commonwealth of Independent States has the largest measure for both vulnerability and government response than any other region.

Appendix

Vulnerability

Dimension 1: Civil and political protections

Confidence in judicial system, political stability, weapons access, discrimination (sexuality), displaced persons, 2016 global slavery index government response, political rights measure

Dimension 2: Social health and economic rights

Financial inclusion (borrowed money), financial inclusion (received wages), cell phone subscription, social safety net, undernourishment, tuberculosis, water improved access

Dimension 3: Personal security

Financial inclusion (availability of emergency funds), violent crime, women's physical security, GINI coefficient, discrimination (intellectual disability), discrimination immigrants, discrimination minorities

Dimension 4: Refuge populations and conflict

Impact of terrorism, internal conflict, refugees residents

Government Responses

Milestone 1: Survivors of slavery are supported to exit slavery and empowered to break cycle of vulnerability.

Information is distributed to the public about how to identify and report cases of modern slavery; this information is distributed systematically and at regular intervals; there has been an increase in number of members of public reporting cases of modern slavery.

Milestone 2: Effective criminal justice responses are in place in every jurisdiction

Conventions on the abolition of slavery, the slave trade, and institutions and practices similar to slavery; protocol to prevent, suppress and punish trafficking in persons especially women and children; supplementing the United Nations Convention against international organized crime; abolition of forced labor convention; domestic workers convention; worst forms of child labor.

Milestone 3: Effective and measurable national action plans are implemented and fully funded in every country

National coordination body exists involving both government and NGOs; national action plan exists with the clear indicators and allocation of responsibilities; government routinely uses the national action plan as a framework for reporting its actions; activities in the national action plan are fully funded; independent entity to monitor the implementation and effectiveness of the national action plan exists; the government is involved in a regional response; agreements exist between the government and countries of origin and/or destination to collaborate on modern slavery issues; the government cooperates with the government of the home country to facilitate reparation; (negative) foreign victims are detained and/or deported; agreements exist between countries on labor migration.

Milestone 4: Laws, policies and programs address attitudes, social systems and institutions that create vulnerability and enable slavery

Government facilitates or funds research on modern slavery; government facilitates or funds research on prevalence or estimation studies of modern slavery; government interventions that aim to address modern slavery are evidence based; awareness campaigns target specific known risks of modern slavery; the government funds labor inspections, affordable health care for vulnerable population exists, public primary education is available for all children regardless of ethno-cultural or religious back ground; national laws criminalize corruption in the public sector;

Milestone 5: Governments stop sourcing goods or services linked to modern slavery.

Conclusion We measured the similarity between the averages of the vulnerabilities and the averages of the government responses as well as the complements of the vulnerability averages and the government response averages. We used the average of the four vulnerabilities and the average of the four government responses. We considered the main routes of trafficking to the United States. The eccentricity of the origin country was determined. The eccentricity provides us with a number which measures the susceptibility of human trafficking in the paths of the origin country to the U.S. We found that Somalia has the highest eccentricity. We also considered the flow of human traffic from countries to the United States with respect to the size of the flow. The ideal situation is that the flow be less than or equal to the minimum of the vulnerabilities (or of the government responses) of the two countries involved. This is not always the case so a quasi-fuzzy incidence graph is obtained. Measures on how much the flow should be reduced to obtain a fuzzy incidence graph were obtained. In future research, we plan to pursue these ideas by replacing the minimum t-norm with an arbitrary t-norm and the conorm maximum with an arbitrary conorm.

References

1. S. Acharjee, J.N. Mordeson, Soft statistics w.r.t. utility and application in human trafficking. New Math. Nat. Comput. **13**, 289–310 (2017)
2. Central Intelligence Agency, Field Listing: Trafficking in Persons (2014). http://www.cia.gov/library/publications/the-world-factbook/fields/2196.html
3. Global Slavery Index (2016). http://www.globalslaveryindex.org/findings
4. http://www.economics-human-trafficking.org/data-and-reports, 3P Anti-trafficking Policy Index - Data and Reports (2014)
5. http://www.state.gov/j/tip/3p/index.htm, U.S. Department of State, Diplomacy in action, 3Ps: Prosecution, Protection, and Prevention (2011)
6. http://www.unodc.org/unodc/en/human-trafficking/whatishumantrafficking, United Nations Office on Drugs and Crime
7. P. Majumdar, S.K. Samanta, On similarity measures of fuzzy soft sets. Int. J. Adv. Soft Comput. Appl. **3**(2), 1–8 (2011)
8. D.S. Malik, S. Mathew, J.N. Mordeson, Fuzzy Incidence Graphs: Applications to Human Trafficking (submitted)
9. J.N. Mordeson, S. Mathew, R. Borzooei, Vulnerability and Government Response to Human trafficking: Vague fuzzy incidence graphs (submitted)
10. J.N. Mordeson, S. Mathew, Human trafficking: source, transit, destination designations. New Math. Nat. Comput. **3**, 209–218 (2017)
11. S.S. Rajaram, S. Tidball, Nebraska Sex Trafficking Survivors Speak - A Qualitative Research Study, Submitted to: The Women's Fund of Omaha (2016)
12. Trafficking in Persons: Global Patterns, Appendices-United Nations Office on Drugs and Crime, Citation Index (2006)
13. Trafficking in Persons: Global Patterns, The United Nations Office for Drugs and Crime, Trafficking in Persons Citation Index
14. United Nations Office on Drugs and Crime (UNODC) Trafficking in Persons Global Patterns (2006)

Chapter 7
Human Trafficking: Policy Intervention

7.1 Introduction

"In [8], causal linkages between commercial sex and human trafficking were examined. It was determined that a three-link chain of necessary conditions existed: a population vulnerable to trafficking, a capable trafficking organization, and a sex market, was identified. All three links are required for trafficking into commercial sex. Thus, trafficking can be examined by policy intervention at any link. Prospects for policy success at the three points of intervention were compared. It was shown in [8] that a strategy of suppressing sex markets is least likely to be successful in reducing human trafficking.

In the first seven sections, we present a mathematical model of the work developed in [8]. The work is based on [14]. We use five methods to derive five different linear equations to measure the success of policy intervention. The methods are the analytical hierarchy process (AHP), [17, 18], the Guiasu method, [3], the Dempster rule of combination method, [3, 6], the Yen method, [21], and the set-valued statistical method, [2]. The equations are based on expert opinion of the relative importance of the factors involved in sex trafficking. The following factors are listed below and their descriptions are given in Sect. 7.7."

ST: Sex Trafficking
V_1: Vulnerable Population
$F_1^{(1)}$: Personal Security
$F_{11}^{(1)}$: Connection with stable families and inclusive social organizations
$F_{12}^{(1)}$: Individual Characteristics
$F_{13}^{(1)}$: Information about trafficking
$F_{14}^{(1)}$: Education
$F_{15}^{(1)}$: Crisis Intervention
$F_{16}^{(1)}$: Training Law Enforcement
$F_2^{(1)}$: Economic Status
$F_{21}^{(1)}$: Economic Opportunities
$F_{22}^{(1)}$: Program to support economic migrants
V_2: Viable Trafficking (Trafficking Organizations)

© Springer International Publishing AG 2018
J. N. Mordeson et al., *Fuzzy Graph Theory with Applications to Human Trafficking*, Studies in Fuzziness and Soft Computing 365,
https://doi.org/10.1007/978-3-319-76454-2_7

$F_1^{(2)}$:   Improving law enforcement agencies
$F_2^{(2)}$:   Extent that specific organizations and individuals are prosecuted
$F_3^{(2)}$:   More robust law enforcement
V_3:   Sex Market
$F_1^{(3)}$:   Suppression
$F_2^{(3)}$:   Legalization

The model can be used to analyze policy intervention of human trafficking in cities, states, regions or countries. We apply the model to a particular state in the USA. We accomplish this by obtaining expert opinion concerning how well the expert feels the state is doing intervening in the factors making up human trafficking. The opinions are given linguistically using the terms very low, low, medium, high, very high. These terms are replaced with the numbers 0.1, 0.3, 0.5, 0.7, 0.9, respectively. Consequently, the analysis can draw upon methods from fuzzy logic such as the Dempster-Shafer theory. Accurate data concerning the flow of human trafficking does not exist. In some cases, its description is replaced by linguistic terms, [19, 20]. We introduce to the study of human trafficking the five methods mentioned above because they can be used in this context. Based on expert opinion, the equations referred to above indicate that the state in question is currently doing an average job in policy intervention.

7.2   Analytic Hierarchy Process

Each expert, $E_j, j = 1, \ldots, n$, assigns a number w_{ij}, to each factor $G_i, i = 1, \ldots, m$, with respect to its importance with respect to the overarching goal. As in [11], the row average w_i of each row i of the matrix $W = [w_{ij}]$ is determined. These row averages make up the coefficients of the linear equation of the overarching goal, the dependent variable, in terms of the independent variables, the factors. In our application, we have three experts, E_1, E_2, and E_3. The following tables give their opinion of the relative importance of the factors.

	E_1	E_2	E_3	Row Avg
V_1	3	3	3	3
V_2	1	3	2	2
V_3	2	3	3	$\frac{8}{3}$
Col Total	6	9	8	$\frac{23}{3}$

Dividing the entries in the row average column by the column total yields the weights (coefficients) of the variables (factors) in the following equation.

$$ST = \frac{9}{23}V_1 + \frac{6}{23}V_2 + \frac{8}{23}V_3. \tag{7.1}$$

The following tables provide the experts' rankings of the factors of V_1, V_2, and V_3.

V_1	E_1	E_2	E_3	Row Avg
$F_1^{(1)}$	8	5	10	$\frac{23}{3}$
$F_2^{(1)}$	3	3	8	$\frac{14}{3}$
Col Total	11	8	18	$\frac{37}{3}$

V_2	E_1	E_2	E_3	Row Avg
$F_1^{(2)}$	3	3	3	3
$F_2^{(2)}$	2	3	1	2
$F_3^{(2)}$	1	2	2	$\frac{5}{3}$
Col Total	6	8	6	$\frac{20}{3}$

V_3	E_1	E_2	E_3	Row Avg
$F_1^{(3)}$	2	3	1	2
$F_2^{(3)}$	1	1	1	1
Col Total	3	4	2	3

We determine the following equations in a similar manner as we did for ST.

$$V_1 = \frac{23}{37} F_1^{(1)} + \frac{14}{37} F_2^{(1)},$$

$$V_2 = \frac{9}{20} F_1^{(2)} + \frac{6}{20} F_2^{(2)} + \frac{1}{4} F_3^{(2)},$$

$$V_3 = \frac{2}{3} F_1^{(3)} + \frac{1}{3} F_2^{(3)}. \tag{7.2}$$

We next present the experts' rankings of the factors making up Personal Security $F_1^{(1)}$ and Economic Status $F_2^{(1)}$.

$F_1^{(1)}$	E_1	E_2	E_3	Row Avg
$F_{11}^{(1)}$	9	4	10	$\frac{23}{3}$
$F_{12}^{(1)}$	10	4	9	$\frac{23}{3}$
$F_{13}^{(1)}$	7	10	4	$\frac{21}{3}$
$F_{14}^{(1)}$	5	10	4	$\frac{19}{3}$
$F_{15}^{(1)}$	4	8	5	$\frac{17}{3}$
$F_{16}^{(1)}$	6	8	4	$\frac{18}{3}$
Col Total	41	44	36	$\frac{121}{3}$

$F_2^{(1)}$	E_1	E_2	E_3	Row Avg
$F_{21}^{(1)}$	2	2	7	$\frac{11}{3}$
$F_{22}^{(1)}$	1	8	6	$\frac{15}{3}$
Col Total	3	10	13	$\frac{26}{3}$

From these tables, we obtain the following equations.

$$F_1^{(1)} = \frac{23}{121} F_{11}^{(1)} + \frac{23}{121} F_{12}^{(1)} + \frac{21}{121} F_{13}^{(1)}$$
$$+ \frac{19}{121} F_{14}^{(1)} + \frac{17}{121} F_{15}^{(1)} + \frac{18}{121} F_{16}^{(1)},$$
$$F_2^{(1)} = \frac{11}{26} F_{21}^{(1)} + \frac{15}{26} F_{22}^{(1)}. \tag{7.3}$$

7.3 Guiasu and Dempster Rule of Combination Methods

We next use the AHP table is used to create the Guiasu table by dividing each entry in the AHP by the column total of the column it's in. The row averages of the Guiasu table are taken and the results give the coefficients equation of the goal in terms of its causal factors.

Because the columns of the Guiasu table add to 1, we can consider the entries in the Guiasu table basic probability assignments for each expert, [12]. We assume that our universe contains two focal elements for each E_i, $i = 1, 2, 3$. Then by [10], Theorem 3.1, p. 195], the combined probability assignment is given by the formula

$$m_{123}(\{G_i\}) = m_1(\{G_i\})m_2(\{G_i\})m_3(\{G_i\})/[1 - m_1(\{G_i\}) - m_2(\{G_i\})$$
$$- m_3(\{G_i\}) + m_1(\{G_i\})m_2(\{G_i\}) + m_1(\{G_i\})m_3(\{G_i\})$$
$$+ m_2(\{G_i\})m_3(\{G_i\})],$$

where i ranges over the number of causal factors G_i of the overarching goal in question. To find the coefficients of the independent variables for the linear equation for the Dempster rule of combination, we divide in each entry of the Dempster column by the column total.

	E_1	E_2	E_3	Guiasu	Dempster
V_1	$\frac{1}{2}$	$\frac{1}{3}$	$\frac{3}{8}$	$\frac{29}{72}$	$\frac{3}{13}$
V_2	$\frac{1}{6}$	$\frac{1}{3}$	$\frac{2}{8}$	$\frac{18}{72}$	$\frac{1}{31}$
V_3	$\frac{1}{3}$	$\frac{1}{3}$	$\frac{3}{8}$	$\frac{25}{72}$	$\frac{3}{23}$
Col Total				1	$\frac{3647}{9269}$

Guiasu: $ST = \frac{29}{72} V_1 + \frac{18}{72} V_2 + \frac{25}{72} V_3.$
Dempster: $ST = \frac{2139}{3647} V_1 + \frac{299}{3647} V_2 + \frac{1209}{3647} V_3.$

V_1	E_1	E_2	E_3	Guiasu	Dempster
$F_1^{(1)}$	$\frac{8}{11}$	$\frac{5}{8}$	$\frac{15}{19}$	$\frac{3581}{5016}$	$\frac{50}{53}$
$F_2^{(1)}$	$\frac{3}{11}$	$\frac{3}{8}$	$\frac{4}{9}$	$\frac{1435}{5016}$	$\frac{3}{53}$
Col Total				1	1

V_2	E_1	E_2	E_3	Guiasu	Dempster
$F_1^{(2)}$	$\frac{1}{2}$	$\frac{3}{8}$	$\frac{1}{2}$	$\frac{33}{72}$	$\frac{3}{8}$
$F_2^{(2)}$	$\frac{1}{3}$	$\frac{3}{8}$	$\frac{1}{6}$	$\frac{21}{72}$	$\frac{3}{8}$
$F_3^{(2)}$	$\frac{1}{6}$	$\frac{1}{4}$	$\frac{1}{3}$	$\frac{18}{72}$	$\frac{1}{31}$
Col Total				1	$\frac{1552}{1984}$

V_3	E_1	E_2	E_3	Guiasu	Dempster
$F_1^{(3)}$	$\frac{2}{3}$	$\frac{3}{4}$	$\frac{1}{2}$	$\frac{46}{72}$	$\frac{6}{7}$
$F_2^{(3)}$	$\frac{1}{3}$	$\frac{1}{4}$	$\frac{1}{2}$	$\frac{26}{72}$	$\frac{1}{7}$
Col Total				1	1

Guiasu: $V_1 = \frac{3581}{5016} F_1^{(1)} + \frac{1435}{5016} F_2^{(1)}$, $V_2 = \frac{33}{72} F_1^{(2)} + \frac{21}{72} F_2^{(2)} + \frac{18}{72} F_3^{(2)}$, $V_3 = \frac{23}{36} F_1^{(3)} + \frac{13}{36} F_2^{(3)}$.

Dempster: $V_1 = \frac{50}{53} F_1^{(1)} + \frac{3}{53} F_2^{(1)}$, $V_2 = \frac{744}{1552} F_1^{(2)} + \frac{744}{1552} F_2^{(2)} + \frac{64}{1552} F_3^{(2)}$, $V_3 = \frac{6}{7} F_1^{(3)} + \frac{1}{7} F_2^{(3)}$.

$F_1^{(1)}$	E_1	E_2	E_3	Guiasu	Dempster
$F_{11}^{(1)}$	$\frac{9}{41}$	$\frac{1}{11}$	$\frac{5}{18}$	$\frac{4775}{24354}$	$\frac{45}{4961}$
$F_{12}^{(1)}$	$\frac{10}{41}$	$\frac{1}{11}$	$\frac{1}{4}$	$\frac{1055}{5412}$	$\frac{1}{110}$
$F_{13}^{(1)}$	$\frac{7}{41}$	$\frac{5}{22}$	$\frac{1}{9}$	$\frac{4133}{24354}$	$\frac{35}{4659}$
$F_{14}^{(1)}$	$\frac{5}{41}$	$\frac{5}{22}$	$\frac{1}{9}$	$\frac{3737}{24354}$	$\frac{25}{4721}$
$F_{15}^{(1)}$	$\frac{4}{41}$	$\frac{2}{11}$	$\frac{5}{36}$	$\frac{6791}{48708}$	$\frac{40}{10363}$
$F_{16}^{(1)}$	$\frac{6}{41}$	$\frac{2}{11}$	$\frac{1}{9}$	$\frac{1783}{12177}$	$\frac{1}{211}$
Col Total				1	

$F_2^{(1)}$	E_1	E_2	E_3	Guiasu	Dempster
$F_{21}^{(1)}$	$\frac{2}{3}$	$\frac{1}{5}$	$\frac{7}{13}$	$\frac{274}{585}$	$\frac{7}{19}$
$F_{22}^{(1)}$	$\frac{1}{3}$	$\frac{4}{5}$	$\frac{6}{13}$	$\frac{311}{585}$	$\frac{12}{19}$
Col Total				1	1

Guiasu: $F_1^{(1)} = 0.18 F_{11}^{(1)} + 0.19 F_{12}^{(1)} + 0.17 F_{13}^{(1)} + 0.15 F_{14}^{(1)} + 0.14 F_{15}^{(1)} + 0.15 F_{16}^{(1)}$, $F_2^{(1)} = \frac{274}{585} F_{21}^{(1)} + \frac{311}{585} F_{22}^{(1)}$.

Dempster: $F_1^{(1)} = \frac{9}{40}F_{11}^{(1)} + \frac{9}{40}F_{12}^{(1)} + \frac{8}{40}F_{13}^{(1)} + \frac{5}{40}F_{14}^{(1)} + \frac{4}{40}F_{15}^{(1)} + \frac{5}{40}F_{16}^{(1)}$, $F_2^{(1)} = \frac{7}{19}F_{21}^{(1)} + \frac{12}{19}F_{22}^{(1)}$.

The above results suggest that when there are exactly two focal sets for the experts, then the two Dempster combinations add to one. We now prove this result. Our results are for a universe X of any cardinality, but with exactly the same two focal elements, say A and B with $A \cap B = \emptyset$ for each basic probability assignment. The following result gives a formula for combining k basic probability assignments on X for arbitrary $k \in \mathbb{N}$, the positive integers. Let $n \in \mathbb{N}$, $n \geq 3$. We use the following notation. Let $C_j^n(A) = \sum_{1 \leq i_1 < i_2 < \cdots < i_j \leq n} m_{i_1}(A)m_{i_2}(A) \cdots m_{i_j}(A)$, $j = 2, 3, \ldots, n - 1$.

Theorem 7.3.1 ([10]) *(i) If n is even, then $m_{12\ldots n+1}(A) = \prod_{i=1}^{n+1} m_i(A)/[1 - \sum_{i=1}^{n+1} m_i(A) + \sum_{j=2}^{n}(-1)^j C_j^{n+1}(A)]$.*

(ii) If n is odd, then $m_{12\ldots n+1}(A) = \prod_{i=1}^{n+1} m_i(A)/[1 - \sum_{i=1}^{n+1} m_i(A) + \sum_{j=2}^{n}(-1)^j C_j^{n+1}(A) + 2\prod_{i=1}^{n+1} m_i(A)]$.

Proof It follows that

$$
\begin{aligned}
K_{12} &= m_1(A)m_2(B) + m_1(B)m_2(A) \\
&= m_1(A)(1 - m_2(A)) + (1 - m_1(A))m_2(A) \\
&= m_1(A) + m_2(A) - 2m_1(A)m_2(A).
\end{aligned}
$$

Thus,

$$
m_{12}(A) = m_1(A)m_2(A)/[1 - m_1(A) - m_2(A) + 2m_1(A)m_2(A)].
$$

(i) Let n be even. Induction hypothesis: Assume

$$
m_{12\ldots n}(A) = \prod_{i=1}^{n} m_1(A)/den,
$$

where

$$
den = 1 - \sum_{i=1}^{n} m_i(A) + C_2^n(A) - C_3^n(A) + \cdots
$$

$$
+ (-1)^j C_j^n(A) + \cdots - C_{n-1}^n(A) + 2\prod_{i=1}^{n} m_i(A).
$$

Now $K_{12\ldots n+1} = m_{12\ldots n}(A) + m_{n+1}(A) - 2m_{12\ldots n}(A)m_{n+1}(A)$ and

$$m_{12\ldots n+1}(A) = \left[\prod_{i=1}^{n+1} m_i(A)/den)/1 - \prod_{i=1}^{n} m_i(A)/den - m_{n+1}(A) \right.$$

$$+ 2 \prod_{i=1}^{n} m_i(A) m_{n+12}(A)/den \Big]$$

$$= \prod_{i=1}^{n+1} m_i(A)/ \left[den - \prod_{i=1}^{n} m_i(A) - den \bullet m_{n+1}(A) \right.$$

$$+ 2 \prod_{i=1}^{n+1} m_i(A) \Big].$$

Now

$$den \bullet m_{n+1}(A) = m_{n+1}(A) - \sum_{i=1}^{n} m_i(A) m_{n+1}(A) + C_2^n(A) m_{n+1}(A)$$

$$- C_3^n(A) m_{n+1}(A) + \cdots + (-1)^j C_j^n(A) m_{n+1}(A)$$

$$+ \cdots - C_{n-1}^n(A) m_{n+1}(A) + 2 \prod_{i=1}^{n+1} m_i(A).$$

Substituting this into the previous equation yields

$$m_{12\ldots n+1}(A) = \prod_{i=1}^{n+1} m_i(A)/ \left[1 - \sum_{i=1}^{n+1} m_i(A) + \sum_{j=1}^{n-1}(-1)^j C_j^m(A) + \prod_{i=1}^{n} m_i(A) \right.$$

$$+ \sum_{i=1}^{n} m_i(A) m_{n+1}(A) - \sum_{j=2}^{n-1}(-1)^j C_j^n(A) m_{n+1}(A) \Big]$$

$$= \prod_{i=1}^{n+1} m_i(A)/ \left[1 - \sum_{i=1}^{n+1} m_i(A) + \sum_{j=2}^{n}(-1)^j C_j^{n+1}(A) \right],$$

the desired result.

(ii) From the case $n = 2$ and from (i), we obtain

$$m_{123}(A) = \prod_{i=1}^{3} m_i(A)/ \left[1 - \sum_{i=1}^{3} m_i(A) + C_2^3(A) \right].$$

Let n be odd. Induction hypothesis: Assume

$$m_{12\ldots n}(A) = \prod_{i=1}^{n} m_i(A)/den,$$

where

$$den = 1 - \sum_{i=1}^{n} m_i(A) + C_2^n(A) - C_3^n(A) + \cdots + (-1)^j C_j^n(A) + \cdots - C_{n-1}^n(A).$$

Now $K_{12\ldots n+1} = m_{12\ldots n}(A) + m_{n+1}(A) - 2m_{12\ldots n}(A)m_{n+1}(A)$ and

$$m_{12\ldots n+1}(A) = m_{12\ldots n}(A)m_{n+1}(A)/[1 - K_{12\ldots n+1}]$$

$$= \left(\prod_{i=1}^{n+1} m_i(A)/den \right) / \left(1 - \prod_{i=1}^{n} m_i(A)/den - m_{n+1}(A) \right.$$

$$\left. +2 \prod m_i(A)m_{n+1}(A)/den \right)$$

$$= \prod_{i=1}^{n+1} m_i(A) / \left[den - \prod_{i=1}^{n} m_i(A) - den \bullet m_{n+1}(A) \right.$$

$$\left. +2 \prod_{i=1}^{n+1} m_i(A) \right].$$

Now

$$den \bullet m_{n+1}(A) = m_{n+1}(A) - \sum_{i=1}^{n} m_i(A)m_{n+1}(A) + C_2^n(A)m_{n+1}(A)$$

$$-c_3^n(A)m_{n+1}(A) + \cdots + (-1)^j C_j^n(A)m_{n+1}(A)$$

$$+ \cdots - C_{n-1}^n(A)m_{n+1}(A).$$

Substituting this into the above equation yields

$$m_{12\ldots n+1}(A) = \prod_{i=1}^{n+1} m_i(A) / \left[1 - \sum_{i=1}^{n+1} m_i(A) + \sum_{j=2}^{n-1} (-1)^j C_j^n(A) \right.$$

$$+ \prod_{i=1}^{n} m_i(A) + \sum_{i=1}^{n} m_i(A)m_{n+1}(A)$$

$$\left. - \sum_{j=2}^{n-1} (-1)^j C_j^n(A)m_{n+1}(A) + 2 \prod_{i=1}^{n+1} m_i(A) \right]$$

$$= \prod_{i=1}^{n+1} m_i(A) / \left[1 - \sum_{i=1}^{n+1} m_i(A) + \sum_{j=2}^{n} (-1)^j C_j^{n+1}(A) \right.$$

$$\left. +2 \prod_{i=1}^{n+1} m_i(A) \right],$$

the desired result.   ■

Theorem 7.3.2 ([14]) *Suppose there exist exactly the same two focal points A and B for each expert, with $A \cap B = \emptyset$. Let $n + 1$ denote the number of experts. Then $m_{12...n+1}(A) + m_{12...n+1}(B) = 1$.*

Proof We use the notation m_i for $m_i(A)$ $i = 1, \ldots, n+1$. Then $m_i(B) = 1 - m_i, i = 1, \ldots, n + 1$. It is easily shown that if n is even, then

$$\prod_{i=1}^{n+1}(1 - m_i) = 1 - \sum_{i=1}^{n+1} m_i + \cdots + \sum_{j=2}^{n}(-1)^j C_j^{n+1} - m_1 m_2 \cdots m_{n+1}$$

and that if n is odd, then

$$\prod_{i=1}^{n+1}(1 - m_i) = 1 - \sum_{i=1}^{n+1} m_i + \cdots + \sum_{j=2}^{n}(-1)^j C_j^{n+1} + m_1 m_2 \cdots m_{n+1}.$$

(i) Suppose n is even. Let $den = 1 - \sum_{i=1}^{n+1} m_i + \sum_{j=2}^{n}(-1)^j C_j^{n+1}$. Then $\prod_{i=1}^{n+1}(1 - m_i) = den - m_1 m_2 \cdots m_{n+1}$. Thus, we have

$$
\begin{aligned}
&m_{12...n+1}(A) + m_{12...n+1}(B) \\
&= \frac{m_1 m_2 \cdots m_{n+1}}{den} + \frac{(1 - m_1)(1 - m_2) \cdots (1 - m_{n+1})}{den} \\
&= \frac{m_1 m_2 \cdots m_{n+1}}{den} + \frac{den - m_1 m_2 \cdots m_{n+1}}{den} \\
&= 1.
\end{aligned}
$$

(ii) Suppose n is odd. Let $den = 1 - \sum_{i=1}^{n+1} m_i + \sum_{j=2}^{n}(-1)^j C_j^{n+1} + 2\prod_{i=1}^{n+1} m_i$. Then $\prod_{i=1}^{n+1}(1 - m_i) = den - m_1 m_2 \cdots m_{n+1}$.

$$
\begin{aligned}
&m_{12...n+1}(A) + m_{12...n+1}(B) \\
&= \frac{m_1 m_2 \cdots m_{n+1}}{den} + \frac{(1 - m_1)(1 - m_2) \cdots (1 - m_{n+1})}{den} \\
&= \frac{m_1 m_2 \cdots m_{n+1}}{den} + \frac{den - m_1 m_2 \cdots m_{n+1}}{den} \\
&= 1.
\end{aligned}
$$

■

We next give a necessary and sufficient condition for the weights of the AHP method and the Guiasu method to coincide. We also develop a method to determine the weights using Dempster-Shafer theory. Let $A = [a_{ij}]$ the $m \times n$-matrix whose entries are the weights of the m factors as to their importance determined by the n experts. That is, a_{ij} is the weight of the ith factor given by expert j with $i = 1, \ldots, m$; $j = 1, \ldots, n$. Let N denote the $m \times n$-matrix determined from A by normalizing the columns of A, i.e.,

$$N = \left[\frac{a_{ij}}{\sum_{k=1}^{m} a_{kj}} \right].$$

Then the weights of analytic hierarchy process are given by

$$A_i = \frac{\frac{1}{n} \sum_{j=1}^{n} a_{ij}}{\sum_{i=1}^{m} \frac{1}{n} \sum_{j=1}^{n} a_{ij}}$$

$$= \frac{\sum_{j=1}^{n} a_{ij}}{\sum_{i=1}^{m} \sum_{j=1}^{n} a_{ij}}$$

for $i = 1, \ldots, m$. The weights G_i of the Guiasu method are given by

$$G_i = \frac{1}{n} \sum_{j=1}^{n} \frac{a_{ij}}{\sum_{k=1}^{m} a_{kj}}$$

for $i = 1, \ldots, m$.

Theorem 7.3.3 Let $C_j = \sum_{k=1}^{m} a_{kj}$, $j = 1, \ldots, n$. For $i = 1, \ldots, m$, $A_i = G_i$ if and only if

$$\frac{a_{i1}}{C_1 + \cdots + C_n} + \cdots + \frac{a_{in}}{C_1 + \cdots + C_n} = \frac{a_{i1}}{nC_1} + \cdots + \frac{a_{in}}{nC_n}.$$

Proof We have that

$$A_i = G_i, \quad i = 1, \ldots, m$$

$$\Leftrightarrow \frac{\sum_{j=1}^{n} a_{ij}}{\sum_{i=1}^{m} \sum_{j=1}^{n} a_{ij}} = \frac{1}{n} \sum_{j=1}^{n} \frac{a_{ij}}{\sum_{k=1}^{m} a_{kj}}, \quad i = 1, \ldots, m$$

$$\Leftrightarrow \frac{a_{i1} + \cdots + a_{in}}{C_1 + \cdots + C_n} = \frac{a_{i1}}{nC_1} + \cdots + \frac{a_{in}}{nC_n}, \quad i = 1, \ldots, m$$

$$\Leftrightarrow \frac{a_{i1}}{C_1 + \cdots + C_n} + \cdots + \frac{a_{in}}{C_1 + \cdots + C_n} = \frac{a_{i1}}{nC_1} + \cdots + \frac{a_{in}}{nC_n},$$

$$i = 1, \ldots, m.$$

■

Corollary 7.3.4 *If $C_i = C_j$, $i, j = 1, \ldots, n$, then $A_i = G_i$ for $i = 1, \ldots, m$.*

Proposition 7.3.5 $C_i = C_j$, $i, j = 1, \ldots, n$ *if and only if for all $j = 1, \ldots, n$ we have that $C_i + \cdots + C_n = nC_j$.*

Proof The proof follows by routine solution of the linear system of equations, $C_i + \cdots + C_n = nC_j$, $j = 1, \ldots, n$. ∎

7.4 Yen Method

In [21], Yen developed an approach that addressed the issue of managing imprecise and vague information in evidential reasoning by combining Dempster-Shafer theory with fuzzy set theory. We apply Yen's method to arrive at the degree of belief of certain subsets of a set $X = \{x_1, \ldots, x_k\}$. We assume that we have n subsets of X, A_j, $j = 1, \ldots, n$, which are focal elements of a function m of the power set of X into the closed interval $[0, 1]$, i.e., $m(A_j) > 0$, $j = 1, \ldots, n$.

In the Guiasu table of weights, we divide each element of the column by the column's maximal entry. This yields a matrix from which we derive a linear equation for the overarching goal.

It is shown in [12] that if $m(A_j) = \frac{1}{n}$, $j = 1, \ldots n$, then the belief function $\text{Bel}(1_{\{x_i\}})$ is the average of the entries in the ith row, where $1_{\{x_i\}}$ is the characteristic function of $\{x_i\}$ in its universe. The following table is obtained from the Guiasu table by dividing each entry in the column by the maximal number in that column.

	E_1	E_2	E_3	Row Total
V_1	1	1	1	3
V_2	$\frac{1}{3}$	1	$\frac{2}{3}$	2
V_3	$\frac{2}{3}$	1	1	$\frac{8}{3}$
Col Total				$\frac{23}{3}$

$$ST = \frac{9}{23}V_1 + \frac{6}{23}V_2 + \frac{8}{23}V_3$$

The following tables and equations are obtained in a similar manner as that above.

V_1	E_1	E_2	E_3	Row Total
$F_1^{(1)}$	1	1	1	3
$F_2^{(1)}$	$\frac{3}{8}$	$\frac{3}{5}$	$\frac{4}{5}$	$\frac{31}{20}$
Col Total				$\frac{91}{20}$

V_2	E_1	E_2	E_3	Row Total
$F_1^{(2)}$	1	1	1	3
$F_2^{(2)}$	$\frac{2}{3}$	1	$\frac{1}{3}$	2
$F_3^{(2)}$	$\frac{1}{3}$	$\frac{2}{3}$	$\frac{2}{3}$	$\frac{5}{3}$
Col Total				$\frac{20}{3}$

V_3	E_1	E_2	E_3	Row Total
$F_1^{(3)}$	1	1	1	3
$F_2^{(3)}$	$\frac{1}{2}$	$\frac{1}{3}$	1	$\frac{11}{6}$
Col Total				$\frac{29}{6}$

$$V_1 = \frac{60}{91} F_1^{(1)} + \frac{31}{91} F_2^{(1)}, \; V_2 = \frac{9}{20} F_1^{(2)} + \frac{6}{20} F_2^{(2)} + \frac{5}{20} F_3^{(2)}, \; V_3 = \frac{18}{29} F_1^{(3)} + \frac{11}{29} F_2^{(3)}.$$

$F_1^{(1)}$	E_1	E_2	E_3	Row Total
$F_{11}^{(1)}$	$\frac{9}{10}$	$\frac{2}{5}$	1	$\frac{23}{10}$
$F_{12}^{(1)}$	1	$\frac{2}{5}$	$\frac{9}{10}$	$\frac{23}{10}$
$F_{13}^{(1)}$	$\frac{7}{10}$	1	$\frac{2}{5}$	$\frac{21}{10}$
$F_{14}^{(1)}$	$\frac{1}{2}$	1	$\frac{2}{5}$	$\frac{19}{10}$
$F_{15}^{(1)}$	$\frac{2}{5}$	$\frac{4}{5}$	$\frac{1}{2}$	$\frac{17}{10}$
$F_{16}^{(1)}$	$\frac{3}{5}$	$\frac{4}{5}$	$\frac{2}{5}$	$\frac{18}{10}$
Col Total				$\frac{121}{10}$

$F_2^{(1)}$	E_1	E_2	E_3	Row Total
$F_{21}^{(1)}$	1	$\frac{1}{4}$	1	$\frac{9}{4}$
$F_{22}^{(1)}$	$\frac{1}{2}$	1	$\frac{6}{7}$	$\frac{33}{14}$
Col Total				$\frac{129}{28}$

$$F_1^{(1)} = \frac{23}{121} F_{11}^{(1)} + \frac{23}{121} F_{12}^{(1)} + \frac{21}{121} F_{13}^{(1)} + \frac{19}{121} F_{14}^{(1)} + \frac{17}{121} F_{15}^{(1)} + \frac{18}{121} F_{16}^{(1)},$$
$$F_2^{(1)} = \frac{63}{129} F_{21}^{(1)} + \frac{66}{129} F_{22}^{(1)}.$$

7.5 Set Valued Statistical Method

We next use the set-valued statistical method [2] to determine the coefficients of the linear equation expressing a goal, say G, in terms of its factors, $G_1, G_2, \ldots, G_m$. Let $\mathcal{G}$ denote a finite set. In our case, $\mathcal{G} = \{G_1, G_2, \ldots, G_m\}$ is a set of factors making up $\mathcal{G}$. Let E denote a set of subjects. In our case, $E = \{E_1, \ldots, E_n\}$ is the set of experts. The problem is to find the degree of membership value μ of the $G_1, \ldots, G_m$ in the set of factors making up $\mathcal{G}$, i.e., the coefficients of the linear equation in our application.

First choose an integer q, $1 \le q \le m$, and then an E_j in order to carry out the following steps:

(1) Select $r_1 = q$ elements from $\mathcal{G}$ such that they are the first group of elements best fit to μ by E_j. This yields a subset $\mathcal{G}_1^{(j)} = \{G_{i_1}, G_{i_2}, \ldots, G_{i_q}\}$ of $\mathcal{G}$.
(2) Select $r_2 = 2q$ elements from G which includes the q elements already chosen in step (1) in such a way that all $2q$ elements are considered better fit to μ than other members of $\mathcal{G}$ by E_j. This yields the following subset of $\mathcal{G}$, $\mathcal{G}_2^{(j)} = \{G_{i_1}, G_{i_2}, \ldots, G_{i_q}, G_{i_q+1}, \ldots, G_{i_{2q}}\}$.
(3) Continuing this process, we construct the sth subset $\mathcal{G}_s^{(j)} = \mathcal{G}_1^{(j)} \cup \mathcal{G}_2^{(j)} \cup \cdots \cup \mathcal{G}_{s-1}^{(j)} \cup \{G_{i_{s-1}+1}^{(j)}, \ldots, G_{i_{sq}}^{(j)}\}$.

This process is continued until we obtain a positive integer t such that $m = tq + u$, where $1 \le u \le q$.

We next calculate $m(E_i)$, the average frequency of E_i using the following formula

$$m(E_i) = \frac{1}{n(t+1)} \sum_{s=1}^{t+1} \sum_{j=1}^{n} \chi_{\mathcal{G}_s^{(j)}}(E_i), \ i = 1, \ldots, m,$$

where $\chi_{\mathcal{G}_s^{(j)}}$ is the characteristic function of the set $\mathcal{G}_s^{(j)}$ in $\mathcal{G}$, $j = 1, \ldots, n$.

A generalization of this method using different sizes of q at each step can be found in [2]. In our application, we let $q = 1$. This yields the following tables and equations.

	E_1	E_2	E_3	Row Total
V_1	3	2	2.5	$\frac{15}{2}$
V_2	1	2	1	4
V_3	2	2	2.5	$\frac{13}{2}$
Col Total				18

$ST = \frac{15}{36} V_1 + \frac{4}{18} V_2 + \frac{13}{36} V_3.$

V_1	E_1	E_2	E_3	Row Total
$F_1^{(1)}$	2	2	2	6
$F_2^{(1)}$	1	1	1	3
Col Total				9

V_2	E_1	E_2	E_3	Row Total
$F_1^{(2)}$	3	2.5	3	$\frac{17}{2}$
$F_2^{(2)}$	2	2.5	1	$\frac{11}{2}$
$F_3^{(2)}$	1	1	2	4
Col Total				18

V_3	E_1	E_2	E_3	Row Total
$F_1^{(3)}$	2	2	1.5	$\frac{11}{2}$
$F_2^{(3)}$	1	1	1.5	$\frac{7}{2}$
Col Total				9

$$V_1 = \frac{2}{3} F_1^{(1)} + \frac{1}{3} F_2^{(1)}, \; V_2 = \frac{17}{36} F_1^{(2)} + \frac{11}{36} F_2^{(2)} + \frac{2}{9} F_3^{(2)}, \; V_3 = \frac{11}{18} F_1^{(3)} + \frac{7}{18} F_2^{(3)}.$$

$F_1^{(1)}$	E_1	E_2	E_3	Row Total
$F_{11}^{(1)}$	5	1.5	6	$\frac{25}{2}$
$F_{12}^{(1)}$	6	1.5	5	$\frac{25}{2}$
$F_{13}^{(1)}$	4	5.5	2	$\frac{23}{2}$
$F_{14}^{(1)}$	2	5.5	2	$\frac{19}{2}$
$F_{15}^{(1)}$	1	3.5	4	$\frac{17}{2}$
$F_{16}^{(1)}$	3	3.5	2	$\frac{17}{2}$
Col Total				63

$F_2^{(1)}$	E_1	E_2	E_3	Row Total
$F_{21}^{(1)}$	2	1	2	5
$F_{22}^{(1)}$	1	2	1	4
Col Total				9

$$F_1^{(1)} = \frac{25}{126} F_{11}^{(1)} + \frac{25}{126} F_{12}^{(1)} + \frac{23}{126} F_{13}^{(1)} + \frac{19}{126} F_{14}^{(1)} + \frac{17}{126} F_{15}^{(1)} + \frac{17}{126} F_{16}^{(1)},$$
$$F_2^{(1)} = \frac{5}{9} F_{21}^{(1)} + \frac{4}{9} F_{22}^{(1)}.$$

7.6 Testing Policy Intervention

In this section, two experts, E_1' and E_2', give their opinion on how well policy intervention is working to suppress sex trafficking in the particular state in question. These opinions are given in the following tables

	V_1	$F_1^{(1)}$	$F_{11}^{(1)}$	$F_{12}^{(1)}$	$F_{13}^{(1)}$	$F_{14}^{(1)}$	$F_{15}^{(1)}$	$F_{16}^{(1)}$	$F_2^{(1)}$	$F_{21}^{(1)}$	$F_{22}^{(1)}$
E_1'	0.5	0.5	0.5	0.5	0.5	0.5	0.3	0.5	0.3	0.5	0.3
E_2'	0.5	0.5	0.5	0.3	0.5	0.5	0.5	0.5	0.3	0.5	0.3

	V_2	$F_1^{(2)}$	$F_2^{(2)}$	$F_3^{(2)}$
E_1'	0.5	0.5	0.3	0.3
E_2'	0.5	0.7	0.1	0.5

	V_3	$F_1^{(3)}$	$F_2^{(3)}$
E_1'	0.3	0.3	0.3
E_2'	0.1	0.1	0.1

As in [14], we determine how well the state is intervening based on the five different methods. We substitute the experts' evaluation into the equations to determine a measure of how well the state is doing in policy intervention with respect to the factors. The use of Dempster-Shafer theory and fuzzy logic yields a number between 0 and 1 which agrees with the linguistic rankings used to develop the equations. In the following equations, we get two values for V_1. The first value is determined from Eqs. (7.2) and (7.3) while the second (followed by $*$) is determined from Eq. (7.2) and the experts' opinions. We also have two equations for ST determined by the two V_1 values.

Analytic Hierarchy Process

$$E_1' : F_1^{(1)} = \frac{23}{121}(0.5) + \frac{23}{121}(0.5) + \frac{21}{121}(0.5) + \frac{19}{121}(0.5) + \frac{17}{121}(0.3) + \frac{18}{121}(0.5) = 0.47$$

$$F_2^{(1)} = \frac{11}{26}(0.5) + \frac{15}{26}(0.3) = 0.38$$

$$V_1 = \frac{23}{37}(0.47) + \frac{14}{37}(0.38) = 0.44$$

$$V_1 = \frac{23}{37}(0.5) + \frac{14}{37}(0.3) = 0.42 *$$

$$V_2 = \frac{9}{20}(0.5) + \frac{3}{10}(0.3) + \frac{1}{4}(0.3) = 0.39$$

$$V_3 = \frac{2}{3}(0.3) + \frac{1}{3}(0.3) = 0.3$$

$$ST = \frac{9}{23}(0.44) + \frac{6}{23}(0.39) + \frac{8}{23}(0.3) = 0.38$$

$$ST = \frac{9}{23}(0.42) + \frac{6}{23}(0.39) + \frac{8}{23}(0.3) = 0.37 *$$

$$E_2' : F_1^{(1)} = \frac{23}{121}(0.5) + \frac{23}{121}(0.3) + \frac{21}{121}(0.5) + \frac{19}{121}(0.5) + \frac{17}{121}(0.5) + \frac{18}{121}(0.5) = 0.46$$

$$F_2^{(1)} = \frac{11}{26}(0.5) + \frac{15}{26}(0.3) = 0.38$$

$$V_1 = \frac{23}{37}(0.46) + \frac{14}{37}(0.38) = 0.43$$

$$V_1 = \frac{23}{37}(0.5) + \frac{14}{37}(0.3) = 0.42 *$$

$$V_2 = \frac{9}{20}(0.7) + \frac{3}{10}(0.1) + \frac{1}{4}(0.5) = 0.47$$

$$V_3 = \frac{2}{3}(0.1) + \frac{1}{3}(0.1) = 0.1$$

$$ST = \frac{9}{23}(0.43) + \frac{6}{23}(0.47) + \frac{8}{23}(0.1) = 0.33$$

$$ST = \frac{9}{23}(0.42) + \frac{6}{23}(0.47) + \frac{8}{23}(0.1) = 0.32*$$

Guiasu Method

$$E_1' : F_1^{(1)} = 0.18(0.5) + 0.19(0.5) + 0.17(0.5) + 0.15(0.5) + 0.14(0.3) + 0.15(0.5) = 0.44$$

$$F_2^{(1)} = \frac{274}{585}(0.5) + \frac{311}{585}(0.3) = 0.39$$

$$V_1 = \frac{3581}{5016}(0.44) + \frac{1435}{5016}(0.39) = 0.43$$

$$V_1 = \frac{3581}{5016}(0.5) + \frac{1435}{5016}(0.3) = 0.44 *$$

$$V_2 = \frac{33}{72}(0.5) + \frac{21}{72}(0.3) + \frac{18}{72}(0.3) = 0.39$$

$$V_3 = \frac{23}{36}(0.3) + \frac{13}{36}(0.3) = 0.30$$

$$ST = \frac{29}{72}(0.43) + \frac{18}{72}(0.39) + \frac{25}{72}(0.3) = 0.37$$

$$ST = \frac{29}{72}(0.44) + \frac{18}{72}(0.39) + \frac{25}{72}(0.3) = 0.38 *$$

$$E_2' : F_1^{(1)} = 0.18(0.5) + 0.19(0.3) + 0.17(0.5) + 0.15(0.5) + 0.14(0.5) + 0.15(0.5) = 0.46$$

$$F_2^{(1)} = \frac{274}{585}(0.5) + \frac{311}{585}(0.3) = 0.39$$

$$V_1 = \frac{3581}{5016}(0.46) + \frac{1435}{5016}(0.38) = 0.44$$

$$V_1 = \frac{3581}{5016}(0.5) + \frac{1435}{5016}(0.3) = 0.44 *$$

$$V_2 = \frac{33}{72}(0.7) + \frac{21}{72}(0.1) + \frac{18}{72}(0.5) = 0.48$$

$$V_3 = \frac{23}{36}(0.1) + \frac{13}{36}(0.1) = 0.1$$

$$ST = \frac{29}{72}(0.44) + \frac{18}{72}(0.48) + \frac{25}{72}(0.1) = 0.33$$

$$ST = \frac{29}{72}(0.44) + \frac{18}{72}(0.48) + \frac{25}{72}(0.1) = 0.33*$$

Dempster Rule of Combination

$$E_1' : F_1^{(1)} = \frac{9}{40}(0.5) + \frac{9}{40}(0.5) + \frac{8}{40}(0.5) + \frac{5}{40}(0.5) + \frac{4}{40}(0.3) + \frac{5}{40}(0.5) = 0.44$$

$$F_2^{(1)} = \frac{7}{19}(0.5) + \frac{12}{19}(0.3) = 0.37$$

$$V_1 = \frac{50}{53}(0.44) + \frac{3}{53}(0.37) = 0.44$$

$$V_1 = \frac{50}{53}(0.5) + \frac{3}{53}(0.3) = 0.49 *$$

$$V_2 = \frac{744}{1552}(0.5) + \frac{744}{1552}(0.3) + \frac{64}{1552}(0.3) = 0.39$$

$$V_3 = \frac{6}{7}(0.3) + \frac{1}{7}(0.3) = 0.3$$

$$ST = \frac{2139}{3647}(0.44) + \frac{299}{3647}(0.39) + \frac{1209}{3647}(0.3) = 0.39$$

$$ST = \frac{2139}{3647}(0.49) + \frac{299}{3647}(0.39) + \frac{1209}{3647}(0.3) = 0.42 *$$

$$E_2' : F_1^{(1)} = \frac{9}{40}(0.5) + \frac{9}{40}(0.3) + \frac{8}{40}(0.5) + \frac{5}{40}(0.5) + \frac{4}{40}(0.5) + \frac{5}{40}(0.5) = 0.46$$

$$F_2^{(1)} = \frac{7}{19}(0.5) + \frac{12}{19}(0.3) = 0.37$$

$$V_1 = \frac{50}{53}(0.46) + \frac{3}{53}(0.38) = 0.45$$

$$V_1 = \frac{50}{53}(0.5) + \frac{3}{53}(0.3) = 0.49 *$$

$$V_2 = \frac{744}{1552}(0.7) + \frac{744}{1552}(0.1) + \frac{64}{1552}(0.5) = 0.40$$

$$V_3 = \frac{6}{7}(0.1) + \frac{1}{7}(0.1) = 0.1$$

$$ST = \frac{2139}{3647}(0.45) + \frac{299}{3647}(0.40) + \frac{1209}{3647}(0.1) = 0.33$$

$$ST = \frac{2139}{3647}(0.49) + \frac{299}{3647}(0.40) + \frac{1209}{3647}(0.1) = 0.35 *$$

Yen Method

$$E_1' : F_1^{(1)} = \frac{25}{121}(0.5) + \frac{23}{121}(0.5) + \frac{21}{121}(0.5) + \frac{19}{121}(0.5) + \frac{17}{121}(0.3) + \frac{18}{121}(0.5) = 0.47$$

$$F_2^{(1)} = \frac{63}{129}(0.5) + \frac{66}{129}(0.3) = 0.40$$

$$V_1 = \frac{60}{91}(0.47) + \frac{31}{91}(0.40) = 0.44$$

$$V_1 = \frac{60}{91}(0.5) + \frac{31}{91}(0.3) = 0.44 *$$

$$V_2 = \frac{9}{20}(0.5) + \frac{6}{20}(0.3) + \frac{5}{20}(0.3) = 0.47$$

$$V_3 = \frac{18}{29}(0.3) + \frac{11}{29}(0.3) = 0.3$$

$$ST = \frac{9}{23}(0.44) + \frac{6}{23}(0.47) + \frac{8}{23}(0.3) = 0.40$$

$$ST = \frac{9}{23}(0.44) + \frac{6}{23}(0.47) + \frac{8}{23}(0.3) = 0.40 *$$

$$E'_2 : F_1^{(1)} = \frac{25}{121}(0.5) + \frac{23}{121}(0.3) + \frac{21}{121}(0.5) + \frac{19}{121}(0.5) + \frac{17}{121}(0.5) + \frac{18}{121}(0.5) = 0.46$$

$$F_2^{(1)} = \frac{63}{129}(0.5) + \frac{66}{129}(0.3) = 0.40$$

$$V_1 = \frac{60}{91}(0.46) + \frac{31}{91}(0.38) = 0.43$$

$$V_1 = \frac{60}{91}(0.5) + \frac{31}{91}(0.3) = 0.44 *$$

$$V_2 = \frac{9}{20}(0.7) + \frac{6}{20}(0.1) + \frac{5}{20}(0.5) = 0.47$$

$$V_3 = \frac{18}{29}(0.1) + \frac{11}{29}(0.1) = 0.1$$

$$ST = \frac{9}{23}(0.43) + \frac{6}{23}(0.47) + \frac{8}{23}(0.1) = 0.33$$

$$ST = \frac{9}{23}(0.44) + \frac{6}{23}(0.47) + \frac{8}{23}(0.1) = 0.33*$$

Set-Valued Statistical Method

$$E'_1 : F_1^{(1)} = \frac{25}{126}(0.5) + \frac{25}{126}(0.5) + \frac{23}{126}(0.5) + \frac{19}{126}(0.5) + \frac{17}{126}(0.3) + \frac{17}{126}(0.5) = 0.47$$

$$F_2^{(1)} = \frac{5}{9}(0.5) + \frac{4}{9}(0.3) = 0.40$$

$$V_1 = \frac{2}{3}(0.47) + \frac{1}{3}(0.4) = 0.45$$

$$V_1 = \frac{2}{3}(0.5) + \frac{1}{3}(0.3) = 0.43 *$$

$$V_2 = \frac{17}{36}(0.5) + \frac{11}{36}(0.3) + \frac{8}{36}(0.3) = 0.39$$

$$V_3 = \frac{11}{18}(0.3) + \frac{7}{18}(0.3) = 0.3$$

$$ST = \frac{15}{36}(0.45) + \frac{8}{36}(0.39) + \frac{13}{36}(0.3) = 0.38$$

$$ST = \frac{15}{36}(0.43) + \frac{8}{36}(0.39) + \frac{13}{36}(0.3) = 0.37 *$$

$$E'_2 : F_1^{(1)} = \frac{25}{126}(0.5) + \frac{25}{126}(0.3) + \frac{23}{126}(0.5) + \frac{19}{126}(0.5) + \frac{17}{126}(0.5) + \frac{17}{126}(0.5) = 0.46$$

$$F_2^{(1)} = \frac{5}{9}(0.5) + \frac{4}{9}(0.3) = 0.40$$

$$V_1 = \frac{2}{3}(0.46) + \frac{1}{3}(0.38) = 0.44$$

$$V_1 = \frac{2}{3}(0.5) + \frac{1}{3}(0.3) = 0.43 *$$

$$V_2 = \frac{17}{36}(0.7) + \frac{11}{36}(0.1) + \frac{8}{36}(0.5) = 0.47$$

$$V_3 = \frac{11}{18}(0.1) + \frac{7}{18}(0.1) = 0.1$$

$$ST = \frac{15}{36}(0.44) + \frac{8}{36}(0.47) + \frac{13}{36}(0.1) = 0.32$$
$$ST = \frac{15}{36}(0.43) + \frac{8}{36}(0.47) + \frac{13}{36}(0.1) = 0.32 *$$

It can be seen that the results of the two experts are similar (except for V_3) as are the results between the five groups.

7.7 Descriptions

The following descriptions are shortened versions of those appearing in [8].

V_1: Vulnerable Populations

$F_1^{(1)}$: Personal Security

A lack of personal security makes a person more subject to kidnapping by a trafficking organizations.

$F_{11}^{(1)}$: Connection with stable families and inclusive social organizations

When people have connections with stable families and inclusive social organizations such as schools, religious groups and social service agencies, they are less subject to exploitation by traffickers.

$F_{12}^{(1)}$: Individual Characteristics

People with more assertive personalities and better "street smarts" are more capable of protecting themselves. Young people alone in society with weaker self-confidence are more vulnerable.

$F_{13}^{(1)}$: Information about trafficking

Programs for family support, provision of information about practices of traffickers,

$F_{14}^{(1)}$: Education

Programs to educate the general public about sex trafficking are important in the combating of sex trafficking.

$F_{15}^{(1)}$: Crisis Intervention

Crisis intervention to assist people who lack reliable links to protective social organizations are practices that are followed and could be enhanced

$F_{16}^{(1)}$: Training Law Enforcement

Training for law enforcement officers that would help them to recognize vulnerable individuals and guide them to appropriate social services is a tactic for improving personal security.

$F_2^{(1)}$: Economic Status

$F_{21}^{(1)}$: Economic Opportunities

Raising standards of living and improving economic opportunities for a vulnerable popular is of extreme benefit. Economics opportunities would take people out of the scope of marketing organizations that exploit individuals intent on migrating.

It is often the case that economic development itself requires labor migration, notably from rural to urban areas.

$F_{22}^{(1)}$:   Program to support economic migrants

A policy to suppress trafficking would be an officially sanctioned and monitored program to support economics migrants. Migrants could pursue economic objectives but would not be as vulnerable to trafficking organizations. Such programs promote ties between the migrants and stable social organizations.

V_2:   Trafficking Organizations

$F_1^{(2)}$:   Improving law enforcement agencies

Law enforcement agencies have a responsibility to suppress trafficking organizations. They can accomplish this through increased resources, training, and intelligence.

$F_2^{(2)}$:   Extent that specific organizations and individuals are prosecuted

The capacity for traffickers to move people coercively into sex markets is reduced.

$F_3^{(2)}$:   More robust law enforcement

More robust law enforcement against trafficking raises the cost engaging in the practice which creates a disincentive for traffickers that manage to evade prosecution.

V_3:   Sex Markets

$F_1^{(3)}$:   Suppression

Suppression imposes increased costs on sex providers, which will shift the sex market equilibrium toward a reduced quantity and increased price.

$F_2^{(3)}$:   Legalization

Price would decrease and the skills of human trafficking organizations in evading detection would become essentially valueless. The information that legitimate sex workers could provide about trafficking organizations and the commercial interests they have in the trade could be a force that assists law enforcement agencies in the suppression of human trafficking into sex.

An interesting problem would be to use techniques of Dempster-Shafer theory and fuzzy logic to study the underground commercial sex economy. In [5], the authors provided such an estimate in seven major cities. They provided a guide to the understanding of the structure of this underground economy. Another possible research project would be the examination of the regulation of prostitution and its effect on sex trafficking, [7].

7.8   Local Look at Human Trafficking

The results in the remaining sections are based on [9].

"We use expert opinion to determine linear equations that aid in deciding how well a local area in the state of Nebraska, USA, is doing in combating sex trafficking. The mathematical model used here can be expanded to larger regions such as states or countries. Our results are in agreement with those of Shared Hope International. We also use outranking relations to determine experts' preferences [9]."

A 2015 survey reported that at least 47 school girls in Nebraska are trafficked a year. The survey's authors estimate that the true number is at least double this amount (Hampton & Bell). Because of the city's location on two interstates, the city is also part of regional and national trafficking networks. According to the Friends of Tamar [4], of the survivors identified in Nebraska in 2014,

99% reported at least one physical health problem;

97% reported psychological health problems (80% reported depression, 61.5% reported PTSD);

95% had experienced some form of violence or abuse while being trafficked;

81.6% reported forced sex

84.3% of survivors also reported substance abuse while being trafficked.

Nebraska received an average grade in the Protection Innocence Challenge, a state-by-state report, funded by Shared Hope International, that examines 41 key legislative components related to sex trafficking. Our model, based on expert opinion, shows that the area in question has an average performance in combating sex trafficking. Our model only uses data that is in the eyes of the experts. Pertinent data is currently not available. The local organizations devoted to combating sex trafficking have dedicated and committed staffs. Nebraska's ratings should continue to rise.

We develop a mathematical model of sex trafficking in the particular area that later can be extended to a state or a regional or a country level to determine how well the region in question is doing in combating sex trafficking. Our approach is based on the paper by Rajaram and Tidball, [16] and on [1]. The approach can be used to place cities, regions and/or states in well-defined tiers. We also hope that an extended model can aid in the beginning of empirical research in Nebraska, aid in the development of a sex trafficking network that includes *source, destination, transit* designations for cities, regions and/or states, and in creating a highly organized Index or Foundation in Human Trafficking.

We use the Analytic Hierarchy Process (AHP), [17, 18], the Guiasu method, [3], and Dempster's rule of combination, [6, 13] to derive linear equations which can be used to determine how well the area in Nebraska is doing in combating sex trafficking. Our presentation is based on [9].

7.9 The Model

Sex Trafficking (ST) is examined through three main factors, Prevention (P_3), Protection of Survivors (P_2), and Prosecution (P_3). Expert opinion is used to obtain an equation based on the relative importance of P_1, P_2, and P_3, say for example,

$$ST = \frac{3}{8}P_1 + \frac{3}{8}P_2 + \frac{1}{4}P_3. \qquad (7.4)$$

Then each of P_1, P_2, and P_3 is broken into components. These components or factors are also broken into subfactors. Equations are developed using expert opinion (or data) with respect to the relative importance of the factors and subfactors. Explanations of the factors and subfactors can be found in Sect. 7.13.

P_1:	Prevention
$F_1^{(1)}$:	Public Awareness
$F_{11}^{(1)}$:	Parents/Schools/Youth
$F_{12}^{(1)}$:	General Public
$F_{13}^{(1)}$:	Professionals
$F_2^{(1)}$:	Early Screening and Detection
$F_{21}^{(1)}$:	Healthcare Workers
$F_{22}^{(1)}$:	Law Enforcement Personnel
$F_{23}^{(1)}$:	Social Service Providers
$F_{24}^{(1)}$:	Hotel Employees
P_2:	Protecting Survivors
$F_1^{(2)}$:	Immediate Needs
$F_{11}^{(2)}$:	Safe House
$F_{12}^{(2)}$:	Counseling by Trauma Experts
$F_{13}^{(2)}$:	Medical Care
$F_2^{(2)}$:	Beyond Immediate Needs
$F_{21}^{(2)}$:	Trauma Therapy and Medical Care
$F_{22}^{(2)}$:	Support Systems
$F_{23}^{(2)}$:	Transitional Housing and Childcare
$F_{24}^{(2)}$:	Job Training and Job Placement
$F_{25}^{(2)}$:	Life Skills
$F_{26}^{(2)}$:	Inter-agency Social Services and Medical Collaboration
$F_{27}^{(2)}$:	Expunge Records
P_3:	Prosecution
$F_1^{(3)}$:	Stiffer Penalties
$F_2^{(3)}$:	Better Follow Up Investigation of Complaint
$F_3^{(3)}$:	Rehabilitation Program for Buyers
$F_4^{(3)}$:	Registry for Sex Trafficking Offenders
$F_5^{(3)}$:	Shaming Strategies

In the following, E_1, E_2, and E_3 are experts. Their rankings of the factors and subfactors as to their relative importance in combating sex trafficking are given in the tables immediately below.

It is shown in [13] that the procedure used in the following is equivalent to the Analytic Hierarchy Approach (AHP).

P_1 : Prevention

P_1	E_1	E_2	E_3	Row Avg
$F_1^{(1)}$	7	8	8	$\frac{23}{3}$
$F_2^{(1)}$	8	6	7	$\frac{21}{3}$
Col Total	15	14	15	$\frac{44}{3}$

We obtain the following equation by taking the row averages and dividing by the column total to obtain the coefficients $\frac{23}{44}$ and $\frac{21}{44}$.

$$P_1 = \frac{23}{44} F_1^{(1)} + \frac{21}{44} F_2^{(1)}. \tag{7.5}$$

We next determine equations for $F_1^{(1)}$ and $F_2^{(1)}$.

$F_1^{(1)}$	E_1	E_2	E_3	Row Avg
$F_{11}^{(1)}$	7	7	6	$\frac{20}{3}$
$F_{12}^{(1)}$	6	8	7	$\frac{21}{3}$
$F_{13}^{(1)}$	8	7	7	$\frac{22}{3}$
Col Total	21	22	20	$\frac{63}{3}$

$$F_1^{(1)} = \frac{20}{63} F_{11}^{(1)} + \frac{21}{63} F_{12}^{(1)} + \frac{22}{63} F_{13}^{(1)}. \tag{7.6}$$

$F_2^{(1)}$	E_1	E_2	E_3	Row Avg
$F_{21}^{(1)}$	6	7	6	$\frac{19}{3}$
$F_{22}^{(1)}$	6	7	6	$\frac{19}{3}$
$F_{23}^{(1)}$	8	7	7	$\frac{22}{3}$
$F_{24}^{(1)}$	6	6	6	$\frac{18}{3}$
Col Total	26	27	25	$\frac{78}{3}$

$$F_2^{(1)} = \frac{19}{78} F_{21}^{(1)} + \frac{19}{78} F_{22}^{(1)} + \frac{22}{78} F_{23}^{(1)} + \frac{18}{78} F_{24}^{(1)}. \tag{7.7}$$

We next substitute Eqs. (7.6) and (7.7) into Eq. (7.5) to obtain

$$P_1 = \frac{23}{44}\left(\frac{20}{63}F_{11}^{(1)} + \frac{21}{63}F_{12}^{(1)} + \frac{22}{63}F_{13}^{(1)}\right) \tag{7.8}$$
$$+ \frac{21}{44}\left(\frac{19}{78}F_{21}^{(1)} + \frac{19}{78}F_{22}^{(1)} + \frac{22}{78}F_{23}^{(1)} + \frac{18}{78}F_{24}^{(1)}\right).$$

P_2 : Protection

P_2	E_1	E_2	E_3	Row Avg
$F_1^{(2)}$	10	9	8	9
$F_2^{(2)}$	8	8	8	8
Col Total	18	17	16	17

$$P_2 = \frac{9}{17}F_1^{(2)} + \frac{8}{17}F_2^{(2)}. \tag{7.9}$$

We next obtain equations for $F_1^{(2)}$ and $F_2^{(2)}$.

$F_1^{(2)}$	E_1	E_2	E_3	Row Avg
$F_{11}^{(2)}$	8	8	8	$\frac{24}{3}$
$F_{12}^{(2)}$	6	6	5	$\frac{17}{3}$
$F_{13}^{(2)}$	6	7	7	$\frac{20}{3}$
Col Total	20	21	20	$\frac{61}{3}$

$$F_1^{(2)} = \frac{24}{61}F_{11}^{(2)} + \frac{17}{61}F_{12}^{(2)} + \frac{20}{61}F_{13}^{(2)}. \tag{7.10}$$

$F_2^{(2)}$	E_1	E_2	E_3	Row Avg
$F_{21}^{(2)}$	6	7	7	$\frac{20}{3}$
$F_{22}^{(2)}$	6	7	7	$\frac{20}{3}$
$F_{23}^{(2)}$	5	6	7	$\frac{18}{3}$
$F_{24}^{(2)}$	8	7	7	$\frac{22}{3}$
$F_{25}^{(2)}$	8	7	7	$\frac{22}{3}$
$F_{26}^{(2)}$	6	6	6	$\frac{18}{3}$
$F_{27}^{(2)}$	6	6	6	$\frac{18}{3}$
Col Total	45	46	47	$\frac{138}{3}$

$$F_2^{(2)} = \frac{20}{138} F_{21}^{(2)} + \frac{20}{138} F_{22}^{(2)} + \frac{18}{138} F_{23}^{(2)} + \frac{22}{138} F_{24}^{(2)} + \frac{22}{138} F_{25}^{(2)}$$
$$+ \frac{18}{138} F_{26}^{(2)} + \frac{18}{138} F_{27}^{(2)}. \tag{7.11}$$

We next obtain an equation for P_2 by substituting Eqs. (7.10) and (7.11) into Eq. (7.9).

$$P_2 = \frac{9}{17} \left(\frac{24}{61} F_{11}^{(2)} + \frac{17}{61} F_{12}^{(2)} + \frac{20}{61} F_{13}^{(2)} \right)$$
$$+ \frac{8}{17} \left(\frac{20}{138} F_{21}^{(2)} + \frac{20}{138} F_{22}^{(2)} + \frac{18}{138} F_{23}^{(2)} + \frac{22}{138} F_{24}^{(2)} \right.$$
$$\left. + \frac{22}{138} F_{25}^{(2)} + \frac{18}{138} F_{26}^{(2)} + \frac{18}{138} F_{27}^{(2)} \right). \tag{7.12}$$

P_3 : Prosecution

P_3	E_1	E_2	E_3	Row Avg
$F_1^{(3)}$	8	8	8	$\frac{24}{3}$
$F_2^{(3)}$	8	8	8	$\frac{24}{3}$
$F_3^{(3)}$	5	6	5	$\frac{16}{3}$
$F_4^{(3)}$	5	5	6	$\frac{16}{3}$
$F_5^{(3)}$	5	6	6	$\frac{17}{3}$
Col Total	31	33	33	$\frac{97}{3}$

$$P_3 = \frac{24}{97} F_1^{(3)} + \frac{24}{97} F_2^{(3)} + \frac{16}{97} F_3^{(3)} + \frac{16}{97} F_4^{(3)} + \frac{17}{97} F_5^{(3)}. \tag{7.13}$$

We next substitute equations (7.8),(7.12), and (7.13) into (7.4) to obtain an equation for ST.

$$ST = \frac{3}{8} \left[\frac{23}{44} \left(\frac{20}{63} F_{11}^{(1)} + \frac{21}{63} F_{12}^{(1)} + \frac{22}{63} F_{13}^{(1)} \right) \right.$$
$$\left. + \frac{21}{44} \left(\frac{19}{78} F_{21}^{(1)} + \frac{19}{78} F_{22}^{(1)} + \frac{22}{78} F_{23}^{(1)} + \frac{18}{78} F_{24}^{(1)} \right) \right]$$
$$+ \frac{3}{8} \left[\frac{9}{17} \left(\frac{24}{61} F_{11}^{(2)} + \frac{17}{61} F_{12}^{(2)} + \frac{20}{61} F_{13}^{(2)} \right) \right.$$
$$+ \frac{8}{17} \left(\frac{20}{138} F_{21}^{(2)} + \frac{20}{138} F_{22}^{(2)} + \frac{18}{138} F_{23}^{(2)} + \frac{22}{138} F_{24}^{(2)} \right.$$

$$+ \frac{22}{138} F_{25}^{(2)} + \frac{18}{138} F_{26}^{(2)} + \frac{18}{138} F_{27}^{(2)} \Big) \Big]$$

$$+ \frac{1}{4} \Big[\frac{24}{97} F_1^{(3)} + \frac{24}{97} F_2^{(3)} + \frac{16}{97} F_3^{(3)}$$

$$+ \frac{16}{97} F_4^{(3)} + \frac{17}{97} F_5^{(3)} \Big]. \tag{7.14}$$

The following table represents expert opinion on how well the area in question is combating sex trafficking. Very well is represented by the number 1, well by 0.8, average by 0.6, poor by 0.4, and very poor by 0.2.

	$F_{11}^{(1)}$	$F_{12}^{(1)}$	$F_{13}^{(1)}$	$F_{21}^{(1)}$	$F_{22}^{(1)}$	$F_{23}^{(1)}$	$F_{24}^{(1)}$	$F_{11}^{(2)}$	$F_{12}^{(2)}$	$F_{13}^{(2)}$
E_1'	0.4	0.4	0.6	0.5	0.5	0.6	0.2	0.4	0.4	0.6
E_2'	0.2	0.4	0.8	0.5	0.6	0.6	0.4	0.5	0.5	0.5

	$F_{21}^{(2)}$	$F_{22}^{(2)}$	$F_{23}^{(2)}$	$F_{24}^{(2)}$	$F_{25}^{(2)}$	$F_{26}^{(2)}$	$F_{27}^{(2)}$	$F_1^{(3)}$	$F_2^{(3)}$	$F_3^{(3)}$	$F_4^{(3)}$	$F_5^{(3)}$
E_1'	0.4	0.6	0.6	0.2	0.4	0.4	0.2	0.6	0.2	0.2	0.2	0.2
E_2'	0.4	0.4	0.4	0.2	0.2	0.2	0.8	0.5	0.5	0.1	0.1	0.1

The numbers in the above table are substituted into Eq. (7.14) to obtain the number stating how well the area is doing with respect to combating sex trafficking. We can also determine how well categories such as P_1, P_2, P_3, and the $F_i^{(j)}$ are doing with respect to sex trafficking in the area.

	$F_1^{(1)}$	$F_2^{(1)}$	$F_1^{(2)}$	$F_2^{(2)}$	P_1	P_2	P_3	ST
E_1'	0.472	0.459	0.466	0.397	0.466	0.434	0.305	0.414
E_2'	0.481	0.529	0.5	0.362	0.503	0.435	0.301	0.427

7.10 Dempster's Rule of Combination

The AHP table below is used to create the Guiasu table by dividing each entry in the AHP table by the column total of the column it's in. The row averages of the Guiasu table are taken and the results give the coefficients of the linear equation of ST in terms of the G_i, $i = 1, \ldots, 9$.

	E_1	E_2	E_3	AHP
$F_1^{(1)}$	7	8	8	23/3
$F_2^{(1)}$	8	6	7	21/3
$F_1^{(2)}$	10	9	8	27/3
$F_2^{(2)}$	8	8	8	24/3
$F_1^{(3)}$	8	8	8	24/3
$F_2^{(3)}$	8	8	8	24/3
$F_3^{(3)}$	5	6	5	16/3
$F_4^{(3)}$	5	5	6	16/3
$F_5^{(3)}$	5	6	6	17/3
Col Total	64	64	64	192/3

	E_1	E_2	E_3	Guiasu	Dempster
$G_1 = F_1^{(1)}$	7/64	8/64	8/64	0.120	0.0025
$G_2 = F_2^{(1)}$	8/64	6/64	7/64	0.109	0.0020
$G_3 = F_1^{(2)}$	10/64	9/64	8/64	0.141	0.0043
$G_4 = F_2^{(2)}$	8/64	8/64	8/64	0.125	0.0029
$G_5 = F_1^{(3)}$	8/64	8/64	8/64	0.125	0.0029
$G_6 = F_2^{(3)}$	8/64	8/64	8/64	0.125	0.0029
$G_7 = F_3^{(3)}$	5/64	6/64	5/64	0.083	0.0007
$G_8 = F_4^{(3)}$	5/64	5/64	6/64	0.083	0.0007
$G_9 = F_5^{(3)}$	5/64	6/64	6/64	0.089	0.0009
Col Total	1	1	1	1	0.0198

Because the columns of the AHP table add to the same number, the coefficients for the ST equation for the AHP and Guiasu are equal, [11]. Hence,

$$ST = 0.120 F_1^{(1)} + 0.109 F_2^{(1)} + 0.141 F_1^{(2)} + 0.125 F_2^{(2)}$$
$$+ 0.125 F_1^{(3)} + 0.125 F_2^{(3)} + 0.083 F_3^{(3)} + 0.083 F_4^{(3)} + 0.089 F_5^{(3)}$$

is the ST equation for both the AHP and Guiasu methods.

We consider the entries in the Guiasu table basic probability assignments for each expert, [6]. For example, $m_1(G_1) = \frac{7}{64}$, $m_2(G_1) = \frac{8}{64}$, and $m_3(G_1) = \frac{8}{64}$. We assume that our universe contains two focal elements one being $\{G_1\}$ for E_1, E_2, and E_3. Then by [10, Theorem 3.1, p. 195], the combined basic probability assignment is

$$m_{123}(\{G_i\}) = m_1(\{G_i\})m_2(\{G_i\})m_3(\{G_i\})/[1 - m_1(\{G_i\}) - m_2(\{G_i\})$$
$$-m_3(\{G_i\}) + m_1(\{G_i\})m_2(\{G_i\}) + m_1(\{G_i\})m_3(\{G_i\})$$
$$+m_2(\{G_i\})m_3(\{G_i\})],$$

$i = 1, \ldots, 9$. Thus, it follows that $m_{123}\{\{G_1\}) = 0.0025$. We divide the elements in the column titled Dempster by its column total to obtain the coefficients for the ST equation. For example, $0.0025/0.0198 = 0.126$. Thus,

$$ST = 0.126F_1^{(1)} + 0.101F_2^{(1)} + 0.217F_1^{(2)} + 0.146F_2^{(2)}$$
$$+0.146F_1^{(3)} + 0.146F_2^{(3)} + 0.035F_3^{(3)} + 0.035F_4^{(3)} + 0.045F_5^{(3)}$$

is the equation for ST determined by the Dempster Rule of Combination method. The following table gives the experts' opinion on how well Omaha-Lincoln is doing with respect to combating sex trafficking with respect to the subfactors.

	$F_1^{(1)}$	$F_2^{(1)}$	$F_1^{(2)}$	$F_2^{(2)}$	$F_1^{(3)}$	$F_2^{(3)}$	$F_3^{(3)}$	$F_4^{(3)}$	$F_5^{(3)}$
E_1'	0.5	0.4	0.4	0.4	0.6	0.2	0.2	0.2	0.2
E_2'	0.6	0.5	0.4	0.2	0.5	0.5	0.1	0.1	0.1

Substituting these values into the ST equations yield the following degree to which Omaha-Lincoln is doing with respect to combating sex trafficking.

	ST	Guiasu	Dempster
E_1'	0.3620		0.3879
E_2'	0.3709		0.3990

7.11 Preference Relations

We now consider a procedure developed in [15]. It's purpose is to determine the combined preference of the three experts. Let G_i be defined as in the previous table, $i = 1, \ldots, 9$. For each expert E_k, $k = 1, 2, 3$, define the fuzzy relation ρ_k on $\{G_i \mid i = 1, \ldots, 9\}$ as follows:

$$\rho_k(G_i, G_j) = \begin{cases} (G_i - G_j + 0.5) \wedge 1 & \text{if } G_i \geq G_j, \\ 1 - [(G_j - G_i + 0.5) \wedge 1] & \text{if } G_i < G_j, \end{cases}$$

$i, j = 1, 2, \ldots, 9$.

Let $r_{ij}^k = \rho_k(G_i, G_j)$, $k = 1, 2, 3$ and $i, j = 1, 2, \ldots, 9$. Let $R_k = [r_{ij}^k]$, $k = 1, 2, 3$. This provides three matrices which represent the preferences of the three experts. We do not display them here because they are easy to derive. For $k = 1, 2, 3$, let

$$a_{ij}^k = \begin{cases} 1 & \text{if } r_{ij}^k > 0.5, \\ 0 & \text{otherwise}, \end{cases}$$

and

$$r_{ij} = \begin{cases} \frac{1}{3} \sum a_{ij}^k & \text{if } i \neq j, \\ 0 & \text{otherwise}. \end{cases}$$

Let $R = [r_{ij}]$. Then

$$
R = \begin{array}{c}
\\
G_1 \\ G_2 \\ G_3 \\ G_4 \\ G_5 \\ G_6 \\ G_7 \\ G_8 \\ G_9
\end{array}
\begin{array}{cccccccccc}
G_1 & G_2 & G_3 & G_4 & G_5 & G_6 & G_7 & G_8 & G_9 \\
\left[\begin{array}{ccccccccc}
0 & \frac{2}{3} & 0 & 0 & 0 & 0 & 1 & 1 & 1 \\
\frac{1}{3} & 0 & 0 & 0 & 0 & 0 & \frac{2}{3} & 1 & \frac{2}{3} \\
\frac{2}{3} & 1 & 0 & 1 & 1 & 1 & 1 & 1 & 1 \\
\frac{1}{3} & \frac{2}{3} & 0 & 0 & 0 & 0 & 1 & 1 & 1 \\
\frac{1}{3} & \frac{2}{3} & 0 & 0 & 0 & 0 & 1 & 1 & 1 \\
\frac{1}{3} & \frac{2}{3} & 0 & 0 & 0 & 0 & 1 & 1 & 1 \\
0 & 0 & 0 & 0 & 0 & 0 & 0 & \frac{1}{3} & 0 \\
0 & 0 & 0 & 0 & 0 & 0 & \frac{1}{3} & 0 & 0 \\
0 & 0 & 0 & 0 & 0 & 0 & \frac{1}{3} & 0 & 0 \\
\end{array}\right]
\end{array}
$$

The matrix R is the combined preference of the three experts.

7.12 Outranking Relations

In order to reflect decision maker preferences, we use relational systems involving those in the following definition. In the following, T is a t-norm, N is a negation operator, and V is a conorm, [6].

Definition 7.12.1 Let $S, I, \succ, R$, and $\sim$ be relations on a set X. Let $x, y \in X$.

S : **Outranking relation**. The statement $x S y$ means "x is not worse than y" and reflects the presence of arguments strong enough to support this assertion.

I : **Indifference relation**. The statement $x I y$ means "x and y are indifferent (i.e., roughly equivalent)" and reflects the presence of arguments strong enough to support both the assertions $x S y$ and $y S x$.

$\succ$: **Preference relation**. The statement $x \succ y$ means "x is preferred (at least weakly) to y" and reflects both the presence of arguments strong enough to support this assertion xSy and the absence of similar arguments to support ySx.

R : **Incomparability relation**. The statement xRy means "x and y are incomparable" and reflects the absence of arguments strong enough to support at least one of the assertions xSy or ySx.

$\sim$: **Non-comparability relation**. The statement $x \sim y$ means " we cannot discriminate between x and y, x and y are either incomparable or indifferent" : it reflects the absence of arguments strong enough to support at least one of the two assertions $x \succ y$ or $y \succ x$.

In the following, $\wedge$ denotes minimum.

Definition 7.12.2 Let S be an outranking relation on a set X. Let (N, T, V) be a DeMorgan triple, [6]. Define the relations $I(S)$, $R_N(S)$, $\succ_{N,T}$, and $\sim_{N,T}$ on X as follows: for all $x, y \in X$,
Indifference index:

$$I(S)(x, y) = S(x, y) \wedge S(y, x),$$

Incomparability index:

$$R_N(x, y) = N(S(x, y)) \wedge N(S(y, x)),$$

Preference index:

$$\succ_{N,T} (S)(x, y) = T(S(x, y), \ N(S(y, x))),$$

Non-preference index:

$$\sim_{N,T} (S)(x, y) = N[\succ_{N,T} (S)(x, y)] \wedge N[\succ_{N,T} (S)(y, x)].$$

The following relations are determined by the t-norm minimum and the standard negation operator, i.e., for all $a \in [0, 1]$, $N(a) = 1 - a$. We let the outranking relation S equal the relation R of the previous section.

$$
I(S) = \begin{array}{c} \\ G_1 \\ G_2 \\ G_3 \\ G_4 \\ G_5 \\ G_6 \\ G_7 \\ G_8 \\ G_9 \end{array}
\begin{array}{c} G_1\ G_2\ G_3\ G_4\ G_5\ G_6\ G_7\ G_8\ G_9 \\
\left[\begin{array}{ccccccccc}
0 & \frac{1}{3} & 0 & 0 & 0 & 0 & 0 & 0 & 0 \\
\frac{1}{3} & 0 & 0 & 0 & 0 & 0 & 0 & 0 & 0 \\
0 & 0 & 0 & 0 & 0 & 0 & 0 & 0 & 0 \\
0 & 0 & 0 & 0 & 0 & 0 & 0 & 0 & 0 \\
0 & 0 & 0 & 0 & 0 & 0 & 0 & 0 & 0 \\
0 & 0 & 0 & 0 & 0 & 0 & 0 & 0 & 0 \\
0 & 0 & 0 & 0 & 0 & 0 & 0 & \frac{1}{3} & 0 \\
0 & 0 & 0 & 0 & 0 & 0 & \frac{1}{3} & 0 & 0 \\
0 & 0 & 0 & 0 & 0 & 0 & 0 & 0 & 0
\end{array}\right]
\end{array}
$$

We see that there is very little indifference between the G_i, $i = 1, \ldots, 9$.

$$
R(S) = \begin{array}{c} \\ G_1 \\ G_2 \\ G_3 \\ G_4 \\ G_5 \\ G_6 \\ G_7 \\ G_8 \\ G_9 \end{array}
\begin{array}{c} G_1\ G_2\ G_3\ G_4\ G_5\ G_6\ G_7\ G_8\ G_9 \\
\left[\begin{array}{ccccccccc}
1 & \frac{1}{3} & \frac{1}{3} & \frac{2}{3} & \frac{2}{3} & \frac{2}{3} & 0 & 0 & 0 \\
\frac{1}{3} & 1 & \frac{2}{3} & \frac{2}{3} & \frac{2}{3} & 0 & \frac{1}{3} & 1 & \frac{1}{3} \\
\frac{1}{3} & \frac{2}{3} & 1 & 0 & 0 & 0 & 0 & 0 & 0 \\
\frac{2}{3} & \frac{2}{3} & 0 & 1 & 1 & 1 & 0 & 0 & 0 \\
\frac{2}{3} & \frac{2}{3} & 0 & 1 & 1 & 1 & 0 & 0 & 0 \\
\frac{2}{3} & 0 & 0 & 1 & 1 & 1 & 0 & 0 & 0 \\
0 & \frac{1}{3} & 0 & 0 & 0 & 0 & 1 & \frac{2}{3} & 1 \\
0 & 1 & 0 & 0 & 0 & 0 & \frac{2}{3} & 1 & \frac{2}{3} \\
0 & \frac{1}{3} & 0 & 0 & 0 & 0 & 1 & \frac{2}{3} & 1
\end{array}\right]
\end{array}
$$

The previous matrix gives the degree of incomparability of the G_i, $i = 1, \ldots, 9$. Note that $G_3 = F_1^{(2)}$ has 0 intensity of incomparability with most of the other G_i, as one would expect from the following matrix.

$$
\succ (S) = \begin{array}{c} \\ G_1 \\ G_2 \\ G_3 \\ G_4 \\ G_5 \\ G_6 \\ G_7 \\ G_8 \\ G_9 \end{array}
\begin{array}{c} G_1\ G_2\ G_3\ G_4\ G_5\ G_6\ G_7\ G_8\ G_9 \\
\left[\begin{array}{ccccccccc}
0 & \frac{2}{3} & 0 & 0 & 0 & 0 & 1 & 1 & 1 \\
\frac{1}{3} & 0 & 0 & 0 & 0 & 0 & \frac{2}{3} & 1 & \frac{2}{3} \\
\frac{2}{3} & 1 & 0 & 1 & 1 & 1 & 1 & 1 & 1 \\
\frac{1}{3} & \frac{2}{3} & 0 & 0 & 0 & 0 & 1 & 1 & 1 \\
\frac{1}{3} & \frac{2}{3} & 0 & 0 & 0 & 0 & 1 & 1 & 1 \\
\frac{1}{3} & \frac{2}{3} & 0 & 0 & 0 & 0 & 1 & 1 & 1 \\
0 & 0 & 0 & 0 & 0 & 0 & 0 & \frac{1}{3} & 0 \\
0 & 0 & 0 & 0 & 0 & 0 & \frac{1}{3} & 0 & 0 \\
0 & 0 & 0 & 0 & 0 & 0 & \frac{1}{3} & 0 & 0
\end{array}\right]
\end{array}
$$

The previous matrix shows that $G_3 = F_1^{(2)}$ is considered to be preferred to the other G_i with intensity 1 in all but one case.

The following matrix displays the non-preference of the G_i with other G_j. Note that $G_3 = F_3^{(2)}$ has non-preference intensity 0 with the other G_i which is in agreement with the matrices above.

$$\sim (S) = \begin{array}{c} \\ G_1 \\ G_2 \\ G_3 \\ G_4 \\ G_5 \\ G_6 \\ G_7 \\ G_8 \\ G_9 \end{array} \begin{array}{c} G_1\ G_2\ G_3\ G_4\ G_5\ G_6\ G_7\ G_8\ G_9 \\ \left[\begin{array}{ccccccccc} 1 & \frac{1}{3} & \frac{1}{3} & \frac{1}{3} & \frac{1}{3} & \frac{1}{3} & 0 & 0 & 0 \\ \frac{1}{3} & 1 & 0 & \frac{1}{3} & \frac{1}{3} & \frac{1}{3} & \frac{1}{3} & 0 & \frac{1}{3} \\ \frac{1}{3} & 0 & 1 & 0 & 0 & 0 & 0 & 0 & 0 \\ \frac{1}{3} & \frac{1}{3} & 0 & 1 & 1 & 1 & 0 & 0 & 0 \\ \frac{1}{3} & \frac{1}{3} & 0 & 1 & 1 & 1 & 0 & 0 & 0 \\ \frac{1}{3} & \frac{1}{3} & 0 & 1 & 1 & 1 & 0 & 0 & 0 \\ 0 & \frac{1}{3} & 0 & 0 & 0 & 0 & 1 & \frac{1}{3} & 1 \\ 0 & 0 & 0 & 0 & 0 & 0 & \frac{1}{3} & 1 & \frac{1}{3} \\ 0 & \frac{1}{3} & 0 & 0 & 0 & 1 & \frac{1}{3} & 1 \end{array} \right] \end{array}$$

7.13 Descriptions

$F_1^{(1)}$: Public Awareness

Survivors stress the need for the general public to become more knowledgeable about sex trafficking. There is a lack of understanding of the difference between prostitution versus sex trafficking. Consequently women are often blamed, stigmatized and held criminally responsible. The lack of awareness fuels the proliferation and hidden nature of the problem. It thus becomes quite difficult for women to escape sex trafficking and to obtain the social and material assistance needed.

$F_{11}^{(1)}$:   Parents/Schools/Youth

Both teachers and students need to be educated. Middle school children need to be educated because even children at this age are trafficked.

$F_{12}^{(1)}$:   General Public

Public awareness is considered essential to address the lack of public awareness in order to reduce the stigma associated with sex trafficking and to provide a safe environment for women and girls get help.

$F_{13}^{(1)}$:   Professionals

For early screening and detection of sex trafficking survivors, law enforcement personnel, social service providers, and hotel employees should be sensitive and understanding.

$F_1^{(2)}$:   Immediate Needs

Survivors need a safe house away from the reach of the trafficker, counseling by trauma specialists, and medical care.

$F_{11}^{(2)}$:   Safe Houses

Women have short term-needs away from their pimp, one that allows them to work through the immediate aftermath of their experiences,

$F_{12}^{(2)}$:   Counseling by Trauma Experts

Survivors may suffer from severe and desolate isolation, serious mental disability, Post-Traumatic Stress Disorder, depression, and anxiety

$F_{13}^{(2)}$:   Medical Care

Many women are to be dependent on drugs as a way to maintain control over them. Thus, substance abuse treatment may be necessary. Some women have been abused physically. Hence, some might even need dental and/or facial work. Some may suffer from disease.

$F_2^{(2)}$:   Beyond Immediate Needs

Survivors require on-going therapy and medical care.

$F_{21}^{(2)}$:   Trauma Therapy and Medical Care.

Survivors require on-going trauma therapy and medical care.

$F_{22}^{(2)}$:   Support Systems.

Survivors need support groups including peer-to-peer support groups and trusted friends and family.

$F_{23}^{(2)}$:   Transitional Housing and Childcare.

A living place is needed for survivors to adjust and prepare for normal life.

$F_{24}^{(2)}$:   Job Training and Job Placement.

Survivors need help preparing them to be successful in the continuation of their life.

$F_{25}^{(2)}$:   Life Skills

Survivors need training in writing a resume, managing a budget.

$F_{26}^{(2)}$:   Inter-agency Social Services and Medical Collaboration.

Ensurance that both referral protocols and necessary services are in place to adequately meet the needs of survivors as they reenter normal society.

$F_{27}^{(2)}$:   Expunge Records

It is extremely important for survivors to have their records expunged of any criminal charges.

$F_1^{(3)}$:   Stiffer Penalties

The prosecution of perpetrators is needed to reduce demand. Penalties should be commensurate with brutality of the offence.

$F_2^{(3)}$:   Better Follow Up Investigation of Complaint

Law enforcement needs to be diligent in conducting follow-up investigations, rather than unjustly blaming women for prostituting themselves.

$F_3^{(3)}$:   Rehabilitation Program for Buyers

Survivors felt that the buyers should be made to go through a rehabilitation program.

$F_4^{(3)}$:   Registry for Sex Trafficking Offenders.

Some survivors suggest creating a sex trafficking registry so that actions of buyers and traffickers would be made public.

$F_5^{(3)}$:   Shaming Strategies
Some survivors felt that buyers and traffickers should be publicly shamed.

We developed a mathematical model of sex trafficking in the Omaha-Lincoln area. The model can be extended to a state or a regional or a country level to determine how well the region in question is doing in combating sex trafficking. An extended model can aid in the beginning of empirical research in Nebraska, aid in the development of a sex trafficking network that includes *source, destination, transit* designations for cities, regions and/or states, and in creating an organized Index or Foundation in Human Trafficking.

References

1. R. Davis, *The Dark Journey, University of Nebraska Omaha Magazine* (2016), pp. 34–37
2. M. Grabisch, H.T. Nguyen, E.A. Walker, *Fundamentals of Uncertainty Calculi With Applications To Fuzzy Inference.* Theory and Decision Library Series B: Mathematical and Statistical Methods (Kluwer Academic Publishers, Dordrecht, 1995)
3. S. Guiasu, Reaching a verdict by weighting evidence, in *Advances in Fuzzy Set Theory and Technology,* ed. by P.P. Wang, vol. II (Bookswright Press, Durham, 1994), pp. 167–180
4. http://www.thefriendsoftamar/magdalane-omaha
5. B. Khan, M. Downey, M. Dank, K. Dombrowski, *A Method for Determining the Size of the Underground Cash Economy for Commercial Sex in Seven US Cities* ed. by S. Cunningham, M. Shah. The Oxford Handbook of The Economics of Prostitution (Oxford University Press, Oxford, 2016), pp. 348–366
6. G.J. Klir, B. Yuan, *Fuzzy Sets and Fuzzy Logic*: *Theory and Applications* (Prentice Hall Inc., Upper Saddle River, 1995)
7. S. Lee, P. Persson, Human Trafficking and Regulating Prostitution, Law and Economics Research Paper Series Working Paper No. 12-08, New York University School of Law, NYU Center for Law, Economics and Organization, 1–43, July 2012
8. E. Lutz, R. Lotspeich, Sex markets and human trafficking: cause-effect and policy interventions. Prot. Proj. J. Hum. Rights Civ. Soc. **2**, 124–199 (2009)
9. J.N. Mordeson, S. Mathew, Local look at human trafficking. New Math. Nat. Comput. **13**, 327–340 (2017)
10. J.N. Mordeson, T.D. Clark, A.D. Grieser, M.J. Wierman, An inductive approach to determining causality in comparative politics: a fuzzy set alternative. New Math. Nat. Comput. **3**, 191–202 (2007)
11. J.N. Mordeson, T.D. Clark, M.J. Wierman, A. Pham, A fuzzy mathematical model of cooperative threat reduction. Adv. Fuzzy Sets Syst. **6**, 153–169 (2010)
12. J.N. Mordeson, H.C. Wething, T.D. Clark, A fuzzy mathematical model of nuclear stability. New Math. Nat. Comput. **6**, 119–140 (2010)
13. J.N. Mordeson, M.J. Wierman, T.D. Clark, A. Pham, M.A. Redmond, *Linear Models in the Mathematics of Uncertainty.* Studies in Computational Intelligence, vol. 463 (Springer, Berlin, 2013
14. J.N. Mordeson, M. Mallenby, S. Mathew, S. Acharjee, Human trafficking: policy intervention. New Math. Nat. Comput. **13**, 341–358 (2017)
15. P. Perny, B. Roy, The use of fuzzy outranking relations in preference modeling. Fuzzy Sets Syst. **49**, 33–53 (1992)
16. S.S. Rajaram, S. Tidball, *Nebraska Sex Trafficking Survivors Speak - A Qualitative Research Study, Submitted to: The Women's Fund of Omaha* (2016)

17. T.L. Saaty, Relative measurement and its generalization in decision making: Why pairwise comparisons are central in mathematics for the measurement of intangible factors: the analytic hierarchy process. RACSAM Rev. R. Span. Acad. Sci. A. Math. **102**(2), 252–319 (2008)
18. T.L. Saaty, L.G. Vargas, *Models, Methods, Concepts and Applications of the Analytic Hierarchy Process*. International Series in Operations Research and Management Science (Kluwer Academic Publishers, Boston, 2001)
19. Trafficking in Persons: Global Patterns, The United Nations Office for Drugs and Crime, Trafficking in Persons Citation Index
20. United Nations Office on Drugs and Crime (UNODC) Trafficking in Persons Global Patterns (2006)
21. J. Yen, Generalizing the Dempster-Shafer theory to fuzzy sets, in *Fuzzy Measure Theory*, ed. by Z. Wang and G.J. Klir (Plenum Press, New York, 1992), pp. 257–283

Index

© Springer International Publishing AG 2018
J. N. Mordeson et al., *Fuzzy Graph Theory with Applications to Human Trafficking*, Studies in Fuzziness and Soft Computing 365,
https://doi.org/10.1007/978-3-319-76454-2

Printed in the United States
By Bookmasters